NUCLEAR REVOLUTION: A CLARION CALL

BOOK 3

Books by Chuck and Tom Paprocki

History, Ideology and Revolution: A Trilogy

Book One: The Untold Story of Western Civilization

Volume One: The Age of Mothers - The Matriarchy
Volume Two: The Age of Warriors
Volume Three: The Age of Intellectual Priests
Volume Four: The Age of Merchant Capitalists
Volume Five: Pax Americana

Book Two: Universal Ideology: The Thought of P. R. Sarkar

Book Three: Nuclear Revolution: A Clarion Call

Books by Chuck Paprocki

Meetings with Spiritual Masters: A Memoir

NUCLEAR REVOLUTION: A CLARION CALL

BOOK 3

Chuck and Tom Paprocki

Inner World Publications
San German, Puerto Rico
www.innerworldpublications.com

Published in the United States by
Inner World Publications
P.O. Box 1613, San German, Puerto Rico, 00683

Library of Congress Control Number:
2020948546

ISBN: 978-1-881717-78-2

Cover Design: Tom Paprocki

Cover Illustration: The cover illustration is from chuckieneedles@adventuretattoo.com. It depicts an illuminated planet Earth that results from a Nuclear Revolution created by an enlightened humanity who have united to meet everyone's basic needs of food, shelter, clothing, health care, and education in an environmentally sustainable manner. The revolution requires physical and intellectual force tempered by a universal spiritual bearing.

Dedication

To the Memory of Prabhat Rainjan Sarkar,
our Spiritual Guide

This book is being published in memory of Chuck Paprocki for his many years of dedicated service for the betterment of all God's creatures, human and nonhuman.

Chuck Paprocki
March 21, 1945 - December 10, 2020

Contents

List of Figures

Introduction

O human being! Be established in the radiance of godliness and the splendor of valor and chivalry, because yours is a path of revolution. Yours is not a path of extra caution and hesitant movement. You are travelers on a difficult path. You have to march ahead, with head held high and with your chest forward. You have not a moment to stagger or look behind. We are near the last stage of the Capitalist Era. If an impact is created it, will help suffering humanity. It is the most opportune moment for creating an all-round revolution.

If you do service, if you teach people and encourage them to develop their rationality, they they will revolt against an inhuman system. Disorganized and irrational people are incapable of launching a revolution.

To teach love, we have to feel love.
To feel love, we have to see all as divine
To see all as divine we have to contemplate the divine within everyone,
And dance with the divinity. P. R. Sarkar

THIS IS THE THIRD book of a trilogy on *History, Ideology and Revolution*. In our history, The *Untold Story of Western Civilization*, we examined the roots and development of the major socio- and geo-sentiments of religious exclusivism, sexism, racism, national chauvinism, and classism. We witnessed how these sentiments were passed along from generation to generation and how they came to constitute our subconscious programming. We have seen how this programming keeps us from uniting as human beings to confront the social, economic, and environmental crises that are boiling over across the world. Arguably, the most dangerous of these crises is the rapid heating of the earth's climate and the consequent mass migration of life-forms trying to adapt to the destruction of their food supplies and habitats. The pollution of air, water, and soil continues to worsen, and the planetary life support systems continue to degrade. Another great crisis is the financial crisis that each day causes more unemployment and underemployment and puts the world's people deeper in debt to a handful of multinational bankers and corporate capitalists, all the while abandoning billions of families and neighborhoods to depression and destitution. These twin crises have spawned other crises in their wake.

There is the "family crisis," in which loneliness, divorce, pornography, drug addiction, and domestic violence become the norm. There is also the "social crisis" where sexism, racism, ageism, homophobia, islamophobia, religious bigotry, and a pseudo culture based on consumerism continue to divide us as human beings. This social crisis is so all-encompassing that it deflects our attention from the money lords who operate their "divide and conquer" strategy from behind a curtain, while billions of people are unable to meet their basic needs. As the unrolling of these crises fill our minds and hearts with fear, society continues to polarize. The Right and Left turn on each other as serious enemies. In the US, we divide along a fault line where the extreme Right wants to bring back the history of white supremacy supported by a policy of law and order and autocratic government. The Left understands that there is no going back and calls for greater diversity and democracy. They want the government to take greater responsibility for protecting all the people. Even so, in the United States, both the Left and the Right are controlled by capitalist political parties, and by the capitalist-owned mass media, neither of which are concerned about ending the US imperialist wars waged for the purpose of stealing resources and enslaving the labor of people of other countries. As these multiple crises interact, they give rise to the "refugee crisis" wherein tens of millions of people are forced to flee from wars, ethnic cleansing, and environmental disasters. Few of these desperate people will ever get their lives back.

Faced with these mounting crises, humanity stands confused, divided, and impotent—our existing ideologies and institutions too limited in scope and too divisive in purpose to unite us in an effort to solve our common problems.

Today, these worsening crises are hacking away at the very support pillars of Western Civilization, weakening its structure and contributing to its demise. The global institutions at the center of the world's political-economy have already begun to fail as they become overwhelmed with problems. As the damage increases and havoc is loosed across the planet like the waves of an incoming tsunami, the situation worsens. Without our ability to change direction, in time, the center will fail, leaving everyone to fend for themselves. This, in fact, is Plan A of the superpowers' militaries—surround the remaining resources with armies and fight it out to the last man.[1]

This characterization of our present predicament is not an anti-utopian fantasy, it is based upon the tide of history. It is based upon our tendency toward sexism, racism, classism, religious bigotry, and national chauvinism. Our existing social institutions and organizations lack the tools to stem this tide because they themselves are based upon divisive sentiments and therefore cannot envision a way out of our predicament. Given the inability of existing paradigms and political organizations to sustain humanity and the planet, we need a new paradigm. We need a universal ideology that seeks the welfare of all of life.

In our second book, *Universal Ideology: The Thought of P. R. Sarkar*, we provide an in-depth look at the major components of a universal ideology. These are: (1) an understanding of social psychology and how the psychologies of the warrior, intellectual, and capitalist classes exist in society and how each has dominated in the different ages of Western history, (2) an understanding of human nature and how individuals and social groups are motivated by instincts, sentiments, rationality, and intuition, and the consequences that result from acting from these different mindsets, (3) an understanding of our spiritual nature, characterized by a thirst for limitlessness, and how it gets perverted into the false longing for infinite wealth and power, (4) a progressive economic model that is based upon economic decentralization, economic democracy, and agricultural revolution, and (5) a progressive political model that is based upon the creation of local social boards composed of local moralists who gain the heart of the people by virtue of their selfless service. This universal ideology has the capacity to inform and guide a nuclear revolution whose goal is to establish a universal human society that can live in harmony with a healthy natural world. A universal ideology addresses our physical, mental, and spiritual needs and offers an optimum set of values and activities to bring us toward human unity. It is comprehensive, hopeful, and provides concrete solutions to our present predication.

While a universal ideology provides us with a vision of how to overcome our present crises, it alone is not enough. We must also be willing to take concrete steps to create a better world no matter where we live or in what condition we find ourselves.

In our struggle to create a better world, we must become the heroes and heroines that we mistakenly thought our religions, corporations, and governments would provide. Each of us must become a leader to create a better world.

The books on History and Ideology are meant to orient us in this task. To create a better world requires that we understand the tide of history so that we can understand all the forces with which we are dealing. It also requires that we share a universal ideology that respects all human beings as spiritual equals. Ultimately, however, creating a better world requires a revolution. Not a political revolution or a cultural revolution, but rather a *nuclear revolution* that addresses the basic needs of our entire humanity as well as the needs of the natural world.

A nuclear revolution is different from the political and cultural revolutions of the past. A nuclear revolution has never been attempted before. The idea of a nuclear revolution is based upon a new interpretation of history, as well as upon the current social and environmental conditions, the scope of which have never threatened humanity in the past. These conditions include the millions of human refugees who have suffered through escalating global wars. They also include the destruction of human habitats due to climate collapse and the wholesale destruction of plant and animal species.

The underlying fact of human history is that, as a species, we possess an unquenchable thirst for limitlessness. This thirst is actually a spiritual urge that we mistakenly attempt to satisfy by possessing limited finite resources. This is verified by man's historic quest for wealth and power. All the money and power on earth can never satisfy our thirst for limitlessness because we will always want more of that which is finite in scope.

The only thing that can satisfy this ache within us is an infinite entity, that is to say, the Absolute Consciousness that is manifest in all forms and yet also transcends forms. The Absolute Consciousness lies within the heart of every finite entity. Everything that is, is fundamentally a form of this infinite consciousness. It lives within each of us as well as outside of us. Everything that is, is contained within it. We are living in an age when even physicists are reaching the conclusion that the world is actually a field of consciousness.[2]

Human history is a story of man's misguided attempts to achieve the Absolute through the accumulation of material wealth and power.

During Ancient History, warrior kings believed they could satisfy their quest for the infinite by conquering greater areas of land and greater numbers of people. This was the age of empire building. During the Middle Ages, the intellectuals, in the form of a priest class, attempted to quench their thirst for limitlessness by controlling the hearts, minds, and even the souls of human beings by telling them they would live in eternal hell if they did not accept the priests as their masters. Again, their motivation for creating such dogmas was primarily to amass wealth and power for themselves. During Modern History, many merchant capitalists mistakenly try to achieve the absolute by hoarding all the money in the world.

Warriors control by physical force; intellectuals control by mental force; and capitalists control by money. These are the three basic ways to have power in this relative world.

It is in the nature of human society to evolve in a cyclical motion in which the warrior age is followed by an intellectual age, which, in turn, is followed by a capitalist age. The end of the capitalist age will then be followed by a second warrior age.

While every society, in every time and place, is composed of warriors, intellectuals, and merchants, the social and environmental conditions differ in each historical age, allowing the class to come to power that best possesses the means to take advantage of those conditions. In this way, the warriors became the ruling class during the age of Ancient History and so forth.

At the beginning of each age, the new ruling class brings new benefits to the masses of people. These benefits accrue during the ascendency of the new ruling class. Once that class achieves the status quo of its golden age, however, it begins to use its privileged position to exploit others, thereby continuing to disenfranchise more people from the system who, therefore, inevitably build the ranks of its enemies.

As the ruling class enters its decline, a new class begins to emerge, gaining greater credence among the disgruntled masses. Then, the old age begins to shift into a new age.

Today we see the capitalist class becoming ever more exploitive of the world's people. The golden age of this class has passed and its decline has begun.

It is at this point that we introduce the possibility of creating a *nuclear revolution*. A nuclear revolution is global in scope. It transcends national or regional boundaries. It transcends religions and other forms of groupism. The entire human race is involved in it because everyone is faced with the same environmental degradation, economic exploitation, and consequent problems.

Having said this, a nuclear revolution is not solely concerned with replacing an old ruling class with a new one. Rather, a nuclear revolution is more profound than that. It addresses not only the physical and mental conditions of this relative world, but also the spiritual course of our individual and collective lives. It sees through the illusion of trying to achieve the Absolute by the possession of wealth and power and sets humanity on a course to build a united human society that exists in harmony with the natural world. As our quest for unity with all of life increases, our understanding of the Absolute becomes greater.

Unless we address the spiritual nature of human beings, we will not be able to advance as a human society. We will merely continue to rotate in the social cycle without any advancement as human beings.

One can only speak of making progress in relation to having a goal. Without a goal, we can have movement, but no progress. The goal of a nuclear revolution is to move us closer to our spiritual goal, which is to merge with the Absolute Divinity. Its purpose is to speed up our evolutionary process.

The quest for Divinity is not something new in human history. Not to be confused with the piecemeal and divisive dogmas of institutional religions, spiritual masters in many times and places have told us how to advance spiritually. Whether from the East or West, their message is always the same. Love God (the Absolute) and love the Creation. The spiritual science of yoga (which means to yoke individual consciousness with Divine Consciousness) was introduced to humanity by Lord Shiva over seven thousand years ago. This science has grown and been refined over time and today constitutes the spiritual foundation of Hinduism, Jainism, Buddhism, and most other religions of the East. This spiritual science develops human intuition, which is even more subtle than reason, and which allows one to communicate with the Divine Self within. In the West, it was the Jewish Essenes and Jesus Christ who taught people how to communicate with the Divinity within. Jesus told his followers, "The kingdom of heaven is within you." To be able to have a relationship with the "Father," Jesus said one must be accomplished in the art of love. He summed up his teachings by saying, "Thou shalt love the Lord thy God with all thy heart, and with all thy soul, and with all thy mind. This is the first and great commandment. And the second is similar. Thou shalt love thy neighbor as thyself. On these two commandments hang all the laws and the prophets" (Matt. 22:37-40).

The great mystics of all places and all times have spoken this same message, albeit using different words and stories. In the wake of some great masters, world's religions have come along to coat their master's spiritual teachings with rules, dogmas, rituals, and ceremonies. At best, these religions provide different styles of worship of the Divine. At worst, the priests and worshippers of these religions condemn others and hold themselves up as superior. Forgetting the commandment to love, they seize on any means to condemn others and, in doing so, believe that they are solely closer to God. They believe this even as they perpetrate the oppression of women and people of other nations, races, or religions. This shows that many religions are the product of intellectuals who create 'divide and conquer' strategies in order to establish their own power. Spirituality is not religion. It is a way of life whose purpose is to unite human beings through the actions of love and self-sacrifice.

There has never been a better time for human beings to join together under a common spiritual worldview and move forward as a united human society.

Our nuclear revolution can only create a united human society by addressing the need for an equitable economy, a just political system, a beneficent culture, and a new way to interact with each other that is cooperative and supportive.

In this way, a nuclear revolution addresses the physical, psychic, and spiritual dimensions of our humanity. In the physical realm, it seeks to ensure that every person on earth is able to secure one's basic needs of food, shelter, clothing, health care, education and to do so in an environmentally sustainable manner. In the psychic realm, it seeks to educate humanity about our common history, our present condition, and our potential as human beings to achieve a Divine status. In the spiritual realm, it provides each individual with a theory and practice for personal realization, and seeks to develop the optimum conditions in society for each individual to merge with Divinity.

In order to move as quickly as possible in this direction, we must be crystal clear in our minds that no existing political, economic, or military institution can save us from the growing chaos that plagues us. We cannot count on any government at the national or state level to solve our problems or to secure our basic needs. They are the institutions of a bygone age. As such, they are helpless to act in any positive manner. They will continue to resist change, while those of us under their authority go about our daily lives watching life decline abjectly while we dream helplessly of a better world. This is our fate unless we create a nuclear revolution.

It follows that a nuclear revolution requires a new kind of leadership. These leaders must know how to effectively channel their thirst for limitlessness to the attainment of Divine Consciousness. (See our book *Universal Ideology: The Thought of P. R. Sarkar* for a description of how this is possible.)

Having accepted the Divine as their shelter, our new leaders will be free to see the Divine in every human being and in every natural form, and to serve the Divine Oneness with each of their actions. Our new leaders will therefore represent each and every gender, race, age, class, nation, and spiritual path. They will also be able to embody the social psychology of the warrior, intellectual, and merchant, as well as the psychology of the laborer who is not afraid to get his or her hands dirty. In this way, our new leaders will be able to best serve humanity regardless of circumstances.

As leaders of a nuclear revolution, we must be willing to accept our own responsibility in perpetuating the crises that currently plague humanity. In supporting existing political and economic institutions, we contribute to our own demise and to the demise of the planet. As all the great revolutionaries in history have told us, including George Washington, Thomas Paine, Sam Adams, and Thomas Jefferson, when the presiding institutions of power are no longer able to support us as free men and women, we no longer owe them our allegiance. In fact, it is our duty to create new institutions to replace those that have become counter-productive and oppressive.

The responsibility for meeting our basic needs and creating a secure natural environment for ourselves, no matter where we live on this weary planet, now falls on us. You and me, our families, and our neighbors.

The revolution will not begin solely on the international or national levels. It will not be led by mainstream institutions or corporations. It will not be featured on TV, nor will its leaders be singled out for praise or reward by social institutions.

To summarize, the purpose of a nuclear revolution is to create peace and harmony across the planet. It is to speed up our process of evolution to reach our highest human potential at the soonest moment. Our highest human potential is to merge with the Supreme Consciousness as taught to us by the great spiritual masters of all times and all places. The method is simple. Learn to love and act from love. God is One and to merge with God is to see the One in everything.

A nuclear revolution involves a constant fight against age-old divisive sentiments based on gender, race, age, class, nationality, and religion. By fighting against those who gratify their egos by slandering and harassing others based on these sentiments, we will create a beneficent culture that seeks the welfare of all.

A nuclear revolution does not begin by toppling a government, rather it begins in the heart of each individual and moves out into the family and neighborhood where moral men and women begin to cooperate to meet the basic needs of the local people. A nuclear revolution expresses itself in all human interactions, be they personal, interpersonal, cultural, economic, political, or environmental. And fundamentally, a nuclear revolution is based upon universal spirituality.

This book on nuclear revolution is intended to transfer our desires, hopes, and dreams of a better world from our imagination onto solid ground. Its purpose is to educate a revolutionary leadership and to provide it with the knowledge, skills, and tactics to combat our fears and lethargy, improve our lives, and unite us in a common cause with families, neighbors, and well-wishers across the planet. A nuclear revolution will help us stem the destructive tide of history and channel our momentum onto a new path that is based upon our highest ideals, our cooperative spirit, and our trust in each other as human beings. This new path will be cleared by our own hands from the ground up.

Chapter One of this book looks at the values of the capitalist ruling class and how these values have led to the gross inequality of wealth that we experience in the world today. It looks at how this inequality impacts the lives of those who are not in the capitalist class, the destruction of the planet's life support systems, and the multiple crises that have been set in motion by the capitalists' end-game strategy. It concludes by presenting a general perspective that is required by a revolutionary leadership if they are to create a nuclear revolution.

Chapter Two defines what a human society is and how it evolves thru time and space. It exams P. R. Sarkar's theory of the Social Cycle and how western society has progressed through different ages that were controlled by different ruling classes. The chapter also discusses in depth the mechanisms by which society moves and how a revolutionary leadership, in understanding these mechanisms, can create a strategy for change in which success or failure can be articulated and measured in concrete scientific terms. It presents the Wave Theory of P. R. Sarkar, in which he demonstrates that neither motion nor station are absolute states, but relative and intermittent to each other. He also demonstrates that neither matter nor ideas are absolute, but are also relative. Mr. Sarkar also discusses the motion of Absolute Consciousness as it progresses from pure consciousness through the mental and physical vibrational states, including the evolution of species and the place of human beings in this motion. The chapter ends with a discussion of the relationship of the individual to the collective and postulates that because the capacity to become God-realized is an attribute of individuals, our new social institutions should be created to provide every opportunity for God-realization to occur.

Chapters Three through Six look at examples of social revolutions in history and in contemporary society. It looks at revolutions against the the western capitalists in Latin American and the eastern capitalists/communists in Eastern Europe and Hong Kong. The Chapter also looks at the revolutions in South Sudan, Greater Kurdistan. In some of these examples, we also see how revolutionaries go astray from fighting for the welfare of the people to betraying their own people due to a lack a universal moral integrity and how success can fail without a grand strategy for success.

Chapter Seven addresses the current Socialist dilemmas in the US and in the world and calls for a unity of revolutionaries from all progressive organizations and political parties working for social change, particularly those doing concrete work on the local level to improve their communities. Currently, the socialists have no viable economic models on the ground, with the possible exception of Cuba and some of the work being done in greater Kurdistan. Another problem with socialism, as it exists, is that socialists do not distinguish between religions and spirituality. This is a significant problem that leaves socialists with an incomplete view of life, stunts their understanding of human potential, and alienates them from great numbers of people.

Chapter Eight unpacks the components of a nuclear revolution and demonstrates how it is significantly different from the political, economic, and cultural revolutions in human experience. It explains how a nuclear revolution is based upon universal spirituality, a universal morality, and the organization of people at the local level. It reviews how revolutionaries need to not only fight for political, economic, and cultural changes, but also for personal enlightenment, positive interpersonal relations, and the security of the natural environment. A nuclear revolution is holistic. It seeks the ultimate balance between our physical, mental, and spiritual propensities, between the individual and the collective, and between human beings and the natural world.

Chapter Nine examines the conditions that are required for a nuclear revolution to occur. These include: (1) the presence of wide-scale exploitation, (2) a universal ideology, (3) qualified leadership, (4) trained cadre,

(5) a revolutionary organization, and (6) a revolutionary strategy. The chapter then focuses specifically on the creation of a revolutionary leadership, the training of cadre, the building of a revolutionary organization, and the issue of strategy. The chapter then looks at the role of students, youth, women, and people of color in a nuclear revolution.

Chapter Ten looks at the Grand Strategy of a nuclear revolution, consisting of multiple campaigns which are complimentary and must be advanced simultaneously. It describes in detail the Personal Campaign, the Interpersonal Campaign, the Socio-Economic Campaign, the Cultural Campaign, the Environmental Campaign, and the Political Campaign.

Chapter Eleven explores the psychology of a nuclear revolution. Because a nuclear revolution is based upon a polarization strategy, our leadership must understand the principles of taking the opposite stance in battle. This principle applies not only to the physical world, but also to the mental and spiritual realms. The chapter also discusses the need for a universal ideology, holistic thinking, and spiritual devotion, which leads to the greatest results in this world.

Chapter Twelve looks at the components of an intellectual revolution. A nuclear revolution will require both intellectual and physical components. An intellectual revolution is based upon the ideal of Cosmic Ownership, which means we may enjoy and utilize the objects of this world, but we cannot declare that any item is private property. Everything is our common inheritance. The chapters looks at the strategies to mobilize the masses by nonviolent action detailed by Gene Sharp in his book *From Dictatorship to Democracy* and includes tactics for protest and persuasion, noncooperation, and interventions to change the capitalist system. The chapter also proposes that because suffering humanity cannot afford to wait for this slow process of change, the revolutionary leadership may create a revolutionary consciousness among the people and sound a clarion call for physical revolution.

Chapter Thirteen discusses next steps in the process to get from the present capitalist system to nuclear revolution. First and foremost is personal transformation, which includes adopting a universal ideology, and performing spiritual practices like meditation, contemplation, devotion to the Supreme, yogic postures, good diet, etc. Other steps include continuing the fight for environmental protection, using sustainable energy, fighting against all expressions of sexism, racism, capitalist dominance, national chauvinism, and religious bigotry, while organizing at the local level to create People's Economies that meet the basic necessities of local people. The chapter also recommends further study in these areas and provides examples of models to emulate.

Chapter Fourteen discusses the need for defending the revolution and the theory and practice of using physical force against an aggressor. It distinguishes between the need of using violence and force against an enemy vs an intent to harm one's enemies. As a nuclear revolution is created mainly from the local level, so too a defensive force may have to be created from a local militia who are tasked with defending the local People's Economies. The chapter looks at rules of combat, the role of policy, strategy and tactics in Sun Tzu's *The Art of War*, as guidance for developing a defensive strategy.

Chapter Fifteen discusses the relationship between local socio-economies, a united human society, and a one world government. It looks at the components of a universal constitution, legal and corrective systems, and how local self-reliance in meeting basic needs is achieved. It presents a vision of a world that works for all.

Chapter One: Why We Need a Revolution

At present, the worth of life is ascertained with an eye to money. One who possesses wealth commands respect and passes for a noble person. Nowadays, if you have no money, you cannot expect to get honor. The poor people belonging to all other classes have to woo the moneyed people just for the sake of earning a livelihood. Human values are taboo here. Human beings are now but a medium of earning money. The rich have purchased all the brains with their money. They have bidden farewell to all their human values and at the same time they are busy playing a chess game with the lives of the other members of society. Bereft of everything, the people toil around the clock to live from hand to mouth. Those who are at the helm of society are absorbed in strife and suspicion, or in reckoning their loss and profit. They have no desire to think of humanity. In order to gratify their self-interest, they are ready to chew the humans' bones. There is no room for mercy, sympathy, or fellow-feeling. All are self-centered. P. R. Sarkar

THE CASE FOR REVOLUTION begins with an understanding of the utter disregard and the brutal exploitation of the planet and its people by the capitalist elite. Today, in the United States, capitalist organizations like the Trilateral Commission, the pseudo-Christian, Republican Party, and the neo-liberal Democrats lay bare the inhumanity of the capitalist leadership and their sociopathic destruction of people's lives and the natural environment to secure greater amounts of wealth and power for themselves. Our governments are controlled by billionaires and millionaires who themselves are members of the elite capitalist class whose interests they represent. The wealth that they exploit from the worlds' people and natural environment are the main cause of the psychological suffering and impoverishment of the human race.

In the United States, due to stress of growing poverty, some men beat their wives and white supremacists oppress blacks, Native Americans, and Chicanos. Due to increased poverty, white nationalists are rising up across Europe and North America. Due to increased poverty, fundamentalist religions across the world prosper as they spread hatred and division among people.

The so-called communist countries are not a sanctuary from this capitalist mentality. In fact, Russia and China are now controlled by capitalists as well. In Russia, Putin gave the wealth of Russia to the capitalist oligarchs in exchange for their support of his government. In China, the leaders of the communist party created a capitalist economy with themselves as the principle recipients of the profits.

The Underlying Values of the Capitalists

The capitalist class did not always exist as the ruling class. Like everything else in this relative world, it came into existence, grew, achieved its status quo, and now is in the process of decline and death. The story of the rise of the capitalist class to a position of power began in the Middle Ages during the time when the Catholic Church controlled Europe. With the onset of the Protestant Revolution, the Catholic Church began its decline. The ranks of Protestant revolutionaries were not only comprised of priests and other religious intellectuals, but also of the merchant class, who wanted to overthrow the Church and the landed aristocracy that were forcing them to pay exorbitant taxes in order to continue to do business. When the merchant class was finally able to establish its own power through the revolutions that began in England, the United States, and France, they eliminated the dominance of the Church and aristocracy by instituting their newly established "constitutional" governments.

The separation of Church and State began in Europe and reached full expression in the First Amendment of the American Constitution. Eager to prevent the religious wars that were continuing to ravage Europe, the Founding Fathers of the United States prevented any religion from gaining dominance in the government, but also conceded that "Congress shall make no law respecting an establishment of religion, or prohibiting the free exercise thereof. . . ."[3] While the bourgeois leadership in the United States did not oppose the Protestant religions, they wanted to ensure that no religion would ever again rise to a position of political dominance. While considered to be progressives at the time of their bourgeois revolution, today, due to their predatory and exploitative behavior, the capitalist class has since become regressive and counter-revolutionary.

Today, in the United States, the Republican Party represents the rule of white nationalists whose objectives include the suppression and control of women, the harassment and suppression of people of other races, and the exhibition of national chauvinism, all of which are cloaked in the wholesale divisiveness of a false portrayal of Christian values. As such, the Party is supported by southern evangelical Christians, many of whom were slavers not long ago. These Christian fundamentalists link the national chauvinism of white rule to the Christian religion by claiming that the United States is a white Christian nation and therefore should be ruled by white Christian capitalists. Because the United States is the supreme hegemonic power on the planet, these destructive values impact the entire world.

The idea of the Protestant Work Ethic, which has long since become a fundamental concept in western sociology and economics, continues to support the dominance of the capitalist class. This idea was derived from John Calvin, a Protestant lawyer and churchman who preached that the new merchant class at the time of the Reformation constituted the chosen people of God and that everyone else had better work hard and demonstrate discipline and frugality under penalty of eternal damnation. This new "work ethic" replaced prayer and the intervention of Catholic priests as the means to know and serve God.

Today, this false ideology justifies capitalist exploitation by allowing them to maintain the myth that they received their wealth as a reward from God because of their hard work. As such, they claim, anyone can do the same. They see charity as a sin against mankind because it encourages laziness. This is still the argument used by capitalist Christian Republicans and neo-liberals today to condemn government sponsored social service programs. This divisive mentality stands in complete opposition to everything that Jesus Christ represents. Thus, these so-called

Christians are Christians in name only. The capitalists' claim that they received their wealth because of hard work is also a fabrication because in reality they received their wealth by enslaving their fellow man and confiscating their natural resources for self-gain.

Individualism is another idea at the root of capitalism. At first it was an extremely empowering ideology for people because it challenged the tyrannical hold of the Catholic Church over European society. It allowed for freedom of thought after centuries of being subject to stultifying dogmas. This value has since degenerated into a license for capitalists to exploit others and to justify their individual right to unlimited accumulation even to the detriment of humanity and the natural world.

Capitalism, which was given its religious justification by men like John Calvin, was further justified in the economic realm by people like Adam Smith and David Ricardo. During the late nineteenth century, it was further ingrained into our social consciousness by the development of the theory of Social Darwinism whose proponents argued that Charles Darwin's theory of "survival of the fittest" should apply to human social conditions as well.

It began with Thomas Malthus arguing that the rate of population growth would outstrip the food supply. As such, charity would only exacerbate social problems. Frances Galton, the father of Eugenics, claimed that welfare and social service programs, like mental hospitals, were allowing inferior human beings to survive and impose on the superior human beings. This furthered the Calvinist argument that charity was a social evil while the definition of inferior humans quickly became a definition of people of other races, particularly blacks. Instead of fostering a vision of a cooperative society, such ideas provided the rationale for the capitalist class to exploit the so-called "inferior" people of other races. Of course, white workers were exploited too, but they had privileges, like better paying jobs, if they were willing to work hard for their capitalist owners.

Materialism is the corner stone of capitalism. This philosophy contends that the human mind is only a reflection of the material condition. The denial of spiritual or mental progress reduces people to become little more than laborers and consumers. We have no identity beyond this as far as the capitalists are concerned. The idea that human beings are children of the Divine is completely replaced by the belief that the value of money becomes the quantitative measure of a person's worth. There are now only laborers/consumers and their owners.

This materialist environment gives rise to "value free" science and technology, which essentially has come to mean the immoral or amoral manipulation of matter in the service of wealth accumulation.

These, then, are the "practical" values that the capitalists use to justify their control over nature and humanity.

What Exactly Are We Up Against?

The capitalist mode of production, which reduces all of life into separate commodified objects that depend upon money for their transaction, has allowed a handful of bankers and the owners of multinational corporations, along with their elite military and intellectual guardians, to keep the rest of us in bondage. The capitalist elite (the 1%) who own the means of production—capital, technology, raw materials, and labor—have imposed their values on society and on the global economy, so that we actually know nothing of our true potential as human beings nor do we have any incentive to cooperate with each other to solve our common problems. The capitalist values discussed above are systematically promoted by the mass media so that they become the underlying values in our collective consciousness. These values are also perpetrated by so-called Christian religions that lead us to believe that our reward will be in the "after life" and that it is not only futile, but against the law of God, to challenge the capitalist status quo.

Those who struggle to gain freedom from the exploitive and oppressive control of the capitalists must contend with the military force and the "policy-oriented intellectuals" who protect their system. The US military (internationally) and the police (domestically) are used to keep individuals and groups "in line" who would challenge the capitalist system. This includes organized labor, women, and minorities, as well as the residents of other countries who want to maintain control over their own resources. This is the underlying reality despite the mass media "fake news" that our military and police are primarily fighting to maintain democracy.

The dominance of the values of the capitalist mindset feeds the pseudo culture of consumerism, which is seductively portrayed through TV programs, films, relentless advertisements, etc. In this pseudo culture, the purchase or theft of "things" replaces human interaction and God-realization as the primary justification for human existence.

Although human beings have lived on this planet for nearly two hundred thousand years, it has only been since the Industrial Revolution, which began around AD 1750, that the merchants became the ruling class and have developed the power to push the Earth to its current state of imbalance. Scientists today argue about whether or not our basic life support systems on the planet (arable soil, drinkable water, breathable air, protective ozone, and diverse forests) can withstand another fifty years of destruction before they collapse. Others forecast the fall of Western Civilization and a mass die-off of species by the end of the century.[4] If this is so, in less than three hundred and fifty years, the capitalists' obsessive thirst for profit, power, and material goods will have reduced the Earth to a point beyond repair and destroyed the ability of the human race to meet its basic needs. There are signs all around us that this process has begun and is accelerating. In the time that we have left, we must begin our revolution with speed and clarity if we want our children to have a future.

The Fruits of Capitalism

Let us begin with a simple fact concerning the gross spiritual depravity of modern capitalism. Today a mere eighty-five rich people own as much wealth as 3.5 billion people living on the earth.[5] This is a pernicious imbalance of God-given resources. Let us be clear, the mountain of wealth controlled by the capitalist elite was not earned by hard work. Rather, it was stolen from people through slavery, colonialism, war, and enforced debt. These elite capitalists care nothing about human beings. They have been destroying people's lives and the lives of their children ever since they came to power. Today, their system for extracting profit resembles a giant vacuum cleaner sucking up every penny that we have. Seven out of ten people live in countries where economic inequality has increased in the last thirty years. In the US alone, the wealthiest one percent captured ninety-five percent of all wealth that has been generated since the financial crisis in 2009. At the same time, the bottom ninety percent became even poorer. The American middle class has shrunk.[6] This is the gratitude that the American people received for bailing out the banks. President Trump, through his "Tax Cuts and Jobs Act" and his Covid 19 stimulus money engineered a wholesale transfer of trillions of more dollars to the capitalist class from the public treasury. This wealth transfer has now led politicians to claim that the federal government does not have enough money for health care, social security, and climate change abatement.

Yet the underlying reality is that the personal wealth of individual capitalists, not counting their corporate wealth, includes seven trillion dollars stored in off-shore private bank accounts alone. If this hoarded personal wealth was distributed in a just and moral manner, every single human on Earth would be able to have their basic needs met and live in a healthy environment. Even if we gave the handful of elite capitalists ten times the average wage required to earn a middle-class income, we could still eliminate hunger, homelessness, illiteracy, and meet people's basic needs over the entire planet.[7]

But this is far from the case at present. Even in the US, people cannot find jobs and even if they do, the jobs often do not pay enough to meet basic needs. How many people are unemployed? It is hard to tell because the government's statistics are intentionally misleading. While we are led to believe that the government's unemployment rate is considered to be a good indicator of where the economy stands, what it hides is also quite revealing. Looking at the unemployment rate, for example, will not tell us how long the unemployed have been unable to find work, or how many quit looking for work out of desperation and are no longer counted in the workforce. It does not reveal anything about the vast number of people who are underemployed and who have to work more than one job to make ends meet.[8]

A recent survey from PayScale found that forty-six percent of workers or roughly twenty-two million Americans say they are underemployed. This means that they're working in full time jobs or part time jobs that do not pay enough for them to meet their basic needs.[9]

A recent report from the Brookings Institute found that low-income households and middle-income households now must spend roughly eighty percent of their budget on housing, food, transportation, health care, and clothing. This does not include child care, household furnishings or supplies, education, taxes, car emergencies, etc. For low-income households, forty percent of their budget went to housing alone.[10]

Many Americans are caught in the trap between unemployment and underemployment as this young mother informs us. "There was a point in time where I was on food stamps and I was eligible. As soon as I became employed, I was not eligible for anything and I was just cut off. So you have to pick yourself up, you gained a job, you gained employment, but you're still picking yourself up," she said.

Many people are in this hellish situation. They just earn enough to be ineligible for the safety net programs that are out there. But yet their income from their job is not enough to live on. Even in industrialized east coast states, this is a major problem. For example, 1.2 million people or 37% of households in New Jersey cannot afford basic needs. In 2007, 600,000 households could not afford them. It was 800,000 in 2014 and in 2017 it was 1,200,000.[11]

This problem is not about minimum wages, it is about capitalist exploitation. The mainstream media would have us believe that families who struggle are not working hard enough, but this is not the case.

In a country that is truly "of, by, and for" the people, we would see all incomes growing; we would see people having more money to spend on things beyond their basic needs; and we would see their purchasing capacity increase. People would have savings. Why? Becomes our productivity as a country keeps growing. The problem is that the people do not reap the rewards for their hard work. The profits go strictly to the capitalists.

Even for those of us who have decent jobs, we must grovel in submission to the multinational capitalists. God forbid we complain about our situation unless we want to be fired. To the capitalists, giving a worker his or her fair due would only mean less profit for themselves.

Historically, as the capitalists were able to amass enough capital through exploitation, slavery, and colonialism to own the factories, they were able to control human labor. Today, they control us by ensuring that our wages are set so low that an ever-growing number of us cannot meet our basic needs and are forced into debt to survive. Once in debt, even an entire state or nation becomes the property of the capitalists, including its human beings, its infrastructure, and its natural resources.

Capitalism today pits a small capitalist ruling class (the 1%) against humanity as a whole. The institutions of capitalism, which include the federal governments of the nation-states, their armies, and the multinational corporations, exploit natural and human resources regardless of the consequences to humanity and the environment. The system turns the natural world into raw materials and sells them back to us as consumer goods. In this way, the capitalists commodify and monetize everything, and in so doing, enslave us to a system where we need their jobs in order to provide our families with our basic needs. In manufacturing these products, the capitalists build in depreciation so that additional profits can be made because we are always forced to buy new products. Consequently, we are swimming in junk products as the world's natural resources are plundered. To accommodate this system, landfills grow to over-capacity as our natural world dies from chemical and plastic pollution.

This analysis was written before the onslaught of Covid 19, which has only put the world's people at a further economic disadvantage and allowed the capitalist class to syphon even more wealth from the people and natural world.

Capitalism and the Natural World

The planetary life support systems have already greatly declined. Our ozone layer, which protects us from harmful rays from the sun, has been depleted by man-made chemicals. The loss of the ozone layer causes skin damage, including aging, cancers, eye disorders, and many other conditions.[12] It is also dangerous to plants and animals.[13]

Our air supply is being polluted primarily from the burning of fossil fuels, which the oil and gas industry fights to protect from the increased use of sustainable energy. The World Health Organization reports that the burning of fossil fuels and the out-gassing of household chemicals kills over seven million people a year.[14]

The destruction of drinkable water is also reaching crisis proportions. According to the UN, 75% of planet Earth is covered in water. 97.5% of that is ocean and 2.5% is freshwater. Seventy percent of freshwater is divided into glaciers and ice caps and the remaining 30% into land surface water, such as rivers, lakes, ponds, and groundwater. Most of the freshwater resources are either unreachable or too polluted, leaving less than 1% of the world's freshwater readily accessible for direct human use. Today 70% of this unpolluted freshwater is used for irrigation, 22% for industry, and 8% for domestic use.[15]

Every time an oil and gas company fracks a well, we lose between two to eight million gallons of water that can never be used again because it is now polluted with thousands of deadly chemicals.[16] A well may be fracked multiple times, with each frack increasing the chances of chemical leakage into the soil and nearby water sources. In June, 2017, for example, an explosion at a Halliburton fracking site in Ohio caused thousands of gallons of toxic chemicals to spill into a tributary of the Ohio River, which supplies drinking water for millions of residents. More cancers on the way. Immediately more than seventy thousand fish died.[17]

But this is a small problem compared to the water that is becoming increasingly polluted every year by fertilizers and pesticides, sewage and food processing waste, lead, mercury, and other heavy metals, chemical wastes from industrial discharges, and chemical contamination from hazardous waste sites. Today, polluted, unsafe water is the leading cause of sickness and death, with half of the world's hospital beds filled with people suffering from water-related illnesses.[18]

Our soil is another fundamental element in the planetary life support system. Like water, it is also being rapidly degraded due to a myriad of industrial processes including oil spills, mining, acid rain, agricultural chemicals, industrial waste, illegal dumping, war zones, etc. Much of it is also destroyed by the expansion of human settlements.

Our soil gives birth to all of life. It is made up of countless species that create dynamic and complex ecosystems that feed our plants and in turn all animals and humans. Under our present economic system, half of the topsoil on the planet has been lost in the last one hundred and fifty years. Soil quality is also being destroyed by compaction, loss of soil structure, nutrient degradation, and soil salinity.[19] Consider that it takes one thousand years to produce three centimeters of topsoil and, if current rates of degradation continue, all of the world's top soil could be gone within sixty years.[20]

Our forests are also being destroyed at a rapid rate. Today, forests cover thirty per cent of the Earth's surface and provide food, medicines, and construction materials, as well as produce oxygen for us to breathe. They provide homes to indigenous populations and to more than eighty per cent of all species of animals, plants, and insects on the planet. Every year, we lose over thirty-two billion acres of forest. In addition, the persistent degradation of drylands has led to the desertification of nine billion more acres. According to the World Resources Institute, which measures the state of the world's resources each year, more than eighty percent of the Earth's natural forests have already been destroyed and eight of the eighty-three hundred known species have already become extinct and an additional twenty-two per cent are at risk of extinction.[21] A report by the United Nations in May 2019 found one million species to be at risk of extinction.[22]

Finally, we need to mention the global warming of our atmosphere caused by the burning of fossil fuels. Despite the propaganda put out by right wing politicians, and the think tanks funded by the fossil fuel industry that deny that climate change is real, over four thousand studies published in peer-reviewed scientific journals show that ninety-seven percent or more of actively publishing climate scientists agree that climate-warming over the past century is due to human activities.[23] In addition, most of the leading scientific organizations worldwide have issued public statements endorsing this position, among them the American Association for the Advancement of Science, American Chemical Society, American Geophysical Union, American Medical Association, American Meteorological Society, etc.[24] Very little of this information is covered by mass media.

Climate change also increases incidents of floods, storms, droughts, forest fires, the melting of ice caps, the rise of sea levels, the migration of germs, insects, plants, and animals and so many other phenomena in the natural world. The major floods caused by hurricanes Katrina, Sandy, Harvey, and Maria do not exist in a vacuum. Rather, they reflect a process of increasing weather intensity due to the climate heating up.

It is typical that Exxon-Mobil and Shell Oil, the largest fossil fuel corporations in the US, knew about the dangers of global warming from its own research, yet denied it in their advertising to the public.[25]

The Trump administration accelerated climate change by ending environmental legislation, reinvigorating the fossil fuel industries, and suppressing renewable energies. It withdrew support from the international Paris Agreement and spread confusion among nations at a time when we most needed a unified global action to reduce climate change.

If climate change is a hoax, then Houstonians recently got fifty inches of hoax dumped on their soaking wet heads.[26] The people of Texas were defenseless against Hurricane Harvey having been told by FEMA, insurance companies, and the fossil fuel giants that there was very little chance of flooding when they bought their homes. Now their homes have filled up with rising flood waters contaminated by sewage and deadly chemicals. Because FEMA had declared large portions of southeast Texas to be safe from major flooding, only twenty percent of the people displaced by Harvey had flood insurance.[27]

These victims of corporate capitalism will return to their damaged homes of which tens of thousands will be unsalvageable. Unfortunately, as the climate deniers continue to press their lies on the people, scientists believe that Texans are going to experience another one hundred-year or five hundred-year flood that will strike within the decade, even before the evacuees of Harvey have been resettled. We can only expect that the number of environmental refugees and wrecked lives will continue to grow.

How do meteorologists know that Hurricane Harvey was influenced by climate change? The answer is actually quite simple. Scientists know that the water-holding capacity of the atmosphere increases about seven percent for each 1.8 degrees Fahrenheit. In simple terms, the warmer the air, the more water it holds and thus the bigger the storms. The temperature of the oceans is also rising. It was this increase in heat from the air and the water in the Gulf of Mexico that turned Harvey into the monster that it was, having dumped over two hundred and seventy-five trillion pounds of water on Texas.[28] Caught up with the Hurricane Harvey disaster, Americans did not even realize that at the same time, the people of Mumbai, Ontario, Macau, Niger, Hong Kong, Istanbul, Ireland, and Sierra Leone were also flooded by severe storms.[29]

Such catastrophes result from the capitalist system that runs on fossil fuels, and which the ruling capitalists will fight to maintain even at the deaths of billions of people. They will not stop it, nor will it stop on its own. We, the people, must put an end to it.

By continuing to view the earth's life support systems as nothing more than raw materials, we are literally destroying our home. We extract resources beyond the capacity of the planet to renew itself. As the planetary systems continue to decline, the speed of societal decline also quickens. This downward trend is on a collision course with an expanding human population that will inevitably require more goods and services, more living room, and thus more natural resources for its survival. This collision has already begun. And the tensions one observes in the world today are the results of this crisis.

The entire region of the Middle East, for example, has turned into a raging hell because the US capitalist elite made their plans to control Eurasia after the fall of the Soviet Union.[30] The first step was to control the states having oil reserves. Today the US has initiated wars directly, or through its surrogates, in Iraq, Afghanistan, Libya, Kuwait, Yemen, and Syria. The Trump administration added Iran to the countries to be dominated. The strategy of waging a relentless war on many fronts against so-called "terrorism" has cost millions of people their lives and their futures. But the reason for these wars was never about terrorism, it was about US capitalist control of the oil reserves in the Middle East. The industrialized world and its armies cannot function without oil and the US, Europe, China, and Russia are circling to ensure that their populations have access to the remaining supplies. The people of the Middle East, ravaged by poverty and war, pay the price. There should be no doubt as to why some people are fighting back against the US and its allies. The US government calls these people "terrorists." The American people do not understand that the US army created the terrorist organization ISIS when it attacked Iraq and destroyed the government of Saddam Hussein. In doing so, his army, along with other insurgents, formed ISIS to fight the invaders of their countries. This has meant the death of thousands of American soldiers. According to Global Research, the persistent wars initiated by the US military on behalf of the multinational capitalists have resulted in the deaths of millions of people. It should not surprise us that every death means astonishing profits for the US war industry which is a partner in crime with the fossil fuel industry.[31]

In fact, every industry controlled by the large US-based multinational corporations is in large part a purveyor of death and destruction. It is not just the war industry or the fossil fuel industry or the banking industry that is the enemy of the world's people. It is also the food, medicine, housing, education, energy, and transportation industries. The entire system is cancerous. For example, while grains provide seventy-five percent of the world's diet, companies like Monsanto, DuPont, Cargill, etc., promote an agricultural system that relies on genetically modified seeds, fossil fuel-based fertilizers, pesticides, insecticides, and herbicides that destroy the soil, plant, and animal life, and even the people who eat the food produced in this system. The four leading causes of death in the United States are now food-related diseases—obesity, heart disease, diabetes, and a range of cancers related to agriculture chemicals and toxic food additives.

The CEOs

CEOs are the chief executive officers of the large corporations. They are the frontline capitalist managers whose status is only exceeded by the elite capitalist class that sits on their boards and invests in their companies. Such people will certainly never raise the flag of revolution because they are the gatekeepers whose job it is to increase the wealth of the capitalists and protect them from public condemnation. As such, they are rewarded with immoral amounts of wealth extracted from workers' labor in the form of salaries and bonuses.

In the chart below, we can see how much CEO's make compared to their employees. In the first example, the CEO earns 434 times what an average worker earns. But this only refers to a difference in salaries. If we look at the last column, we can see the Total CEO compensation due to additional bonuses, stock options etc.[32] In this case, the ratio becomes 819:1.

Company Name	CEO	Median Worker	CEO	Pay Ratio	Total CEO
		Annual Pay	Annual Pay	(Cash)	Compensation
		(Cash)	(Cash)		
CVS Health Corp	Larry J. Merlo	$27,900	$12,105,481	434:01	$22,855,374
CBS Corp	Leslie Moonves	$59,900	$23,652,883	395:01	$56,352,801
Walt Disney CO/	Robert A. Iger	$71,400	$26,208,003	367:01	$43,490,567
TJX Companies Inc /DE/	Carol Meyrowitz	$22,400	$7,330,584	327:01	$17,962,232
Twenty-First Century Fox, Inc.	K. Rupert Murdoch	$54,800	$17,047,636	311:01	$22,192,923.
Comcast Corp	Brian L. Roberts	$55,800	$16,819,942	301:01	$27,520,744
L Brands, Inc.	Leslie H. Wexner	$33,900	$9,665,925	285:01	$26,669,306
Honeywell International Inc	David M. Cote	$81,600	$22,767,851	279:01	$33,105,851
Pepsico Inc	Indra K. Nooyi	$61,500	$15,937,828	259:01	$22,189,307
Wynn Resorts Ltd	Stephen A. Wynn	$50,100	$11,930,391	238:01	$20,680,391
Goodyear Tire & Rubber Co	Richard J. Kramer	$54,700	$12,779,784	234:01	$17,772,128
Allergan Plc	Brenton L. Saunders	$92,900	$21,565,325	232:01	$21,565,325
Viacom Inc.	Philippe P. Dauman	$79,400	$18,263,525	230:01	$54,140,509
Cablevision Systems Corp /NY	James L. Dolan	$62,000	$13,649,028	220:01	$24,539,725
Liberty Media & Liberty Interactive	Gregory B. Maffei	$51,100	$10,947,607	214:01	$26,868,931
Wal Mart Stores Inc	C. Douglas McMillon	$24,600	$5,133,256	209:01	$19,404,042
Lennar Corp / NEW/	Stuart A. Miller	$61,700	$12,813,693	208:01	$17,909,693

Time Warner Cable Inc.	Robert D. Marcus	$51,700	$10,699,868	207:01	$18,034,506
Starbucks Corp	Howard Schultz	$28,000	$5,722,076	204:01	$20,091,353
Amtrust Financial Services	Barry Zyskind	$53,800	$10,942,190	203:01	$13,942,205
General Electric Co	Jeffrey R. Immelt	$86,400	$17,434,376	202:01	$26,637,142
Time Warner Inc.	Jeffrey L. Bewkes	$78,200	$15,610,633	200:01	$31,493,211
Ameriprise Financial Inc	James M. Cracchiolo	$50,100	$9,901,300	198:01	$19,372,300
T-Mobile Us, Inc.	John J. Legere	$60,700	$10,782,502	178:01	$24,457,987
Omnicom Group Inc	John D. Wren	$84,300	$13,496,047	160:01	$23,576,047

Despite this travesty of economic justice, CEOs of the five hundred largest multinational corporations got another eight hundred thousand dollars pay raise in 2019.[33] Because the capitalists' single focus is to increase profits, all public corporations have a fiduciary responsibility to increase their stock value despite any damage they might do to their employees or the natural world. This legal requirement prevents even well-meaning chief executives from voluntarily increasing wages or reducing pollution, if doing so threatens profits. This can be seen when American Airlines CEO, Doug Parker, recently tried to raise workers' salaries and was immediately slapped down by Wall Street investors who began to sell the airline's stock in the company. Even in this highly profitable industry, the capitalists found it unacceptable to spread the wealth. They believe in their distorted minds that any increase in corporate wealth is due exclusively to themselves. Profits are their God-given right. This is why JP Morgan condemned the pay raise as a "wealth transfer of nearly $1bn" to workers.[34]

Impact of US Capitalism on the World's People

The poverty and misery of people in any country are the sins of the leaders. True leaders should always be vigilant and think how to work best for the human society; they must be ever cautious that under their guidance the people are not led to darkness, death, and immorality. P. R. Sarkar

For the reasons mentioned above, the people of the world can neither depend on the leaders of multinational corporations nor federal governments to relieve their suffering. Rather, these institutions, which are controlled by a small class of capitalists, continue to exploit the earth and its people, the consequences of which we daily witness around us. Capitalism means deprivation, poverty, and warfare on a massive scale. It means watching our children die from illness and starvation as the world's resources become increasingly scarce.

In an essay titled "Apocalypse," Junot Diaz observed that Haiti had entered the "zombie stage of capitalism," where people, having nothing, move around in a manner that is not quite alive. Many other countries have also entered the zombie stage of capitalism. Africa, Central America, South America, the Middle East, the Far East, the countries of Eurasia, and even Europe and the US all have large populated areas where people, sick with malnutrition and brain fog, due to dire poverty, live their directionless lives, unable to change their conditions. The capitalists suck the life's blood from these people, many of whom, in trying to survive, leave their homelands and their families, only to be denied access to other countries. Forced into refugee camps with thousands of others, they have no hope to rebuild their lives. They have nothing to do but wait to die.[35] Today, tens of millions of human refugees find themselves in this situation and the numbers grow every day. This is not some horror movie fiction. This is reality.

There is no nation or international government to which people can turn for solace. Even the United Nations, whose initial goals were to safeguard peace, protect human rights, and promote economic and social progress, is powerless to heal humanity because it has no authority on its own. It merely represents the interests of the nation-states, and more particularly the capitalist nation-states of the United States and Europe, who remain detached from the suffering of humanity because it does not profit them to stop their exploitation of labor and resources that are the cause of this wide spread human misery.

The Multiple Crises of Capitalism

The exploitation by the international capitalist class continues day by day to spin off multiple crises. In addition to the economic crisis that has manifested in unemployment, war, imperialism, international debt, currency imbalances, the growth of the military industrial complex, etc., there are also other crises.

Massive poverty due to unemployment and underemployment has given rise to a family crisis that has even touched the lives of middle-class Americans. Today, the unbearable stress our families are experiencing, due to their inability to earn enough money to meet their basic needs, has led to a drug and opioid epidemic across the country and the rise of suicide, especially among young men. It has led to an obsession with porn and the heightening of domestic violence. Under such pressure, our children are not cared for in a manner that will allow them to grow up to be healthy and intelligent human beings.

Industrialized societies are now beginning to suffer from the same inbred sentiments that gave rise to fascism, racism, sexism, and ultra-nationalism prior to WWI and WWII. The fear for one's survival leads to greater inequality and this inequality is fed by religious bigots and political opportunists. The mass media fans the flames of mutual distrust between the Left and the Right while the capitalist class remains insulated behind the politicians and military elite who manage their divide-and-conquer strategies for them.

The overlapping of these crises, including the environmental crisis, is increasing the speed of societal collapse. Everyone knows that things are not getting any better, but we do not know what to do about it. We cannot see a way out. Even as our capitalist institutions fail us, socialism and communism have also failed us. Neither can we turn to religious leaders because they are unable to think in terms of the benefit of all. As such, we watch hopelessly as life unravels around us leading to greater human suffering with each passing day.

For all these reasons we need a nuclear revolution that is based upon a Universal Ideology. We already know in our heart of hearts this is true. Yet, we try to block it out because we do not really know what to do and we fear violence. But the violence pointed at us by the capitalist ruling class has been perpetrated for years, decades, and centuries.

While revolution is certainly approaching, it is not yet here. In the United States, the white middle class still remains privileged compared to most of the population of the world. We travel in functioning cars, live in comfortable homes, and have credit cards to keep the debt collectors temporarily at bay. We assuage our pains and frustrations by purchasing more material goods. Even though this behavior does not bring lasting relief, we still follow our conditioned response to buy more.

As long as we are able to purchase temporary relief from our pain, we will not change our behavior. When our ability to buy distracting toys ends, however, our mood will turn sour and we will then focus on our true enemy, those who have been lying, exploiting, and misleading us all along. Let us hope by then that a new revolutionary leadership has emerged to temper the violence and guide us in our quest for greater oneness.

At the time when the middle class takes to the streets in revolution, the masses will also rise. And when the capitalist ruling elite is sufficiently vanquished, the people will begin to look for ways to achieve true happiness in their lives. This is the time for which all progressives must be prepared. The process that began as a trickle will become a torrential flood.

A revolutionary leadership must focus its energies on the future and not be obsessed with the past or the distractions of the present. We must be ready for the wave of change that will occur and be prepared with a concrete solution to meet the people's basic needs. Our leadership must possess a universal ideology in order to serve all the people and to become the beacons of positivity that the people, in their suffering, so desire. We must be as honey to the bee, able to channel grace and compassion in order to lead the people to a better world. As Che Guevara taught us, "The true revolutionary is guided by strong feelings of love."[36]

Erich Fromm, the social psychologist who developed the idea that freedom was a fundamental part of human nature, described in his book, *Art of Loving*, how revolutionary love entails being self-disciplined, being able to concentrate, being patient with others, being concerned for the welfare of others, and being willing to master the art of loving.[37]

While the capitalist process of destroying the American middle class has already begun, it has not reached the stage where a social revolution is possible. But, in the meantime, the numbers of people who are struggling to make ends meet continues to grow.

For the time being, the people, egged on by religious bigots, opportunistic politicians, and media personalities, will continue to blame each other for their troubles and busy themselves in attempts to reform the federal government to improve their living conditions. This is fruitless behavior. Our time is better spent preparing for the coming judgment day when capitalist exploitation is thrown into the dust bin of history. This day is not far off.

It is the purpose of this book to prepare the intellectuals, students, farmers, workers, warriors, and small business men and women of the world from all races and ethnic groups for that eventful day when the masses of people rise up in the streets because they have nothing left to lose. At this time, when society is approaching chaos, our revolutionary leaders, our spiritual warriors, must seize the moment to create positive change. If we are prepared, we will be able to take the reins of society at this time and turn it onto a viable path for the good and well-being of all.

Having painted a comprehensive picture of decadent capitalism, it must be added that it will not serve us to continually denounce capitalism at every opportunity because this only allows the capitalists to become alert and invent more devious ways to exploit the people. Those who lack a constructive ideology will never be able to destroy capitalist exploitation, even if they speak sweet words, or use threats, or create circumstantial pressure.

As revolutionary leaders, we must be willing to fight and die not only for our families and neighbors, but also for the welfare of every human being on the planet. Our approach is not to fight for diminishing global resources, but to organize ourselves, beginning at the local level, to meet our basic needs using local resources in a sustainable manner. While organizing locally to meet our basic needs, we must also be able to defend ourselves and to build bridges with neighboring communities and the world's people. In this way, we will learn from each other and support each other and establish relations with each other to build a world that secures our livelihoods and allows us to live in peace. The success of our revolution will depend upon our having taken the time to personally engage in the path of self-perfection and to have imbibed a universal ideology, while also working to build a local People's Economy that meets the basic needs of the people with whom we interact. The knowledge required to do this is the subject of the second book of this trilogy entitled *Universal Ideology: The Thought of P.R. Sarkar*.

Having said this, our first task in becoming revolutionaries is to understand what constitutes a human society and the dynamics by which it moves and changes. This is the subject of the next chapter.

Chapter Two: What is the Human Society and How Does It Evolve?

The Role of Social Psychologies in Revolution

When capitalist sociologists analyze class relations, they use variations in income levels to determine class distinctions. They speak of the upper class, middle class, and lower class and articulate the differences in life-style that is attributed to each class. Marxist sociologists, on the other hand, define class relations based upon people's relationship to the means of production (capital, factories, machines, raw materials, etc.). The capitalists own the means of production, while the workers are forced to sell their labor to the capitalists in order to meet their basic needs. Marx refers to these two classes as the bourgeoisie and the proletariat.[38] Marx also identified two other minor classes. These are the petty bourgeoisie (small businessmen) and the lumpen proletariat (the poor and unorganized lower classes who are not interested in revolutionary advancement). According to Marx, the capitalists and the workers are forever at odds because in order for the capitalists to increase their profits they must keep the wages of the workers as low as possible.

While these two class analyses are valuable, they do not, in themselves, give us enough information to conduct a nuclear revolution. The class analysis of progressive socialism (PROUT), which is the theory and practice upon which this book is based, is not determined by income categories or relationships to the means of production. Rather, the class analysis of a nuclear revolution identifies four existing classes and one potentially new class based upon social psychology. These classes are identified by how they exercise power in this relative world. To gain a full understanding of the toiling masses, warriors, intellectuals, and merchant/capitalists and their roles in history and contemporary society, see our books on *The Untold Story of Western Civilization* and *Universal Ideology.*

Because the capitalists are the ruling class of the present age, their psychology dominates human society. It is fruitless to believe that the capitalist class will change their psychology and share their wealth with the rest of human society. We may find an individual capitalist here or there who might listen to his conscience, but that is as much as we might expect. This means that the revolution against capitalist exploitation will be led by the radical warriors and intellectuals among the middle and lower classes. It is from these classes that we can expect the rise of a global class of spiritual warriors who will lead the nuclear revolution that will be further defined as we proceed in this book.

In addition to having strategies for working with the four traditional classes, a nuclear revolution requires the creation of an entirely new social class, that of the spiritual warriors. In fact, a nuclear revolution cannot be accomplished without the creation of this new class. It is their very purpose to lead the revolution and guide humanity along the path of a universal ideology. This class will come into existence as local organizers, who live according to the tenets of a universal ideology, begin to associate with each other and build greater networks within and outside their own communities. In time, these local spiritual warriors will gain a class consciousness and, in so doing, set the stage for the creation of a global revolutionary leadership. We will discuss this class of spiritual warriors more in subsequent chapters.

As mentioned in the introduction to this book, during the period of ancient history, the Warrior psychology dominated and the Warriors constituted the ruling class. Their vehicle was the newly developed city-states. During the Middle Ages, the Intellectual psychology dominated and the Intellectuals became the ruling class by using the Catholic Church as a church-state. In modern history the Capitalist psychology dominates and the Capitalists are the ruling class. They use nation-states and multinational corporations as their vehicles for conquest. In the next turn of the Social Cycle, the Capitalist Age will end and a new Warrior Age will begin.[39] Each Age begins when the masses of people support the new ruling class because the class that dominated the old order became too greedy and exploitive of the people. The Capitalist class will cease to be the ruling class as a result of its greed and abuse of power.

Before we look more closely at the dynamics by which a society evolves and how a revolutionary leadership is able to judge the motion of social change, let us look at the idea of the Social Cycle because it provides the larger context for understanding the definition of society and social motivity.

The Social Cycle

A nuclear revolution is based upon a different interpretation of history. Specifically, history is not linear. It is cyclic in nature. It is governed by a Social Cycle in which one class consciousness cedes to the next in a cyclical order. The warrior consciousness of Ancient History ceded to the intellectual consciousness that governed the Middle Ages, which, in turn, ceded to the merchant/capitalist ruling class of Modern History. In like manner, the Capitalist Age will cede to a second Warrior Age and so forth.

The warrior, intellectual, and merchant mentalities evolved in history because they constitute the only means by which to gain and maintain a position of power in society. The warrior mentality controls by the exercise of physical force. The intellectual mentality controls by the the exercise of mental force. And the merchant psychology controls by owning the money of society.

The Social Cycle is the wheel of history. The acceleration of the forward movement of the Social Cycle is called evolution, while the acceleration of the forward movement by the application of great force is called revolution. Conversely, the backward movement of the Social Cycle is called counter-evolution, while an accelerated backward movement, by the use of great force, is called counter-revolution.

Social progress may take place through evolution or revolution. However, to eliminate capitalist exploitation, the application of tremendous force is absolutely essential. A slow evolutionary process will no longer work because the suffering of humanity cannot go on forever. Neither will the earth's systems continue to tolerate being exploited as they are under the capitalist system. Today, human beings are faced with a two-headed monster that is characterized by the breakdown of nation-states due to war and social upheavals, and the breakdown of the natural world due to climate change and the destruction of the planetary life-support systems, including arable soil, potable water, breathable air, a protective ozone layer, and forest habitats.

As a result of capitalist exploitation, those having Warrior and Intellectual mentalities are transformed into the disgruntled slaves of the Capitalists. They have no alternative but to toil at the behest of the Capitalist to meet even their most basic needs. Those Warriors and Intellectuals who are forced into poverty under circumstantial pressure carry a simmering discontent in their hearts. This group is known as the "disgruntled workers" ("viksubdha shudras"). These disgruntled workers—the exploited Warriors and Intellectuals and even some of the capitalist psychology—will, in time, give systematic expression to the frustrations of the laboring masses to end Capitalist exploitation. These are the people from which the leadership of the nuclear revolution will emerge.

The revolution that is set in motion to terminate the Capitalist Era will require the united struggle of the disgruntled workers. The masses of society can never be the actual revolutionaries because they lack sufficient moral stamina, sense of responsibility, and fighting spirit. They are vulnerable to many vices and are not well-established in human values, hence they cannot develop the requisite revolutionary character. Even so, under the leadership of the spiritual warriors, they will rise up when called upon to put an end to all forms of capitalist exploitation. It is only the "disgruntled workers" who, by making themselves into spiritual warriors, can develop into genuine revolutionaries because they have the ability to acquire the moral courage and discipline to oppose exploitation.

The revolution against capitalist exploitation is termed "the revolution of the toiling masses," ("shudra revolution").[40] Although the warriors and intellectuals are turned into disgruntled workers due to capitalist exploitation, immediately after the shúdra revolution, they will revert back to their warrior and intellectual psychologies. Because of the martial character of shúdra revolution, the leadership of society will pass into the hands of the warriors, hence a new Warrior Era will emerge. Even during the Warrior Era, however, the Intellectuals will continue to try to establish their influence through force of intellect and attempt to move society along the path of the Intellectual psychology. Thus, after some time, the Intellectual Era will follow the Warrior Era and, in turn, be followed by a second Capitalist Era. A Shudra Revolution will occur to end each age and usher in a new age. Shudra means toiling masses. By this process we see that the movement of the social cycle and revolution are inseparably related.

We also see that three types of psychology dominate the social cycle in cyclic progression. As each successive ruling class establishes its status quo, it tries to usurp social power and establish its hegemony by trampling on the rights of others. This conflict has been discernible from the dawn of human civilization. Through this clash and cohesion, human beings have tried to find the path of emancipation. Towards the end of an era, the collective psychology undergoes marked deterioration. Moral degeneration and social retardation cause psycho-social stagnation. Exploitation becomes rampant. This sort of unhealthy situation signals the end of an era.[41]

This can be seen in the modern world where capitalist exploitation is rampant almost everywhere. Capitalism is rapidly moving into the final stage of degeneration. In the early part of the capitalist era, society experienced certain advantages, but towards the end, society has become the victim of insatiable rapacity, unbearable hardship, and heartless deprivation. Those countries suffering under the weight of capitalist exploitation are inexorably moving towards shúdra revolution. These revolutions have occurred and will continue to occur against western capitalists and the communist governments that have become capitalist in psychology.

The Social Cycle will always continue. No force can stop it. Not even the revolutionary leadership.

Now that we understand the cyclical motion of human society, let us look more closely at the dynamics by which this cyclical motion occurs.

Definition of Society

A society may be described as the collection of numerous parallel psychic waves that originate due to the psychic tendency of people moving together in unison.[42] Social movement is the active effort to demolish a static structure and construct a dynamic structure in its place, and the characteristic of this movement is rhythmic or systaltic, rather than linear. This is to say that when a person or group has the requisite energy, it is able to move. When it runs out of energy, it pauses. In stopping, it gains energy for the next pulse forward until it again runs out of energy. Thus, the movement is systaltic, characterized by alternating periods of motion and station. When, the person or group is no longer able to gather energy in its state of pause for the next pulse forward, it dies.

The periods of motion and station may also be characterized as alternate periods of expansion and contraction. It is the nature of liberals to want to expand the society, while it is the nature of conservatives to want to contract it. Conservatives want to maintain the status quo or revert to the past. Liberals what to change the status quo by expanding its capacity. Thereby the liberals look to the future for solutions while the conservatives look to the past. The conservatives dominate politically during periods of contraction, while the liberals dominate during the periods of expansion. These periods of contraction and expansion alternate. The more powerful the period of contraction, the more powerful will be the period of expansion that follows, and vice-versa. For example, in the United States, the expansive period of the Obama administration was followed by the contractive period of the Trump administration. It is a natural movement to expand and contract. Thus, the capitalist society requires both these periods to maintain its existence. As such, it requires both liberals and conservatives to function. There is nothing revolutionary here in that both liberals and conservatives seek to maintain the capitalist system status quo.

In considering the motion of society, one must also consider its speed. In comparison to the speed of social change over the last few centuries, the speed of change has greatly accelerated in the twenty-first century. In the near future, changes in society will take place with even greater speed. Speed in social movement is related to the direction of change. For example, in the growth or expansion of a system, speed is slower at the onset but faster as it gains in maturity. Similarly, in the contraction or death of a system, speed is slower as it declines from its peak but gathers greater speed the further down the slope it proceeds. Today the domination by capitalists is in decline and gathering speed as it plummets to its end.

How Does Human Society Evolve?

A careful study of the social history of the world will reveal that until now every attempt at counter-revolution has not only caused enormous psychic and financial suffering and plunged humanity into the mire of gloom and despair, but has also lengthened the period of social contraction. This in turn, in the next phase, has helped to accelerate the speed of the period of social expansion—has inspired the chariot of revolution to advance towards victory with greater momentum. P. R. Sarkar

Society, as with everything else, moves according to natural laws. To understand this movement, revolutionaries in the modern age have developed the methodology of dialectics to explain social movement and to chart revolutionary strategies.

Like all revolutionaries of the past, the leaders of nuclear revolution will also need a method for defining and monitoring social change. They will need to know if they are making progress or not and how fast change is occurring. Are periods of progress longer than periods of contraction?

Our method of analyzing social change is an advance on the methodology of dialectics employed by social scientists of the past. As such, it bears explanation. This new methodology is based on the wave theory of P. R. Sarkar.[43]

Because social science as a field of study emerged at a time in history when the ruling class (the dominant class in society) was the capitalist class, we find that contemporary social scientists largely fall into two main camps—the capitalist sociologists and the Marxist sociologists.

The capitalist sociologists preceded the Marxist sociologists. We may call their methodology "metaphysical idealism." The Marxists, on the other hand, expressly called their methodology "dialectical materialism."

"Metaphysics versus Dialectics" and "Idealism vs. Materialism" constitute the two main contradictions in the analytical methods of the two camps. Let us look now at the distinction between Metaphysics and Dialectics.

Metaphysics vs Dialectics

Metaphysics	Dialectics
1. Nature is studied as an agglomeration of isolated phenomena unconnected and independent of each other.	1. Nature is studied as an integral and connected whole in which phenomena are organically connected, dependent upon, and determined by each other.
2. Nature is immobile and immutable.	2. Nature is a state of continuous movement with some things always developing and some things dying away.
3. There are quantitative changes and there are qualitative changes. These remain distinct.	3. Quantitative changes result in qualitative changes and vice-versa.
4. All things are inherently unified.	4. Internal contractions are inherent in all things. Diversity is the nature of formal reality.
5. Life is harmonious.	5. Life is a disclosure of the contradictions inherent in phenomena; it is a struggle between opposing forces.

In order to present the distinction between these two worldviews without going into too much detail, the chart above will be of help. It provides a quick overview of the typical characteristics of each method of analysis.

In essence, what this chart indicates is that metaphysics stresses the motionlessness of reality (i.e., it fixes or stabilizes it). The dialectical interpretation, on the other hand, stresses the motion or motivity of reality.

Historically, all Marxists have utilized dialectics to study social progress and have refined the methodology according to their own political experiences. Lenin, Stalin, and Mao have all developed treatises on the Dialectical Method.[44]

While the Marxists have developed the methodology of Dialectical Materialism (originally called Historical Materialism) for their social analysis, the capitalist social scientists remain largely metaphysical idealists. The reason for this is that the Marxists, as revolutionaries, use a methodology that stresses process and allows them to analyze and create change in a social system. The defenders of capitalism, on the other hand, prefer a methodology that stresses motionlessness and allows them to justify the status quo.

Materialism and Idealism

Despite the large number of philosophical schools and subtle nuances between philosophies, they all can be said to fall into one of two categories: Idealism or Materialism.[45] The primary distinction between them is the way they answer two fundamental questions: "Of what does reality consist?" and "How does it originate?" To idealists, the objects of mind (ideas) are primary, and matter secondary. To materialists, matter is primary, and ideas are secondary. Ideas are no more than a shadow of matter.

In philosophy, then, idealism refers to those metaphysical philosophies that assert that reality is fundamentally mentally constructed, or otherwise immaterial. Idealism rejects the possibility of knowing anything independent of mind. According to this view, consciousness exists before and is the pre-condition of material existence. Consciousness creates and determines the material and not vice versa. Idealists believe consciousness and mind to be the origin of the material world and aim to explain the existing world according to these principles.

In a sociological sense, idealism emphasizes how human ideas, especially beliefs and values, shape society. As an ontological doctrine, idealism goes further, asserting that all entities are composed of mind or spirit. Idealism thus rejects those empiricists and dualist theories that fail to ascribe priority to the mind.

The earliest arguments that the world is grounded in the mental realm derive from India and Greece. However, during the fifteen hundred year reign of the Catholic Church in Europe, free thought in Western civilization was all but suppressed. It was only after the fall of the Catholic Church, during the eighteenth century, that intellectuals were free to think for themselves again. This gave rise to science and philosophy. While science stressed material conditions and considered only those things "real" as could be grasped by the senses, philosophy tended to argue that only the mind could determine what was real. Beginning with Immanuel Kant, German idealists such as Georg Wilhelm Friedrich Hegel, Johann Gottlieb Fichte, Friedrich Wilhelm Joseph Schelling, and Arthur Schopenhauer dominated nineteenth century philosophy.

These philosophers were metaphysical idealists who held that reality is incorporeal or experiential at its core. Beyond this, idealists disagree on which aspects of the mental realm are more basic. There are subjective idealists, phenomenalists, epistemological idealists, transcendental idealists, objective idealists, etc. The point is that idealists tend to downplay the importance that material reality has on human beings. For example, while Karl Marx was studying at the University, his fellow students were all enamored by Hegel. Hegel had created quite an intellectual stir by developing the idea of dialectics and analyzing reality, not as a fixed entity, but as a process. Marx, appreciated Hegel's dialectic, but reminded his fellow students that while they were arguing over philosophical fine points, the capitalists, who now constituted the new ruling class, were enslaving humanity in their factories, and preventing common people from meeting their basic needs.

This reality caused Karl Marx and his friend Frederick Engels to develop a new analysis of reality that they called Dialectical Historicism.

Later the followers of Marx would change the name to Dialectical Materialism.

Materialism holds that matter is the fundamental substance in nature, and that all things, including ideas and consciousness, are results of material interactions.

Therefore, in Idealism, mind and consciousness are first-order realities to which matter is an expression. In Materialism the converse is true. Here mind and consciousness are by-products of material processes (for example, the biochemistry of the human brain and nervous system) without which they cannot exist. According to this doctrine, material creates and determines consciousness, not vice versa.

Materialist theories are mainly divided into three groups: (1) Naive materialism, which is due to the difficulty that persons, who have had no philosophic training, experience in observing and appreciating the importance of the immaterial facts of consciousness. The pre-Socratics may be classed as naïve materialists; (2) Metaphysical materialism which examines separated parts of the world in a static, isolated environment; and (3) Dialectical materialism which adapts the Hegelian dialectic for materialism, examining parts of the world in relation to each other within a dynamic environment.

Karl Marx and Friedrich Engels extended the concept of materialism to elaborate a materialist conception of history centered on the roughly empirical world of human activity (practice, including labor) and the social institutions that are created, reproduced, or destroyed by that activity.

The Methodology of P. R. Sarkar (Wave Theory)

The methodology of P. R. Sarkar resolves the contradiction between metaphysics and dialectics, as well as the contradiction between idealism and materialism. To Sarkar, neither motion nor station, neither idea nor matter, is permanent. Rather all share a relative existence.

Let us begin our analysis of Sarkar's methodology by looking at his definition of matter.

While discussing the nature of matter in *Subhasita Samgraha, Part VI*, Sarkar says:

> Call the fundamental essence of matter by any name, molecule or atom, the basic cause of matter is nothing but bottled up energy. It is the condensed or phenomenized state of the particles of energy that we call matter. It will be no means correct to take matter as an original substance. To adequately enjoy matter one has got to be acquainted with the energy entity. The greater the acquaintance a man makes with energy and understands matter accordingly, the greater the right he earns to the use of matter as an object of enjoyment.[46]

If, therefore, matter is essentially energy, what does Sarkar say about energy? May we consider it to be an "original substance"? Sarkar says no. As matter is condensed energy, so also energy is a condensation of consciousness. Consciousness is the most subtle entity in the universe, matter and energy are crudifications of it.

In *Human Society, Part II*, Sarkar states the link between Matter and Consciousness:

> The field of comprehension wherein the principle of relativity, consequent upon time, place and person, finds its existential consummation—the inactiveness of that cognizant bearing constitutes the ultimate stability of matter.[47]

This sentence, although difficult to understand, specifically states that what we call the stability of matter to Sarkar is actually the inactiveness of Consciousness. This implies, and indeed Sarkar states this many times in his writings, that the manifest universe, or what we call relative reality, is essentially composed of waves of consciousness undergoing transmutations that modify it into different forms of energy and which achieve their ultimate stability in forms of matter.

Therefore, we can speak about the motionlessness or station of matter only when it is deprived of external stimulation. The motion that we call the movement of Matter is dependent on an "external agency," (i.e.

energy-consciousness). Matter, independent of external activation, is immobile. Thus, we see that in analyzing the material condition, its motivity is dependent upon consciousness.

According to Sarkar, motion and station have a relationship with each other. They are, in fact, alternating characteristics of all finite entities. The change from motion to station or from station to motion constitutes the wave pattern that underlies all forms, whether material or ideal. The graph below demonstrates Sarkar's analysis of a single pulse in the wavelength of an entity.

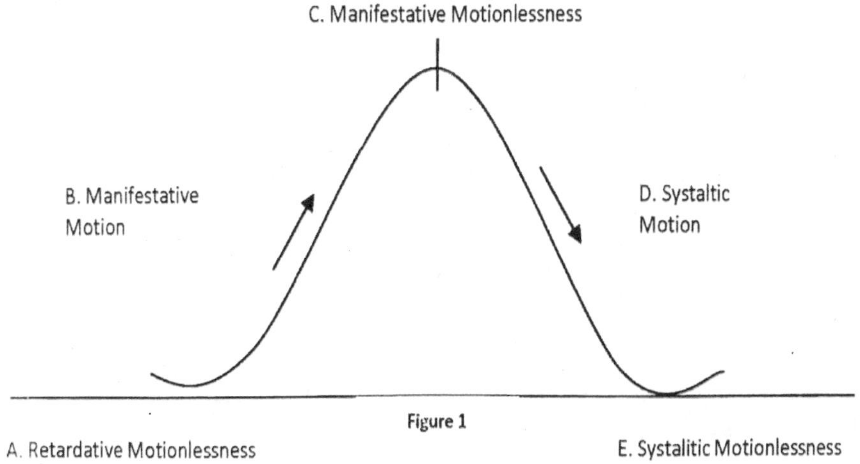

Figure 1

Using the respiratory function as an example of this pulse, Sarkar explains that: (1) inhalation (inspiration) is a manifestative motion proceeding from retardative motionlessness to manifestative motionlessness; (2) the retention of breath at the end of inhalation is the manifestative motionlessness; thereafter (3) the exhalation (expiration) is a systaltic movement towards systaltic motionlessness; and (4) the holding of breath after its complete voidance is a systaltic motionlessness.

Sarkar explains:

> After every such complete breath (i.e., in every post-exhalation pause), there occurs the death of the entity. Then again after gathering vitality for the second time from this death or state of motionlessness, it becomes

> alive again during the next inhalation. If, after this bi-functional respiration (i.e. after one full cycle of inspiration and expiration), it is noticed that the physical mechanism is unable to gather vital force from motionlessness then inhalation becomes an impossibility. This condition is what we colloquially call death.[48]

Looking at the graph below, we can see the relationship of inhalation and exhalation to motion and station.

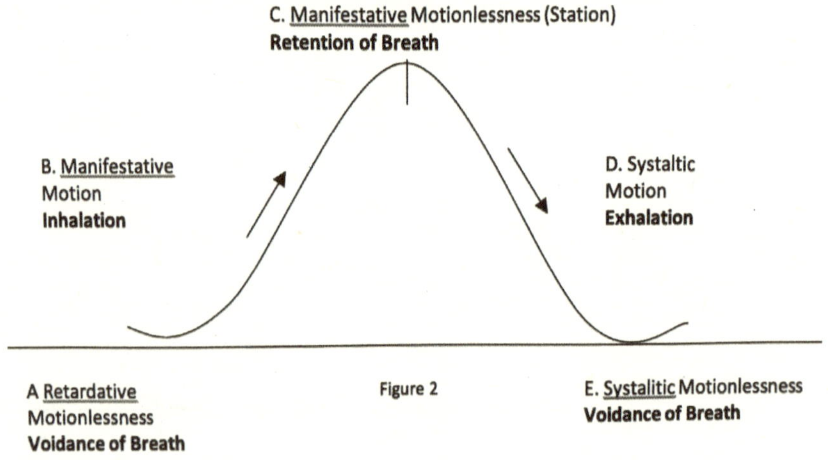

Figure 2

But is death the final stability or motionlessness? Sarkar says no. He explains:

> Relatively speaking, when the observable objects do not seem to change places, we call such a state static. But in such an event the seer-and-the-seen-ness that is on the move in the greater field, passes our crude or subtle comprehension. So such so-called immobilities cannot be adjudged as the final stability. In individual life final stability can only be, when the activity of even the causal mind becomes nil. Nevertheless, we cannot call even the bodiless, or mindless condition of a unit as the final stability, for in that case the seed of activity is compelled to remain active on the body and mind of the Macrocosm by virtue of Its creative potentials. From

> this we may conclude that the destruction of the seed of activity of the mind is verily the ultimate stability.[49]

By this example, Sarkar indicates that motion rather than station is the principle characteristic of the condition of relative reality. It is the primary aspect in the dialectic between motion and station. This being the case, how does Sarkar's theory compare to the dialectician's theory of movement. Visually we may graph a comparison:

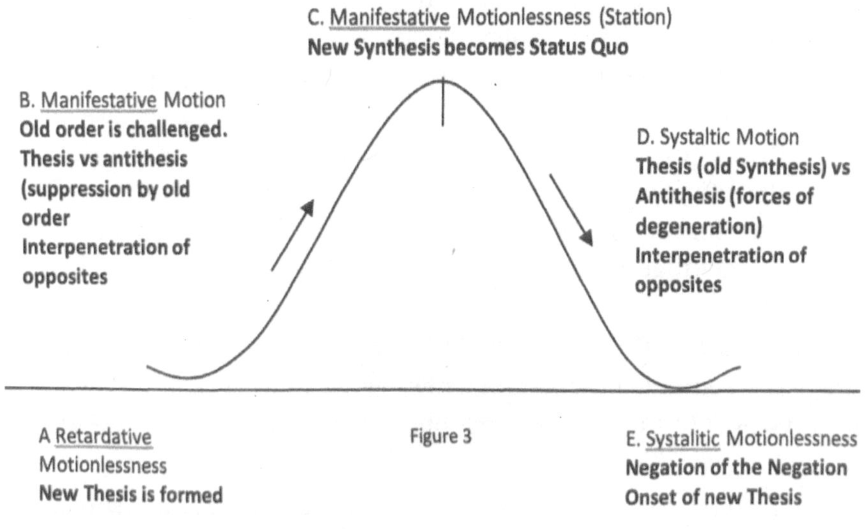

Figure 3

In the above graph we see that the Manifestative Motion (B) is equated with the interpenetration of opposites (i.e., the thesis-antithesis). Here a new movement fights the old status quo to replace it with a new institutional form. At point (C), the Manifestative Motionlessness represents the thesis becoming a new synthesis as it achieves its position as the new status quo. As it tries to maintain its stability in the face of continued motion (change), however, it begins to act contrary to the laws of motion. In this way, it becomes exploitive, less people benefit as the ruling class becomes more oppressive.

Consequently, a new movement (antithesis) develops to challenge the old order. The degeneration of the old order is represented by (D),

Systaltic Motion when the forces of change and internal degeneration bring down the status quo. The Systaltic Motionlessness (E) represents the stability at the end of the conflict, the Negation of the Negation. Here the potential for a new thesis gains vitality to become manifest.

Discussing the vitality of a new system, Sarkar says:

> The manifestative movement, in its former systaltic bearing, acquires the vital energy from motionlessness. The mildness or forcefulness of this vital force depends upon the length of the period of immobility as well as upon the capacity of the structural stamina. The length of the state of motionlessness is adjudged as death only when the old structure is unable to hold and appropriate the vital force of motionlessness. It is in such a case that a newer structure becomes necessary for the manifestative expression of motionlessness. This new structure may either be a newer form of the old one or that of an altogether different one.[50]

Here we find that the "new structure" can be either a reform of the old system or an entirely new system. How can we differentiate a reform from a new system? Is it simply a matter of change in degrees? Mr. Sarkar says:

> That structure can be called new, only when the mind . . . of society fails to establish its consciousness of identity between the older and the newer forms. During the period of cessation there occurs the death . . . of one structure due to the suppression or destruction of its vital energy by another structure preceding the cessation. Such worldly deaths do not come to individuals alone but to the society as well.[51]

According to dialectical materialism, society fails to establish its consciousness of identity between the older and newer forms because quantitative changes have resulted in a qualitative change. For example, heat may be applied to water by an increase in degrees, but eventually this will result in a qualitative change from water to steam.

From this analysis, we can surmise that there is a difference between reformers and revolutionaries. A reformer simply wants to reform the existing system. The revolutionary, on the other hand, wants to create an entirely new system. The new system would entail a change in the ruling class and all institutions of authority. It would represent a change in the political economy and the culture as well. Examples of true revolutions would be the change from Monarchy to church-state to democratic nationalism, or from democratic nationalism (capitalism) to progressive socialism.

The below graph expresses a simple, general movement in which a society (or any form, for that matter) is born, grows, matures, decays, and dies. If we look in greater detail at this process, we would have a graph that indicates the interpenetration of opposites at each stage of development. Theoretically, it would look more like this:

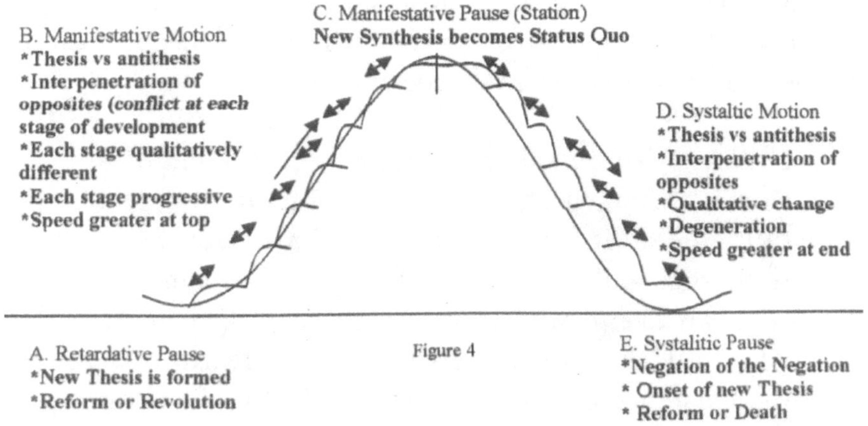

Figure 4

The above graph shows a stylized version of the contradiction between motion and station within the birth, growth, maturation, decline, and death of a system. In reality, the forces at play create infinite wave forms having very complex natures. If one were to graph the movement of basic trends in a society, the interpenetration of opposites may look more like this:

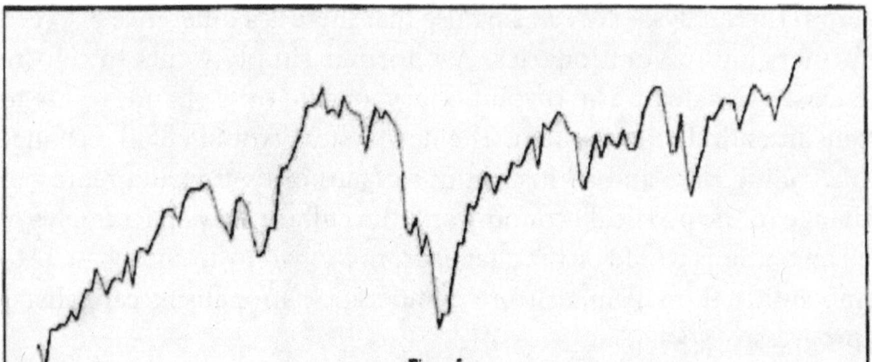

Figure 5

In the above graphs, we can see that Sarkar's wave theory essentially coincides with the dialectician's analysis of movement. His major contribution to dialectics is to demonstrate that motion is not continuous, but rather it is systaltic, or pulsative. The following words therefore caution both the dialectician as well as the metaphysician:

> Nothing can be called stationary in this manifest world, thronged as it is with the plethora of relativities. Had it been so, all would have been essentially one—all would have attained homoeosis. In such an event "all" would have ceased to be all. All would have lost their distinctive features. So the existence of this manifest world finds its consummation in the relativity of movements and not in the absolute motion or motionlessness.[52]

A nuclear revolution then is an act of replacing one age with another by destroying exploitation and bringing about a change in the collective psychology within a short period of time through the application of tremendous force.

As the illustration below shows, it should be the goal of revolutionaries today to create a nuclear revolution as quickly as possible to prevent human society from plummeting into greater misery as the Capitalist system moves ever quicker toward its demise.

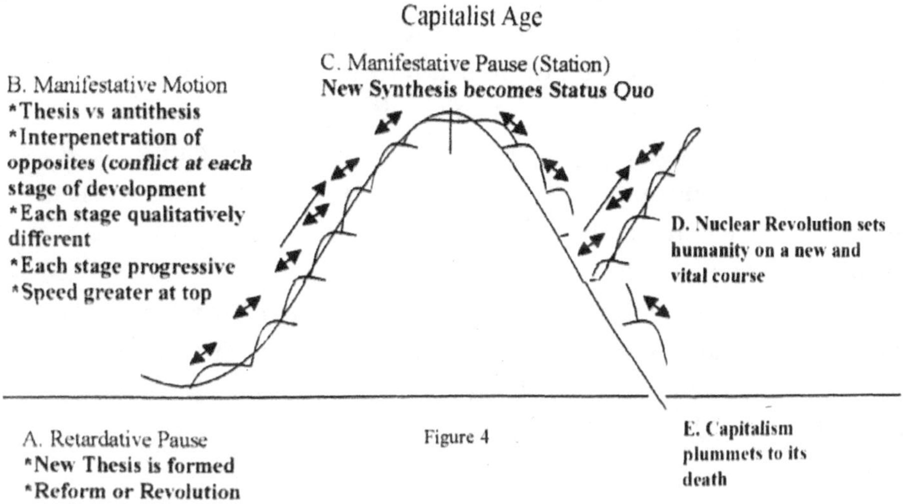

Figure 4

Based upon P. R. Sarkar's wave theory, the flow of civilization based upon the social psychologies would like this:

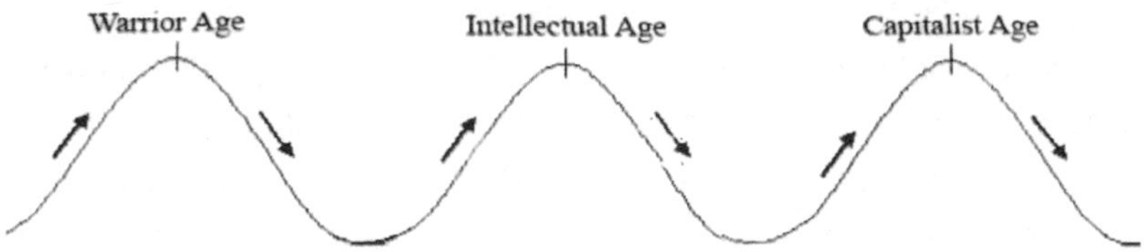

If a reverse movement takes place, and the social cycle moves in the opposite direction by the application of tremendous force, it is called "counter-revolution." This is the process we saw under the Trump administration. We can say this because the Trump administration represented all the major divisive sentiments of history including sexism, racism, class warfare, national chauvinism, and religious exclusivism. It also perpetrated all the behavior responsible for the destruction of the planet, its species, and life-support systems.

As we have said, society moves according to the inexorable laws of the social cycle, yet those who are concerned for the collective welfare will always endeavor to accelerate the speed of social movement by fighting relentlessly against exploitation so that all can move forward together in unison with a minimum of pain and death.

When the psychic balance in social life undergoes a revolutionary change, society will experience a collective upsurge against exploitation and a revolution against the status quo. There will be an effective struggle of the benevolent forces against the malevolent forces and the people will strive for a new collective psychology. The leaders of the nuclear revolution will guide this quest for a new collective psychology by ensuring that it is motivated by a deep-seated spiritual longing.

Therefore, we observe that Sarkar's methodology is neither metaphysical nor dialectical, but rather explains their relationship within a larger synthesis.

Spiritual Science

Having seen how motion and station are relative conditions, now let us look at Sarkar's interpretation of Idealism and Materialism. To Sarkar, again, both are relative positions. To him, it is equally misguided to assume a materialist position and discount an idealist position, as it is to assume an idealist position and discount the materialist position. Relative reality, according to P. R. Sarkar, is a "psycho-physical parallelism." Neither psychic waves, nor material waves, exist in isolation, but continually interpenetrate each other. Sarkar classifies materialism and idealism as "Logician's Philosophy." Neither hold much weight in his eyes. He says:

> Logician's Philosophy at times weaves a web of logic with matter and at the other it endeavors to find the relation between thought and matter and then again at places it plays the blind-man's bluff in the blind alley of thoughts. Nevertheless, the place of Logician's Philosophy should indeed be below Physical Science instead of being between the Spiritual and the Material Philos-

> ophies, for it does not appease the hunger of the body, mind and soul of the living beings, rather a confusion is created due to the mind being clouded with the smoke of logic.[53]

In the above paragraph, Sarkar's use of the term Material Philosophies does not mean the philosophy of Materialism as we have been discussing it. Rather it refers to the philosophy that guides Physical Science and which, in turn, is continually being altered by virtue of scientific investigation. It is not, therefore, like Materialism, a static concept about the essence of reality.

But if both Materialism and Idealism are unacceptable to Sarkar to explain the nature of reality, what then is? It can only be Spiritual Philosophy, to which he makes reference above. Sarkar calls Spiritual Philosophy an Intuitional Science. It is not abstract, but rather based upon the laws of Consciousness and the process by which Consciousness gains expression through waves that ultimately are perceived as material forms. Intuitional Science is the underlying philosophy upon which the oldest religions on earth (i.e., Hinduism, Buddhism, and Jainism) have been built. It is essentially the foundation of all yogic practices. In the East, this Intuitional Science is called Tantra, which means "expansion of mind." Tantra itself is not a religion; rather it is the science of self-realization (Dharma) that gives Eastern religions their spiritual meaning. According to Sarkar, the summum bonum (the greatest good) of human existence is not merely to survive in a peaceful manner as material beings, but rather it is to achieve God-realization (i.e., to transform one's entitative rhythm from animality to humanity to Divinity).

Intuitional Science (Tantra) proposes that the manifest universe is essentially Qualified Consciousness (Saguna Brahma),[54] which emerges out of Unqualified Consciousness (Nirguna Brahma) as a result of an imbalance between two original forces inherent in Unqualified Consciousness. These are the Cognitive Principle (Purusha) and the Operative Principle (Prakriti).

In Nirguna Brahma, Purusha and Prakriti remain in balance. As such, there is no vibration, no form, no expressed duality, only a singular Nuclear Consciousness, the Whole characterized by unity or blissful love.

The manifest universe begins only when Prakriti gains the upper hand over Purusha, with His permission, and begins to bind or qualify Him. Prakriti is said to have three chains (gunas) with which to bind Purusha. These three chains are the sattva guna (sentience), the raja guna (mutation) and tama guna (staticity). Sentience is the subtlest chain and with it, Prakriti (the Operative Principle) binds Purusha (Consciousness) into having a sense of Its own existence. In this way the I-feeling of the cosmic mind is born. This is referred to as the Mahatattva.

With the raja guna (mutative force), Prakriti further binds the I-feeling of Consciousness into having a sense of willfulness, called the Ahamtattva. Thus, the Supreme I-feeling not only has a sense of existence, but also a sense of action, a sense of "I-do." Finally, Prakriti further crudifies Consciousness with the tama guna (static force) into having a mental plate (Citta) or "field of comprehension." These three mutations of Consciousness constitute the vibrational realm of cosmic mind.

It is through a process of continued crudification, that a portion of the Macrocosmic Mental Plate, which Sarkar says is composed of "inter-ectoplasmic atoms," is further transformed into the ethereal factor (Space), and then into the aerial factor, luminous factor, liquid factor, and finally the solid factor respectively. These constitute the five fundamental factors that compose the material universe. This process occurs as the Operative Principle increases pressure on Consciousness and compacts its inter-ectoplasmic atoms into greater degrees of density.

The process of transmutation from Macrocosmic I-feeling to the solid factor is called Saincara, or the process from subtle to crude. The return process, or the process from crude to subtle, is called Pratisaincara. It is characterized by the evolution of unit entities through forms of greater complexity that permit Conscious greater expression, which will ultimately allow each unit consciousness to reunite with the Cosmic Consciousness.

The birth of the solid factor allows for the manifestation of units. Without matter, we cannot speak of individual units. The transition from Saincara to Pratisaincara is triggered when a portion of Consciousness is bound up as the solid factor. Depending upon the interpenetration of external and internal forces, either one of three things can occur: (1) the solid object will remain temporarily inert due to a balance of external and internal forces, (2) if the external force dominates, the object will explode, or (3) if the internal force dominates, the unit will gain a nucleus of vital energy and thereby become a material entity. Material forms are, in effect, atomized units of consciousness.

During the transmutation of the Macrocosmic Consciousness into the Solid Factor, by virtue of the movement from subtle to crude, Prakriti, or the Operative Principle, keeps applying greater and greater pressure on the solid factor until it can no longer maintain its formal integrity. The consequence is the Macrocosmic explosion western scientists call the Big Bang. This is the event that gives birth to the material universe as we perceive it.

During the initial blast, as the gases began to cool, simple atomic structures like helium and hydrogen emerge. This is due to the clash and cohesion of atoms under conditions of hot and cold, attraction and repulsion, etc. In time, the interpenetration of atoms under varied conditions gave rise to additional atomic structures of ever more complicated natures. After billions of years, the complex molecules that have been formed in the bodies of stars produce the DNA molecule. For the first time in the known universe, molecules are able to express a modicum of consciousness that we recognize as intention. "Like atoms" bond with "like atoms" within the DNA molecule. They are not paired accidentally or by force of external pressure. From this simple act, the evolution of species on planet Earth begins. The next step is the evolution of the single cell and then cell colonies, etc.

In this process, Consciousness, which had been bound up in the different mental and physical vibrational states, begins its return journey through more and more inclusive formal expressions and thus greater freedom. Present scientific evidence indicates that the Earth is about 4.5 billion years old and that the oldest life forms to be uncovered are microscopic, one-celled creatures that are about 4.1 billion years old.

Intuitional Science holds that organic forms (marine life, insects, plants, animals) that have the ability to reproduce themselves are composed of the five factors and also possess the vibrational factor of the mental plate. This mental plate, upon which they can reproduce their image, and thus their physical being, is the base of the unit mind. As the process of evolution continues, higher animals have an evolved sense of doing (Ahamtattva). In other words, they have a sense of performing actions. It remains unique to certain individual primates and human beings, however, to inculcate the vibrational state of the sense of I-feeling. Human beings have a sense of "I" that underlies everything that we think, say or do. We have, in other words, a unit mind that mirrors completely the attributes of the Macrocosmic Mind.

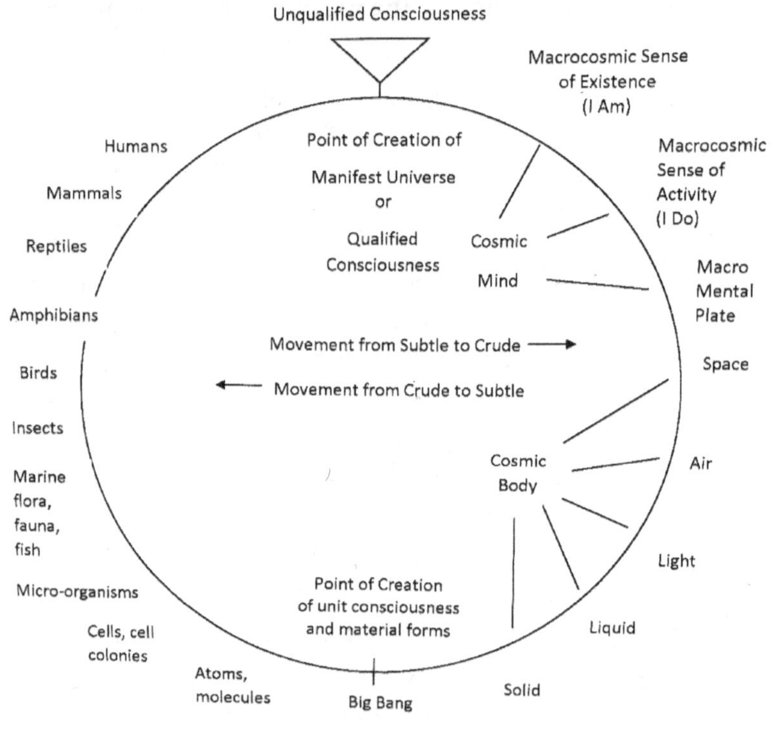

Chart illustrating the Creation Theory

The Mahatattva (literally 'the highest sense') or I-feeling gives each human being a sense of individual distinction within relative reality. This sense of distinction is not well developed, if at all, within the minds of lower evolved species. In other words, where animals have a sense of doing, humans have a sense of I am doing. This results when the Mahatattva (sense of I) permeates the Ahamtattva (sense of doing).

In this way, humans distinguish themselves as individuals from their environment.[55]

The Mahatattva is the highest sense of the human mind, but it is not the pure subjective Self. Rather, the Mahatattva is only a reflection within the human mind of the Qualified Cosmic Consciousness. It is the pure subjective Self that says "I know that I exist." In the East this Self is called the Atman. In the West it is called the Soul.

The highest achievement of human beings is to identify themselves, not with their ego (I am), or their activity (I do), but rather to identify with their Soul—the knower of existence. The Soul or Atman is the perfect reflection of the Cosmic Mind; in fact, we may even say that the Cosmic Mind is only a collective name for all the Atmans. Once evolution proceeds beyond matter and mind, the maintenance of units is impossible; thus there is no difference between Atman and Cosmic Consciousness. Unit subjectivity merges into Cosmic subjectivity. Distinct definitions are no longer possible. Contradiction, wave, and object are no longer possible. Separation is no longer possible. Bliss, the experience of pure unity, is the only reality. This is the highest achievement of human existence.

According to Mr. Sarkar:

> The only true progress for human beings is spiritual progress. The wise will, therefore, concern themselves only with the spiritual sphere. The only concern with the physical and intellectual spheres will be adjusting the base on which spiritual progress will thrive. While concerning themselves with the adjustment in the physical and intellectual spheres, the wise should also continue to consume cruder waves into subtler ones . . . The absence of such consumption or transmutation in the physical and intellectual spheres is bound to lead to retardation. The wise, remembering that the goal of life is the subtlest goal, will make efforts to reach supreme blessedness. This is the only way to progress. There is no other way.[56]

While this may sound sweet to the ears of Idealists, does not this condemn all lower species and virtually all of humanity to never being able to attain their highest self-realization? If we consider environmental conditions and hereditary predispositions, how can this be fair? Most of us are completely absorbed, out of necessity, with just maintaining our lives—with just "reproducing" ourselves as Marx expressed it.

The condition in which we, as humans, find ourselves has led generation after generation, and civilization after civilization, to focus on simply "reproducing" ourselves. This activity has so permeated human history that it has become the supreme objective of human existence. Who will not say that the rich and powerful among us are making greater human progress than the rest of us?

Intuitional Science seems to offer a greater goal than the accumulation of wealth and power, but one which, on the surface of things, appears unattainable. How does one become self-realized within such a short lifetime when one must spend the greatest portion of one's days simply making ends meet?

According to most religions, one achieves spiritual success only after death, if one has led a moral life and believed in God. On the other hand, if one is immoral, and does not believe in God, their future will consist of eternal unspeakable suffering. Fortunately, life is not so simplistic. One does not become God-realized within a single lifetime. Being a good person is not enough. Morality is relative and dependent upon circumstances and group identity. Believing in God is not enough. Simply believing in anything does not allow one to enjoy the object of intention. An effort must be made to attain it. To achieve the Consciousness of the universe, the greatest prize of all, will require a Herculean effort and much grace. The fact is that we are on a long journey, transitioning out of animality, into humanity, and heading for divinity. It may take hundreds or thousands of lifetimes to evolve from an atom to a realized human being.

The unit consciousness evolves through the process of reincarnation. The human being has lived through a myriad of forms including the identities of less evolved species in order to attain a human body. This life, this current identity, is only one life within a long process that extends through time, space, and circumstance as our consciousness evolves in a process to achieve its original Divine state. Evolution is the process by which consciousness gains its freedom, by adopting consecutively higher life forms that allow it greater room for expression. Once one achieves a human form, the mind continues to seek more freedom, to have greater happiness. In this process, one discovers that infinite happiness cannot be achieved by ideating on finite objects. Insofar as finite objects follow the laws of wave theory, they have a beginning, a middle, and an end. The greater that one's attachment is to a finite object, the greater one's pleasure, but so also the greater the pain when the object ceases to give pleasure.

In Intuitional Science, there are no accidents, only incidents. Each unit mind inhabits a physical form complimentary to its propensities or desires. To fulfill its evolutionary potential, a unit consciousness will ultimately require a human body/brain. This is because the human form contains the capacity to imbibe and express all psycho-physical wave lengths, even the subtle wave length of the Mahatattva. Without a human body and brain, the unit consciousness cannot reflect a sense of "I am"; nor can it transcend this feeling and merge with the Cosmic Subjectivity.

The cause of the motion of the unit consciousness is the Attraction for the Great, but it is also the clash and cohesion that occur through the interaction of the body and mind with the conditions of relative reality. Through the interpenetration of thesis and antithesis, in the process of pause and motivity, the mind gains vital energy and is able to achieve greater expressions of understanding.

The process of Pratisaincara (the movement from crude to subtle) is the process in which unit consciousnesses returns to Cosmic Consciousness. It is the process by which units attain God (or Self) Realization.

Discussing the relationship between the unit consciousness and Cosmic Consciousness, Sarkar states:

> Now regarding the relationship of átman [soul] and Paramátman [supreme soul], the átman is the object and the Paramátman is the subject. So Paramátman is the Supreme Subjectivity. He is the Supreme Subject of all other subjects. He is the Supreme Multiple of all other expressed multiplicities. Because He is the Supreme —in Sanskrit "supreme" is parama—He is known as Paramátmá. He is the Supreme Subject.[57]

Wave Theory and Spiritual Science

The Idealist who says that the earth is a product of heaven or that matter is an outward expression of God or mind, perceives reality as the process from subtle to crude. He or she puts emphasis on the process of Saincara. There is a relative truth to this analysis. The problem for the Idealist, however, is that by dwelling solely on this aspect of creation, they rob the individual of physical integrity and allow those who would exploit our material condition to enslave us. The philosophies and theologies of the Idealists have systematically been used by ruling classes to confuse the masses so that they do not understand the true cause of their suffering, which is not the wrath of God, but rather the exploitation of their life force by the ruling elites.

From the perspective of spiritual science, Karl Marx and the materialists begin their explanation of relative reality from the point of view in which Consciousness is bound in the solid factor (i.e., expressed as Matter). They, therefore, stress Pratisaincara, the process from crude to subtle. Beginning from this perspective, Marx concluded that the mind of the human being was a product of his or her material condition. Mind evolved out of the individual's clash and cohesion with the material condition. Therefore, according to Marx, the material condition preceded and determined the content of the human mind.

According to Intuitional Science, this assumption is correct. Mind is a product of the material condition, as Marx explains. The flaw in Marx's thought, however, is that he did not distinguish between Consciousness and the human mind per se. Spiritual science holds that Consciousness precedes the evolution of species and is inherent in all forms. Mind, on the other hand, does not precede the evolution of species, rather it is a concomitant development of material evolution.

Marx made an important contribution to human thought when he pointed out to the intellectuals of his day that the individual mind is a product of its interaction with the material condition. This fundamental understanding had evaded the Idealists and left them solely preoccupied with "mental abstractions."

By bringing this perspective to the attention of human society, Marx created a focus on the material welfare of humanity, especially the masses of humanity that were exploited by ruling elites. Nonetheless, by failing to distinguish between unit mind and Consciousness, Marx inadvertently robbed us of any motivations for living other than providing our material necessities or enjoying their pleasures. In a word, we became reduced solely to economic beings—our future being no more than some vague and static "classless society" that is unable to nourish the human heart or the human thirst for limitlessness. Marx's vision holds no promise of human greatness.

It is unlikely that Marx's intended to confine our potential in such a way. He was a moralist, a humanist, and a champion of the common man. Rather, he saw a deep need to change the dominant philosophy of the intellectual class so that they would get their heads out of the clouds and focus on helping their fellow human beings. The need for this still exists today, as people try to extricate themselves from the rapacious power of the capitalists who enslave them and drain the planet of its resources (including other life forms) in their relentless and mindless quest for profit.

Given this reality, P. R. Sarkar has moved beyond Marx to develop a completely new political economic theory that will revolutionize production in a post-capitalist society. His system of PROUT (Progressive Utilization Theory) can neither be called capitalism nor communism.[58] Neither can it be called idealism nor materialism.

At this point, we simply want to state that, to P. R. Sarkar, both Cosmic Consciousness and the individual mind are real entities and as such can be experienced. Each has its own vibrational motivity. The relationship of the individual mind to the Cosmic Mind can therefore be determined by the relationship of their respective motivities. Mr. Sarkar says:

> When the waves of the unit-mind lag behind those of the Cosmic Mind or move towards the opposite direction—we call it the downfall of the unit. When they move at par with the latter—we call it the natural movement of the unit mind and when they travel faster than the latter we call it advancement or progress.[59]

What is the motivity of the Cosmic Mind to which Mr. Sarkar refers? Theoretically, we can envision it as a straight line. In actuality, it is the field of being. The motivity of the unit mind, however, is a wave that oscillates toward or from this line, thereby gaining contradictory perspectives. This gives rise to the duality of thought.

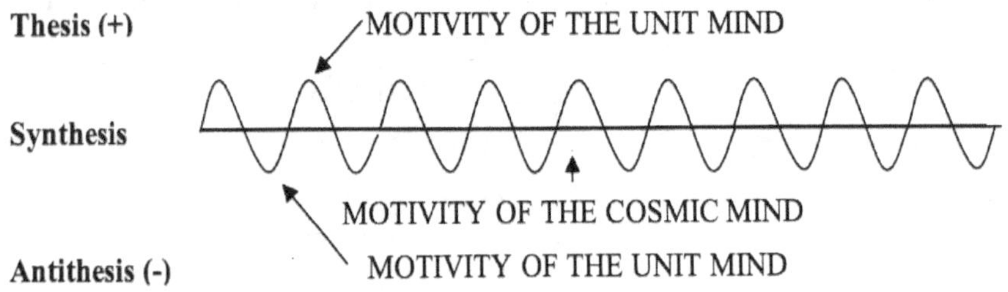

The progress of the human being, therefore, lies in the struggle to resolve experienced contradictions, neither attaching the mind to one polarity or the other, neither seeking pleasure nor avoiding pain. By seeing Divinity in all manifestations, and by performing spiritual practices and selfless service, the mind gains the ability to overcome inherent contradictions and to merge into the holistic consciousness of the One. This occurs when the mind is completely stilled. How an individual is able to achieve that state, without isolating oneself from the material struggles of the environment and society, is the subject of Intuitional Science.

Wave Theory, the Individual and Society

The reason that the Supreme Oneness can be attained by service to one's fellow human beings and the created universe is that the Supreme Oneness exists not only within the individual mind, but also outside the individual mind. As such, by serving the external humanity and the universe, the individual mind expands and begins to merge with the Supreme Oneness.

It must be noted however that selfless service is not an easy path. It requires sacrifice. Let us look now at how wave theory is applied to determine the relationship of the individual to the collective, or in its fullest expression, to society as a whole. According to P. R. Sarkar:

> It is with the assembly of many individuals that a society comes into being. In such a society, although every individual wishes to move as per the momentum of his own Id or samskaras (propensities) yet he just cannot do so cent per cent. Individualism is possible in the subtle and causal spheres, not in the sphere of this crude world. The social order that is created through the conglomeration of several trends, indirectly influences each and every one of them, which gets inclined more or less to keep abreast of or adjust itself to the collective social evolution at least in the worldly spheres.[60]

But if, according to Sarkar, the individual must adjust himself or herself to the collective social evolution in order to maintain physical existence, so too is the collective compelled to adjust itself to the motivity of the individual in like manner. Sarkar puts it this way:

> The vibrational fluctuations of the resultant motivity of the various individual motivities that can well be called the motivity of the social order are somewhat shorter than those of the unit. It is this shortness thereof that paves the way for evolution or revolution.[61]

The shortness of motivity of the social order results from the process of "averaging" that occurs when a collective of entities must all be accounted for. The elongation of the wave of the unit occurs because the unit can move faster than the thought or activity of the mass of society.

Therefore, we see that Sarkar is not willing to hold up and glorify the individual at the expense of the collective as capitalist ideology is want to do, but neither is he willing to favor the collective at the expense of the individual as Marxist ideology is want to do.

For the above reasons, we can determine that the methodology of wave theory is an evolutionary step in the development of methodologies. It allows those who study the motion of any object to have a greater understanding of its process than any theory that has preceded it. It synthesizes longstanding contradictions between metaphysics and dialectics, between materialism and idealism, and between the individual's and the collective's status in society. At the same time, wave theory adds new dimensions to the theory of motion. For example, it observes that motion is not continuous, but rather a pulse that gains its momentum from motionlessness (Consciousness). It also observes that motion is faster at the beginning of a pulse, and at the end of the pulse during the decline. In its manifestative motion, it has more energy at the initiation and slows down as it reaches its peak. In the systaltic motion, it is slower at the peak and speeds up as it descends to its trough. During its manifestive motion, it is progressive in nature, while, in its systaltic motion, it is regressive.

Finally, PROUT Methodology (wave theory) introduces spirituality or Intuitional Science as the ground of being. This revolutionary idea allows human beings the potential for greatness beyond just meeting our basic needs. We are not just material or economic beings; we are spiritual beings. We exist for the purpose of achieving Divinity. As such, our social institutions must be created to assist the individual in this pursuit. Our political, economic, social, and environmental policies and institutions must be geared in this direction. Hence, the very purpose of a nuclear revolution is to facilitate humanity in achieving its optimum potential. Mr. Sarkar says:

> Spirituality is not a utopian ideal but a practical philosophy which can be practiced and realized in day-to-day life, however mundane it be. Spirituality stands for evolution and elevation, and not for superstition in action or pessimism. All fissiparous tendencies and group or

> clan philosophies which tend to create the shackles of narrow-mindedness are in no way connected with spirituality and should be discouraged. That which leads to broadness of unison alone should be accepted. Spiritual philosophy does not recognize any distinctions and differentiations unnaturally made between one human being and another. [It] stands for universal fraternity.[62]

He goes on to say:

> In the present environment many fissiparous tendencies are working to divide humanity into mutually belligerent groups. Spirituality must inculcate [common] sense in human psychology, and develop a natural affinity amongst [the] species of the creation. The approach of spirituality should be psychological and rational, and should offer a touching appeal to the deepest psychic sentiments of human beings. Human beings should appreciate by a rational analysis their relationship with the Cosmic Entity and recognize the most benevolent kindness of the most beloved Entity. Spirituality should lead human beings to the one Cosmic truth from which they have derived their selves and which is the ultimate destiny. That ultimate and absolute ideal is the Cosmic ideal—an ideal beyond the scope of time, place and person. It is the Absolute, without and beyond relativity. It stands with its own luster for all times and for every factor of the Cosmos, may it be a human being or a less-evolved animal. The Cosmic ideal alone can be the unifying force which shall strengthen humanity to smash the bondages and abolish all narrow domestic walls of fissiparous tendencies.[63]

While regional differences will exist and should be allowed to exist because they give color and variety to human existence, a Cosmic Ideal alone can unify us as a species on this small planet. But how do we nurture a Cosmic Ideal that is common to all. To Mr. Sarkar, this will require the realization of both idealistic and materialistic goals. He says:

> The inspiration of Cosmic sentiment will depend upon certain objective physical problems which must be solved on a collective humanitarian basis. In the relative objective sphere, the following few fundamental problems must be attempted and solved. These are: (1) Common Philosophy of Life, (2) Same Constitutional Structure, (3) Common Penal Code, and (4) Availability [supply and purchasing capacity] of the Minimum Essentials of Life.[64]

We will briefly examine these "fundamental problems" in Chapter 12 for they provide us with the basic contradictions to which a revolutionary leadership must attend, even prior to announcing the clarion call for revolution. But for now, let us conclude this chapter with a look at the relationship between our roles as individuals and as social beings.

Relationship between Individual and Social Expressions

Human beings are simultaneously individuals and social beings. Sarkar describes the spiritual path as "a subjective approach with objective adjustment."[65] Because the spiritual path leads to the greatest good (the summum bonum of human progress), and because the spiritual path is a personal path, the individual assumes the role of the primary aspect in the contradiction within human expression. The social being is the secondary aspect. This being the case, the society (collective) should exist to serve individuals, not the other way around. A political economy, regardless of ruling class, should be structured to allow individuals the optimum opportunities to achieve spiritual realization. On this premise, Sarkar, developed his political economic theory called progressive socialism or PROUT. PROUT is an acronym for Progressive Utilization Theory.

Having said this, Sarkar also sets up guidelines to keep the individual from exploiting the collective. Amenities should only be provided to those individuals who contribute to the greatest good.

Sarkar argues:

> If the maximum amenities of meritorious people become excessively high, then the minimum requirements of common people should be immediately increased. For example, if a person with special qualities has a motorbike and an ordinary person has a bicycle, there is a balanced adjustment. But if the person with special qualities has a car, then we should immediately try to provide the common people with motorbikes.[66]

To summarize P. R. Sarkar's description of human beings, he holds that the universe is composed of Consciousness, which is crudified into fundamental factors that give rise to the building blocks of different forms, both mental and physical. Each form is distinguished by its wave characteristics. The unique characteristic of the human mind is that it has the capacity to reflect any wavelength within the entire spectrum of the manifest universe (psycho-physical reality) and thereby is able to overcome its inherited and environmental conditioning that defines it within the evolutionary process. Through overcoming these bondages or attachments, the unit consciousness is eventually able to merge in the pure I-feeling of Cosmic Consciousness. This is true for every human being, although we must also concede that evolution is non-simultaneous. That is to say, not everyone evolves as the same rate of speed, nor does every human being begin at the same time or place. This gives rise to our distinctions as human beings, but it does not deny that, regardless of our state of development as individuals, our goal and our spiritual heritage remain the same. Therefore, it behooves each of us to help each other toward our highest attainment.

In the evolutionary process, each unit mind inhabits a form that is complimentary to its propensities and desires. By gradually extracting oneself from attachment to psycho-physical forms and by concentrating attention on the Absolute Consciousness, one evolves from humanity to divinity. This process of enlightenment does not occur by accident, but by a complicated process of clash and cohesion (the interpenetration of opposites) in which the attraction to matter is powdered down into an attraction for more subtle vibrational states. This occurs in a process over many lifetimes.

The evolutionary phase of creation is called Pratisaincara (the process from crude to subtle). It indicates the unit's return to Pure Consciousness. The return to Pure Consciousness occurs, not only through spiritual practices like meditation on the Supreme Oneness, but also on selfless service to humanity and the created universe. In this process, nuclear revolution provides the optimum means by which to serve humanity and the world.

With this cursory view, the leaders of a nuclear revolution are able to define who we are as human beings and as a human society. We can envision our current position as a human species in relation to our evolutionary trajectory toward our highest state of being. We have identified a new, dynamic method for analyzing the movement of society and for charting our progress in the course of creating a revolution. We can now prioritize the activities of our campaigns in relation to our goals and our direction. Our new methodology will also help us chart our progress and speed of motion.

Chapter Three: What Is a Revolution?

Human society is like a garland which is made from different types of flowers linked by one common thread. The overall beauty of the garland is dependent upon the beauty of each flower. P. R. Sarkar

Revolutions in History

THE MOVEMENTS OF SOCIETY in periods of expansion may be evolutionary or revolutionary. The movement of society in periods of contraction may be counter-evolutionary or counter-revolutionary. The main factor in revolution is the application of tremendous force to move society forward. This application of great force also creates greater speed of movement. Nuclear revolution implies the use of great force to terminate the old capitalist system and create a new, vital system which seeks the benefit of all.

The definition of nuclear revolution is built upon and exceeds the understanding of revolutions as they have existed throughout history. As early as 350 BC, Aristotle, the Greek philosopher, analyzed the topic of revolution. He was interested in the causes and nature of political revolution. He also wanted to know how states are preserved and by what means each state may best be preserved.

Aristotle believed that no matter the type of government, they all gained power by promising justice and equality to the people. Democracy, he said, "arises out of the notion that those who are equal in any respect are equal in all respects; because men are equally free, they claim to be absolutely equal." Whereas, "Oligarchy is based on the notion that those who are unequal in one respect are unequal in all respects; that is, in property, they suppose themselves to be unequal absolutely."[67] The contest between those who espouse democracy and those who espouse oligarchy is the wellspring of revolution.

Aristotle believed that both the democrats and the oligarchs failed to meet a universal standard and that when either side ever believed that their share in the government did not accord with their preconceived ideas, they stirred up revolution.

In defense of a universal standard, which seeks the good of all, Aristotle held that, "Those who excel in virtue have the best right of all to rebel . . . but then they are of all men the least inclined to do so." [68] In a nuclear revolution, we will actualize Aristotle's defense of a universal standard because our moral leadership is inclined to rebel.

In the past, the upheavals that most people call revolutions were either political, cultural, or economic revolutions. When people think about revolution, they usually think about a political revolution in which a government is overthrown. Such revolutions go back to the beginning of Greco-Roman civilization. For example, there was the Athenian Revolution in 508 BC that occurred when a Spartan army invaded Athens to clamp down on its budding democracy. The Athenians drove them out, becoming thoroughly democratic and inspiring scores of nations to follow suit.

In 378 BC, there was the Theban Revolt in which a small ruling class (oligarchy) of Spartans was overthrown when seven conspirators, dressed like women, infiltrated a party attended by the rulers and stabbed them to death. This revolution created a democracy in Thebes.

In Rome, from 73-71 BC, a slave revolt led by the gladiator Spartacus raised an army of sixty thousand and defeated nine Roman armies before it was put down. This had the effect of prompting Roman citizens to prefer a dictator rather than the indecisiveness of a democracy.

Not all revolutions were led by men. In AD 60, Boudicca, who was a wife of a British king, was brutally mistreated by the Romans after her husband's death. She started a revolution, defeated an entire Roman legion and destroyed three cities, including London, before she was defeated. She was never captured and has since become a symbol of resolve against all odds for the British people.

Another example of revolution is the Jewish Revolt against the Roman army from AD 66-70. The Jewish Revolt led to massive civilian death and the destruction of the Temple in Jerusalem (AD 70) as well as to the capture of the fortress Masada (AD 73). The revolt spurred a movement that eventually led to the ascendency of the Jewish priest class and the split between early Christians and Jews.[69]

A final example is the story of revolution in Rome itself. Between 133 BC to AD 14, the tribunate of Tiberius Gracchus was replaced by the accession of Tiberius to the throne. During this time, Rome witnessed the collapse of the Republican political system, twenty years of civil war, and the consolidation of a monarchical system of power under the reign of Emperor Augustus, the first Roman emperor.

After the fall of the Roman Empire, the Catholic Church took over the task of governance and became a multinational church-state. During its rule, it faced schisms and heretical reformers, but it did not face revolution within its ranks until the Protestant revolution in 1517 when Martin Luther published his *Ninety-Five Theses.* After many previous attempts to reform the Church, this event created a massive schism between the conservative hierarchy and the progressive reformers that ultimately led to a hundred years of war between the Catholics and the progressive Protestants across Europe. The revolution ended with the creation of many new Protestant religions. While the *Protestant Reformation*, which actually was a revolution against a Church oligarchy, never replaced the Church hierarchy or its iron rule that had lasted for over fifteen hundred years, the Church's dominance was finally at an end.

With the political power of the Catholic Church severely damaged by the Protestant Reformation, a series of democratic bourgeoisie revolutions occurred throughout Europe beginning in the seventeenth century, immediately following the Protestant revolution. The purpose of these revolutions, which were led by the middle-class merchant classes, was to permanently overthrow the rule of the Catholic Church, the landed aristocracy, and the feudal economy which supported them. During these revolutions, the bourgeoisie (merchant class) ascended to power and created a new form of government under their control. In this development, the power of the church-state was replaced by the power of the nation-state. In England, by 1688, the term "revolution" was well established and was used to describe the bourgeoisie revolution in which James II was replaced by Willian III in what became known as the "*Glorious Revolution*."

In colonial countries like the United States, the aim of the bourgeois revolution was to attain national independence. The American Revolution was a fight by the American bourgeoisie against the ruling English bourgeoisie. At the time, these bourgeoisie revolutions were progressive in nature, insofar as they promised justice and equality, and attempted to better meet the needs of the people.

It was not long, however, after the bourgeois revolutions occurred, that the capitalists' concern for the people's need for justice and equality was subverted into a strict preoccupation with their own personal aggrandizement of wealth and power (see *The Untold Story of Western Civilization*, Volumes 4 and 5 for this history). The merchant class quickly changed from being democrats to oligarchs. Aristotle's definition of revolution and its causes has rung true in the history of Western Civilization.

In their bid for greater riches, the small oligarchy of rich capitalists created the means to separate the people from their wealth and resources. In short, they accomplished this through the perpetration of war, colonialism, slavery, and enforced debt. We can say with certainty that the capitalists did not "earn" their money, they stole it by force and intimidation.

To counter the industrialized capitalist countries' use of force and intimidation (imperialism) against the world's people, the ideas of Marxism, communism, and democratic socialism arose. While the democratic socialists believe that capitalism can be reformed, the Marxists and communists did not believe that capitalism could be reformed and proposed revolution. They believed and still believe that the workers of the world must unite in revolution and free themselves from capitalist oppression in order to create a world run by and for the working class. Thus, in the Marxist view, worker revolutions need to happen in countries all over the world. This ideology set off a series of anti-capitalist revolutions in Russia, China, Vietnam, Korea, Cuba, etc., which were partially or completely successful in overthrowing the capitalist ruling elite of those countries. Finland, Germany, Hungary, Spain, Yugoslavia, and other countries also attempted revolutions but failed.

Despite the communist revolutions, we find that, in most cases, the leadership of those countries has reverted to a capitalist mentality. Today both Russia and China are controlled by capitalist oligarchs. While they may have their differences with the oligarchs of the West, they share the same exploitive mentality.

That being said, today, the United States and the European capitalists have established a global hegemony over the world's people. While local leaders may sometimes struggle to protect the resources of their countries against the international capitalist, they are easily destroyed. In such situations, a ground war against "insurgents" or "terrorists" may continue for years or until the capitalists are able to insert a puppet regime that will allow them to exploit the country's resources and labor without opposition.

As social scientists grapple with the idea of revolution, they have developed more complex definitions. Jack Goldstone, one of the leading authorities on the study of revolutions and long-term social change, defines a revolution as "an effort to transform the political institutions and the justifications for political authority in society, accompanied by formal or informal mass mobilization and non-institutionalized actions that undermine authorities."[70]

Jeff Goodwin, another expert, writes in his book, *No Other Way Out: States and Revolutionary Movements, 1945-1991*, that revolutionary movements are not only a response to economic inequality and social injustice, but are also a response to political repression and violence. He holds that revolutions entail not only mass mobilization and regime change, but also rapid and fundamental social, economic, and cultural change, during or soon after the struggle for state power.[71]

This leads us to acknowledge that the term revolution often denotes great changes that spill over beyond the political sphere. These can cause economic revolutions and cultural revolutions. We can see this in the case of the revolutions in England, France, the United States, Russia, China, etc.

There are also cases where economic or cultural revolutions occur without a political revolution.

For example, the Industrial Revolution, the Green Revolution, and the Technological Revolution denote periods of rapid transformations in the development of machines, agricultural production, and digital communication that do not affect the political structure, but have an impact on the economic structure. Such revolutions, however, do not replace the capitalist economy, they only change its expression. They signify a rapid change within specific sectors of the capitalist economy.

Aside from these sectoral changes in the capitalist global economy, the term revolution is also attributed to cultural developments that may also have a great impact on society. For example, in the United States, there was the Sexual Liberation movement, the Youth movement, and the Civil Rights movement. These revolutionary movements did not challenge the state or the economy, but rather they pitted sections of society against each other. In the oscillations of these social movements, society moves forward, but at a very slow pace.

These economic and cultural revolutions are not revolutions in the true sense. They are, at best, evolutionary movements. They certainly were not nuclear revolutions because they did not result in the change of the collective psychology and the forward movement of the social cycle. Nor where they spiritually motivated.

In the next three chapters, we will explore the dynamics of nine contemporary revolutions, many of which are still being fought at the time of this writing. We note that in every revolution today the oligarchs, whether from so-called capitalist or communism countries, represent the forces of counter-revolution whereas the pro-democracy movements represent the forces of revolution. The struggle between the oligarchs and the people is being played out across the entire planet today. While, on the surface, the revolutions may seem to have nothing in common with each other, it is this underlying reality that ties them all together. The people are all fighting a capitalist mentality regardless of the different political ideologies under which they are being controlled and exploited. To understand this fact, we will take a quick look at the contemporary revolutions in the Cuba, Nicaragua, Chile, Ukraine, Belarus, Bulgaria, Hong Kong, South Sudan, and Kurdistan. In doing so, we will see how the relationship between capitalist oligarchs and pro-democracy advocates play out under different conditions no matter where they occur in the world.

Chapter Four: Contemporary Revolutions Under Western Capitalism

LATIN AMERICA HAD BEEN colonized by Europeans and then Americans since the beginning of the Capitalist Age in the fifteenth century. The contemporary revolutions in Cuba, Chile, and Nicaragua, discussed below, were fought against the predations of American capitalism.

Cuba

The oldest of all contemporary revolutions in Latin America took place in Cuba in 1959 when Fidel Castro led a small guerilla army to overthrow the corrupt capitalist dictatorship of Fulgencio Batista.

Cuba is located in the northern Caribbean where the Caribbean Sea, Gulf of Mexico, and Atlantic Ocean meet. It is less than one hundred and fifty miles from Florida's southern coast.

Havana is the largest city and capital of Cuba. The area of Cuba is 42,800 sq. miles.

From the fifteenth century, it was a colony of Spain until the Spanish–American War of 1898, when Cuba was occupied by the United States and gained nominal independence as a *de facto* United States protectorate in 1902.

In the years immediately following its independence from Spain, the Cuban republic saw significant economic development. For example, Cuba had Latin America's highest per capita consumption rates of meat, vegetables, cereals, automobiles, telephones, and radios, although about one third of the population remained poor and enjoyed relatively little of this consumption.

The economic expansion, however, came at the price of political corruption and a succession of despotic leaders, which did not end until the overthrow of the dictator Fulgencio Batista by the 26th of July Movement, led by Fidel and Raúl Castro Ruz, during the 1956–59 Cuban Revolution.[259]

Batista had been receiving financial, military, and logistical support from the United States government. He had aligned himself with the wealthiest landowners, including US-based multinational corporations, who owned the largest sugar plantations in the country. Batista, therefore, became the caretaker of a stagnating economy that widened the gap between the rich and poor Cubans. To minimize dissent, Batista had suspended the Constitution and revoked most political liberties. He denied the right of workers to strike.[260]

During the 1950s, different organizations, including groups like Castro's, who advocated armed uprising, competed for public support to bring about political change. By this time, Havana was controlled by US-based mafia capitalists and the CIA. The city was rife with gambling casinos, exotic restaurants, department stores, red light district, etc., in order to serve as a destination for rich American tourists. The mafia shared some winnings with the officials of the Cuban government while the Cuban people suffered unseen in their filth and poverty. The tourist trap prevented the American people from seeing the real lives of the Cuban people, just as similar tourist traps prevented Americans from seeing the lives of the people of other Latin American countries.

Castro offered the Cuban people a better life than they had ever known before. He offered them better houses than mud huts, food so their families would not go hungry, better health care so they did not have to live short miserable lives and die young, and better education so children could learn to read and write.

The rural peasants made up half the population of Cuba. Castro built his Revolution on the power of the peasant. It was not based on a struggle between ruling power groups like the military or political organizations or capitalists.

The peasants never cared about such kinds of revolutions because their lives never changed no matter who was in power. Castro made the peasants understand what political power was and how it could be theirs. They flocked to him.

Castro's revolution proved that the gulf between the poor and the rich could be bridged. It proved for all the people of Latin America that better lives were possible.

Castro's ideas did not find advocates among the white Christian people of North America whose mainstream press never reported on the intentions and behaviors of the US-based multinational corporations, but instead lied to them saying that the corporations and CIA were only trying to bring democracy to the less fortunate and uncivilized people in Latin America and the world. And so, Castro invited the communist enemies of the capitalist United States to come to Cuba and support his revolution.

In return for their generous assistance, Castro allowed the Russians to set up a military base in Cuba that targeted the US. At the same time, the US had Guantanamo Bay Naval Base at the southeastern end of Cuba. Since the Cuban Revolution of 1959, the Cuban government has consistently protested against the US presence on Cuban soil and called it illegal under international law, alleging that the base was imposed on Cuba by force.

Since 2002, the naval base has contained a military prison, the Guantanamo Bay detention camp, for alleged unlawful combatants captured in Afghanistan, Iraq, and other places during the War on Terror. Cases of torture of prisoners, and their denial of protection under the Geneva Conventions, have been condemned internationally.[261]

Before the 1959 revolution, the big capitalist land owners had been exploiting Cuban labor and taking resources out of Cuba. For example, just seven companies owned five million acres, or one-half of Cuban farm land. They created products that were not for the Cuban people. The peasants cut sugar cane for one hundred and sixty-one sugar mills that exported the sugar abroad. The Cuban people also worked in the foreign-owned mines that sent chromite, manganese, copper, iron ore, and nickel to the US.

The Russians, in the meantime, were inviting the young people of Cuba to travel to Russia, all expenses paid, to gain a free education and to learn what had to be done to stabilize their country after a revolution.

Revolution of 1956

The armed revolution against Batista's government began in 1956 when Fidel Castro and about eighty supporters sailed a yacht called *Granma* from Mexico and landed in Cuba with the intention of starting a rebellion against the Batista government.

Batista had correctly predicted where the landing would take place and his troops were ready. The landing party was bombarded by helicopters and airplanes. Because the terrain on the coastline provided little cover, the party was an easy target. The revolutionaries moved inland.

Che Guevara gave this account of the landing:

> We reached solid ground, lost, stumbling along like so many shadows or ghosts marching in response to some obscure psychic impulse. We had been through seven days of constant hunger and sickness during the sea crossing, topped by three still more terrible days on land. Exactly 10 days after our departure from Mexico, during the early morning hours of December 5, following a night-long march interrupted by fainting and frequent rest periods, we reached a spot paradoxically known as Alegría de Pío (Rejoicing of the Pious).[262]

Here, the group was attacked again by Batista's army. After losing more men, the survivors continued to the foot of Pico Turquino in the Sierra Maestra to carry out their mission.

Initially, Batista did not know who among the revolutionaries had died during the landing. International media widely reported that Fidel was dead. This was not the case, however. Of the eighty-two revolutionaries who left Mexico, only twenty had survived, including Fidel, his brother Raúl, and Che Guevara.

Throughout 1957, Fidel Castro's small band of revolutionaries operated out of a base in the Sierra Maestra mountains. They staged hit and run attacks on Batista's government. In the beginning, the Cuban army and the Batista government did not take these attacks seriously, but their attitude changed in the spring of 1958 when the revolution started to gain international recognition. When Castro called for a General Strike, Batista decided to destroy Castro's small army. [263]

In May 1958, General Eulogio Cantillo was given the mission to do so. Cantillo planned to use nearly all of the Cuban regular army to surround the Sierra Maestra, set up a blockade to prevent weapons from going in, and then attack from the north with fourteen battalions.

Castro's troops knew the terrain well, and they set up mine fields and built defensive positions along the major routes through which they expected the army to attack. Castro had excellent knowledge of where the army was and what they were planning. He also had the support of local peasants, who assisted in the transmission of information on Cantillo's troops and risked their lives to hide rebel supplies.

The first attack was on June 28, 1958 when Cantillo's troops moved out of the Estrada Palma Sugar Mill. This attack was halted by an ambush from Che Guevara's troops. Armored cars that moved off the road ran into a mine field. The army began to retreat while Che's troops continued to fire on them.

On July 11, the army landed a battalion at the mouth of the La Plata River. The government wanted to surround Castro's mountain defenses at Turquino Peak. The Cuban soldiers (most were new recruits) again were ambushed by Castro's guerrillas and were soon surrounded and immobilized. A second battalion landed and tried to help, but they were halted at the beach. A third battalion was then sent, but they ran into another part of Castro's forces and did not push through the road blocks. After more than a week, on July 21, Battalion eighteen surrendered. They had lost forty dead and thirty wounded, and two hundred and forty became prisoners. Castro's troops lost just three of their own men.

General Cantillo withdrew Battalion seventeen, but he laid a trap if Castro chose to follow the retreating soldiers. This resulted in the Battle of Las Mercedes. For the first time in the fighting, Castro's guerillas were caught in a trap and more than seventy were killed in the first two days of fighting. Those killed included a senior rebel leader, René Ramos Latour.

Castro managed to get his army out of the trap by opening negotiations with General Cantillo and Batista. By August 8, all of Castro's forces had escaped and the government's offensive was over.

By trying and failing to destroy Castro's guerrilla army, the Batista government looked weak and ineffective. The impact upon the morale of the Cuban army was devastating. The majority of the junior officers who had fought so hard over the preceding weeks were disgusted that Cantillo had even stooped to negotiate. Moreover, Castro's masterful maneuver had come at just the time when the regular army, after having fought well for the first time in the campaign, seemed to have all the advantages.

Uplifted by the government's failure, Castro went on the offensive and within four months, he had taken control of Cuba.[264]

Late in 1958, the rebels had broken out of the Sierra Maestra and launched a general popular insurrection. After Castro's fighters captured Santa Clara, a centrally located city with two hundred and fifty thousand people, Batista fled with his family to the Dominican Republic on January 1, 1959. On January 8, 1959, Fidel Castro's forces entered the capital city of Havanna. The liberal Manuel Urrutia Lleó became the provisional president.

After the success of the revolution, Castro nationalized American banks, oil refineries, sugar and coffee plantations along with other American-owned businesses in Cuba.

From 1959 to 1966 Cuban counter-revolutionary forces fought a six-year rebellion in the Escambray Mountains against the new Castro government. The rebel group of insurgents was a mix of former Batista government operatives and soldiers, big farmers who had their land confiscated, and also former allied guerrillas who had fought alongside Castro against Batista during the Cuban Revolution, but who were anti-communist. The end result was the elimination of all insurgents by Cuban government forces in 1965. The Cuban government describes the rebellion as the "War against the Bandits."

The CIA provided aid to the counter-revolutionaries, but withdrew all support after the failed Bay of Pigs Invasion in 1961.

The Bay of Pigs Invasion was a military invasion of Cuba undertaken by the Central Intelligence Agency's counter-revolutionary group called Brigade 2506. It was composed of Cuban exiles who had traveled to the United States after Castro's takeover, but also US military personnel, with the intention of overthrowing Castro's government. Launched from Guatemala and Nicaragua, the invading force was defeated within three days by the Cuban Revolutionary Armed Forces, under the direct command of Castro.

US President Dwight D. Eisenhower had been very concerned at the direction Castro's government was taking, and in March 1960, he allocated $13.1 million to the CIA to plan Castro's overthrow. The CIA proceeded to organize the Bay of Pigs invasion. Eisenhower's successor, John F. Kennedy, approved the final invasion plan on April 4, 1961.

Castro decided to take personal control of the operation and stopped the invasion. As the US involvement became apparent to the world, and with the invasion stalled, Kennedy decided against providing further air cover. As a result, the operation only had half the forces the CIA had deemed necessary. The original plan devised during Eisenhower's presidency had required both air and naval support. On April 20, the invaders surrendered after only three days, with the majority of them being publicly interrogated and put into Cuban prisons. With the failure of the Bay of Pigs invasion, the counter-revolt in Cuba was also soon defeated.

The failed invasion helped to strengthen the position of Castro's leadership and made him a national hero. The animosity between Cuba and the US became more entrenched and the relationship between Cuba and the Soviet Union strengthened.[265]

The US State Department has estimated that thirty two hundred people were executed by Castro's government from 1959 to 1962. According to Amnesty International, however, official death sentences from 1959–87 numbered two hundred and thirty-seven of which all but twenty-one were actually carried out. The majority of those executed were policemen, politicians, and informers of the Batista regime accused of crimes such as torture and murder. The public trials and executions had widespread popular support among the Cuban population.[266]

Castro Reforms

Today Cuba has one of the world's only socialist economies. According to the United Nation's Human Development Index (HDI), Cuba ranks very high in national performance included health care and education.

The HDI is a measure for assessing long-term progress in three areas of human development: a long and healthy life, access to knowledge, and a decent standard of living. A long and healthy life is measured by life expectancy. Knowledge level is measured by mean years of education among the adult population and access to learning and knowledge by expected years of schooling for children of school-entry age. Standard of living is measured by Gross National Income (GNI) per capita expressed in constant 2011 international dollars converted using purchasing power parity (PPP) conversion rates.

Cuba's 2017 HDI of 0.777 is above the average of 0.757 for countries in the high human development group and above the average of 0.758 for countries in Latin America and the Caribbean.

Cuba also has high marks for the status of women in society. It ranks 65 out of 160 countries in the 2017 index. In Cuba, 48.9 percent of parliamentary seats are held by women, and 86.7 percent of adult women have reached at least a secondary level of education compared to 88.9 percent of their male counterparts. For every 100,000 live births, 39 women die from pregnancy related causes; and the adolescent birth rate is 43.6 births per 1,000 women of ages 15-19. Female participation in the labor market is 41.5 percent compared to 67.3 for men.[267]

Cuba is also the only country in the world to meet the conditions of sustainable development put forth by the World Wildlife Fund.[268]

An example of the Castro government's embrace of sustainable development can be seen in its response to the destruction of its agricultural system when the fall of the Soviet Union caused Russia to stop supplying gas and oil products that it desperately needed to run its tractors, grow its crops, and get the food to market. Cuba was also under a US-embargo.

Instead of searching for a new supplier of fossil fuels or caving to the US capitalist multinational corporations, the new government hired six Permaculture experts from Australia to teach the people how to plant intensive organic gardens in the cities. The Australians shared their permaculture vision with key agriculturalists, and members of organizations like the *Antonio Nunez Jimenez Foundation for Nature and Humanity* (FANJ). In a few short years, FANJ began teaching permaculture throughout the country, creating a movement that would grow tremendously using a model of *campesino to campesino* (*farmer to farmer*) knowledge sharing. FANJ wrote a book on permaculture methods for Cuba, *Permacultura Criolla* and developed numerous other methods for spreading the promise of permaculture far and wide.[269]

According to the *Architectural Review*, Havana has become an exemplary model of this new self-provisioning, a precedent that demonstrates both the opportunities and obstacles for the transference of organic urban agriculture to other regions. The city has more than two million people and resembles New Orleans more than any other city in the Caribbean. Havana now provides an example of a systematic approach to rethinking urban landscapes for more productive means. A food production infrastructure is now completely woven into the city fabric, with projects that range in size from backyard gardens to large peri-urban farms. More importantly, the Cuban government bolsters these urban growing efforts with training and support, hosting many dozens of subsidized agricultural stores, three compost production sites, seven artisanal pesticide labs and forty urban veterinary clinics. This combination of top-down state support and ground-up citizen participation has proven wildly successful. Economist Sinan Koont estimates that more than eighty-six thousand acres of land are being used in urban agriculture in Havana alone.[270]

The social progress of the Cuban people came at a steep price for Fidel Castro himself. Over his term of office, his enemies made over six hundred attempts on his life. Most of these were initiated by the US-based mafia and by the CIA, who spent years grooming every kind of operative to kill the Cuban leader. Murder attempts included poisoning him with cyanide pills, dosing his diving suit with fatal botulism, and blowing him up during a speech.[271]

The Obama administration in the US initiated a warming of Cuba–United States relations that began in December 2014. This attempt at peace ended a fifty-four-year stretch of hostility against Cuba. In March 2016, Barack Obama became the first U.S. President to visit Cuba since 1928.

In 2015, the Obama administration announced that Cuba would be removed from the United States State Sponsors of Terrorism list.

In 2017, however, President Trump announced that he was "cancelling" the Obama administration's treaty with Cuba. Business travel again became illegal. In 2019, he reinforced a ban on educational and recreational travel and cut the amount that Cubans in the US could remit to relatives in Cuba to one thousand dollars per person per quarter. These actions deny the human rights of the American and Cuban people to get to know and learn from each other.

Lessons from Cuba

The major lesson to be learned from the revolutions in Latin America is that the United States government, particularly the CIA and military, working in consort with US-based multinational corporations, act repeatedly to overthrow popular socialist governments and pro-democracy movements in order to exploit the local people's resources and labor by putting their own puppets in power.

In Cuba, this exploitation unfolded in the following way. Cuba became a US protectorate after the Spanish American war. This war began after the US seized Cuba in their War of Independence against Spain. Immediately after the war, Cuba's economy flourished. This so-called economic progress, however, came at the cost of political corruption and a succession of despotic leaders who took the side of US corporations at the expense of the Cuban people.

Batista had been receiving financial, military, and logistical support from the United States government and had aligned himself with the wealthiest landowners, including US-based multinational corporations, who owned the largest sugar plantations.

Castro's revolution began by soliciting the help of the poorest of the poor. These were the small farm workers, not the urban workers. He promised the campesinos a life that provided their basic needs of food, shelter, clothing, health care, and education.

The Cuban revolution has lasted because Fidel never forgot the people's needs. Neither multiple assassination attempts, nor the CIA's funding of counter-revolutionary forces, could kill the revolution or its leader.

Fidel's Cuba remains a beacon of light for the third world countries of the world, who also remain grateful to the people of Cuba for the doctors and teachers that they continually send to help the poor people of third world countries.

If there was a flaw in Fidel's government, it was that he was unable to distinguish between religious dogma and a true spiritual path. As a consequence, in order to stamp out the corrupting power of religion, he also did harm to people's spiritual longing. In the future, revolutionaries must be able to distinguish between religious dogmatists and their support of corrupt dictators versus spiritual science and its foundation of universal morality.

As a socialist, Fidel Castro's goal was to meet the basic needs of the Cuban people. To hold off the corrupting power of the CIA, the US military, and the counter-revolutionaries, many of whom had fled to Florida after the revolution, Fidel depended on the Soviet Union for help. It was a survival tactic, but it was not based on a need for long-term dependency. The Cuban people discovered the wisdom in striving for local autonomy, when the Soviet Union collapsed and stopped providing the Cuban people with oil and gas. This destroyed the people's agricultural system, which then led the people to create a new one based upon low fossil fuel inputs. They adapted a system that combined Permaculture and worker's cooperatives and, after two years of belt-tightening, created the first national organic food system on the planet. Now the Cuban people eat locally grown nutritious food at affordable prices that is grown in an environmentally sustainable manner. The new system has also created many good living-wage jobs.

Chile

In 1970, the people of Chile had elected Salvador Allende, a socialist, to be President of the country. This happened in large part because US corporations like ITT, Kennecott, Anaconda, Purina, Bank of America, and others were ravaging the resources of the country and the people wanted it to end. During the election process, ITT's director, John McCone, a former head of CIA, invested heavily in an effort to stop Allende from being elected president. After Allende was elected, McCone changed strategies and developed a plan to conduct a coup d'état. The first step was to spread "economic chaos" in the country to soften up the population. As McCone plotted, Allende began his administration by nationalizing the Anaconda and Kennecott's copper mines. The US government under Nixon swung into action. They cut the US Export-Import Bank credits to Chile and pressured the World Bank and IMF (in which they had the dominant voice) to disapprove all further loans to Chile. They also pressured other private banks to cut off credit. During Allende's first year in office access to capital in international markets dropped from two hundred and twenty million dollars to thirty-five million dollars. The global corporations also joined the campaign by refusing to sell Chile spare parts and machinery even for cash. These actions destroyed the economy of Chile.

The American people had no idea what was going on. Certainly, Chile presented no national security risk to Americans. The people of Chile simply wanted greater freedom to control their own resources and livelihood.

Following an extended period of social unrest and political tension between the capitalist-controlled Congress of Chile and the socialist President Salvador Allende, which was set against a backdrop of economic warfare against Chile that was ordered by US President Richard Nixon, President Allende was overthrown by the armed forces and national police in 1973.

On September 11, 1973, the military, citing a call by the Chilean Congress to end Allende's presidency, illegally staged a coup against him. As the armed forces surrounded La Moneda Palace, Allende gave his last speech vowing not to resign. He died in the assault on the presidential palace during the coup. After Allende's ousting, General Augusto Pinochet, who led the coup d'état, was installed in power along with a military junta that suppressed all public dissent. The United States government, which had worked to create the conditions for the coup, promptly recognized the junta government and supported its consolidation of power.[272]

In the first months of the fascist takeover, the dead and disappeared numbered in the tens of thousands.

What happened in Chile was the United States' way of telling the people of the world not to oppose its control over their own resources.

While the banks and corporations tightened their grip on the world's people and their resources, the US government was telling the American people that our government was fighting to defend democracy and to stop the spread of communism in the world. What they did not tell us, however, was that their definition of a communist was anyone who voiced any opposition to US imperialism. These people not only included dissidents and leftists, but also union workers, peasant leaders, priests and nuns, students, teachers, intellectuals, small business people, in short, anyone who loved their country and their freedom as much as we Americans love ours.

Once the military dictatorship was established in Chile, the US corporations took over the economy. The new government cut tariffs on imports, suppressed union power, and "privatized" hundreds of state-controlled industries. Human services were put under the control of private corporations.

This is the story of how Allende and his socialist government was overthrown and how the people of Chile were forced to endure torture, rape, and death at the hands of the US supported military junta.

Prior to the military coup sponsored by US corporations and the CIA, Chile had been hailed as a beacon of democracy and political stability for decades, a period that had seen the rest of South America plagued by military juntas and dictators. This demonstration of local freedom, however, did not sit well with the US-based multinational corporations who wanted control of Chile's markets, its resources, and labor.

Consequently, during the 1950s and 1960s, the United States CIA had been active in funding radio and print propaganda campaigns aimed at impeding the presidential aspirations of the socialist candidate Salvador Allende. Throughout this time, the United States had also successfully impeded other left-wing parties from gaining power.

In the 1958 presidential election, Jorge Alessandri, a nominal independent with support from the Liberal and Conservative parties, defeated Allende by nearly 33,500 votes to claim the presidency. His laissez-faire policies, supported by the United States, were put forward as the solution to the country's economic problems. Under recommendations from the United States, beginning in 1959, Alessandri steadily reduced tariffs on imported goods, a policy that caused the Chilean market to be overwhelmed by American products. These policies cut local jobs and angered the working people of Chile who demanded higher wages. The repercussions of this massive discontent were felt in the 1961 congressional elections. President Alessandri suffered a terrible defeat, sending a clear message that his laissez-faire policies were not desired. The president had accepted one hundred and thirty million dollars from the US banking industry, the U.S. Treasury Department, the IMF, and the CIA to enforce US policies and make Chile more dependent on the United States.

Salvador Allende was a top contender in the 1964 election. The American capitalists and the CIA plotted to get the Christian Democratic contender Eduardo Frei Montalva elected as president fearing that Alessandri's failures would lead the people to support Allende. Clandestine aid to Frei was put forward through the Latin American Alliance for Progress, which promised "twenty billion dollars in public and private assistance in the country for the next decade."

The US feared the example of a "well-functioning socialist experiment" in a region run by military dictators who were under the thumb of the US. According to the 1975 Church Commission Report,[273] covert United States involvement in Chile in the decade between 1963 and 1973 was extensive and continuous. A declassified file from August 19, 1970 reveals the minutes of high-level officials in the CIA known as the "Special Review Group. It was chaired by Henry Kissinger. The minutes detail correspondences between each of the officials and reveal the underlying interests of the United States in the US-based copper companies in Chile. The election in Chile represented the potential that the US economic dominance in Chile could collapse if its outcome did not align with US interests.

At a September 8, 1970 meeting of the *40 Committee*,[274] which is one of the least-known, most secretive, and most powerful committees of the Government, the chairman of the committee asked for an analysis of where the US/CIA stood in terms of taking action to prevent Allende from becoming President of Chile. William Broe, a high-ranking CIA officer, reported that the election of Eduardo Frei Montalva, who had previously held the office of president, was essential to the situation in Chile. The strategy of the CIA was to support Montalva's candidacy. The

40 Committee asked that the CIA collect information and create more intelligence reports to see what could be further done in Chile. The committee decided it was unlikely they were going to be able to influence the October 24 congressional election to go against Allende. The Director of the CIA, Richard Helms was also concerned about Allende supporters in the Chilean military, as it seemed they would support Allende in the event of a coup. As a result of all this information, the Committee decided they wanted a full analysis of two things: (1) a cost versus benefit analysis of organizing a military (Chilean) coup and (2) a cost versus benefit analysis of organizing future oppositions to Allende to topple his influence. They presented two options for Henry Kissinger: political maneuvering or outright force.

Salvador Allende won the 1970s presidential election by a plurality vote (nearly thirty-seven percent). On September 15, 1970; before Allende took office, Richard Nixon gave the order to overthrow Allende. According to a declassified document from the NSA, the handwritten notes from Richard Helms (CIA director at the time) states:

> 1 in 10 chance perhaps, but save Chile!; worth spending; not concerned; no involvement of embassy; $10,000,000 available, more if necessary; full-time job--best men we have; game plan; make the economy scream; 48 hours for plan of action.[275]

These notes came from a meeting Helms had with President Nixon, indicating the administration's willingness to stage a coup in Chile and the extent to which Nixon was willing to go to succeed.

On November 5, 1970, Henry Kissinger, chair of the 40 Committee, advised President Nixon against peaceful coexistence with the Allende administration and instead advocated one of two positions. The U.S. government, according to Kissinger, had two approaches to fighting socialism as represented by Allende. "Track I" was a State Department initiative designed to thwart Allende by subverting Chilean elected officials within the bounds of the Chilean constitution that excluded the CIA. Track I was expanded to encompass some policies whose ultimate goal was to create the conditions that would encourage a coup. "Track II" was a CIA operation overseen by Henry Kissinger and CIA's director of covert operations, Thomas Karamessines. "Track II" excluded the State Department and Department of Defense. The goal of Track II was to find and support Chilean military officers who would conduct a coup.

A first step in overthrowing Allende was the removing of General René Schneider, the army chief commander.[276] Schneider was a constitutionalist and would oppose a coup d'état. To assist in the planned kidnapping of Schneider, the CIA provided "$50,000 in cash, three submachine guns, and a satchel of tear gas, all approved at headquarters . . . ." [277] The submachine guns were delivered by diplomatic pouch.

A group was formed, led by a retired general, General Roberto Viaux. Viaux, however, was considered to be unstable by the U.S. and was discouraged from attempting a coup alone. The CIA encouraged him to join forces with an active duty general, Camilo Valenzuela, who had also been approached by CIA operatives. They were joined by an Admiral, Hugo Tirado, who had been forced into retirement after the Tacnazo insurrection, which was a military protest against underfunding. On October 22, Viaux went ahead with a plan to kidnap General René Schneider. Schneider drew a handgun to protect himself from his attackers, who mortally shot him. He died in Santiago's military hospital three days later.

In 1970, the US manufacturing company ITT Corporation owned seventy percent of Chitelco, the Chilean Telephone Company, and funded El Mercurio, a Chilean right-wing newspaper. The CIA used ITT as a conduit to finance aid to opponents of Allende's government. On September 28, 1973, ITT's headquarters in New York City, was bombed by the *Weather Underground* for the alleged involvement of the company in the overthrow of Allende.

On September 10, 2001, a suit was filed by the family of Schneider, accusing former US Secretary of State Henry Kissinger of arranging Schneider's 1970 murder because he would have opposed a military coup. CIA documents indicate that while the CIA had sought his kidnapping, his killing was never intended. Kissinger said he had declared the coup "hopeless" and had "turned it off." However, the CIA claimed that no such "stand-down" order was ever received.

The attempted kidnapping and murder of General Schnieder shocked the people of Chile and caused them to rally around the

Chilean Constitution and the Allende presidency. With the Track I strategy in ashes, the US capitalist elite, spearheaded by the Nixon administration and the CIA went immediately into their Track II strategy, which was to use Chilean military operatives to overthrow Allende's government.

On November 25, 1970, Henry Kissinger issued a memorandum that detailed the Covert Action Program that the US would spearhead in Chile. In the memorandum, Kissinger stated that there were five principles of the program. The US would continue to make contacts in the Chilean military, take steps to divide Allende's supporters, cooperate with the media to run anti-Allende propaganda campaigns, support non-communist political parties in Chile, and publish materials that stated that Allende did not adhere to the democratic process and also wanted to form connections with Cuba and the Soviet Union.

In response to US intervention and the murder of his general, Allende nationalized several US corporations and the copper industry.[278]

After, Kissinger's memorandum regarding the Covert Action Program, the US operatives set the stage for Allende's overthrow in its fight against Chilean democracy. The US not only created the conditions for the coup, but helped to consolidate power in the hands of the ruthless junta of General Pinochet after the coup.[279]

On June 29, 1973, Colonel Roberto Souper surrounded the presidential palace with his tank regiment but failed to depose the Allende Government. In August 1973, a constitutional crisis occurred when the Supreme Court publicly complained about the government's inability to enforce the law of the land. On August 22, the Chamber of Deputies (with the Christian Democrats united with the National Party) accused the government of unconstitutional acts and called upon the military to enforce constitutional order.

For months, the Allende administration had feared calling upon the *Carabineros* national police, suspecting them of disloyalty. On August 9, Allende had appointed General Carlos Prats as Minister of Defense, but the general was forced to resign both as defense minister and as the Army commander-in-chief two weeks later when an embarrassing traffic incident caused him to be scandalized in the press.[280] General Augusto Pinochet replaced him as Army commander-in-chief the same day.

Following several more months of political manipulations by the CIA, on September 11, 1973, early in the morning, the Chilean Navy seized Valparaíso, a major city and seaport in Chile. It strategically stationed ships and marine infantry along the central coast and closed radio and television networks. The Province Prefect informed President Allende of the Navy's actions and immediately the president went to the presidential palace with his bodyguards. By 8:00 am, the Army had closed most radio and television stations in Santiago city while the Air Force bombed the remaining active stations. This caused the President to receive incomplete information, and led him to believe that only a sector of the Navy conspired against him and his government.

President Allende and Defense Minister Orlando Letelier were unable to communicate with military leaders. Admiral Montero, the Navy's commander and an Allende loyalist, was rendered incommunicado; his telephone service was cut and his cars were sabotaged before the *coup d'état* to ensure he could not thwart the opposition. Leadership of the Navy was transferred to José Toribio Merino, planner of the *coup d'état* and executive officer to Admiral Montero.

Augusto Pinochet, General of the Army, and Gustavo Leigh, General of the Air Force, did not answer Allende's telephone calls. José María Sepúlveda, the General Director of the *Carabineros* (uniformed police) and Alfredo Joignant the head of the Investigations Police (plain clothes detectives) answered Allende's calls and immediately went to the presidential palace.

When Defense Minister Letelier went to the Ministry of Defense to assess the situation, he was arrested as the first prisoner of the *coup d'état*.

Despite mounting evidence that all branches of the Chilean armed forces were involved in the coup, Allende hoped that some units remained loyal to the government. He was convinced of Pinochet's loyalty, telling a reporter that the *coup d'état* leaders must have imprisoned the general. Only at 8:30 am, when the armed forces declared their control of Chile and Allende was deposed, did the president grasp the magnitude of the military's rebellion.

At approximately 9:00 am, the uniformed police left the presidential palace. At this time, the armed forces controlled Chile, except for the city center of the capital, Santiago. Allende refused to surrender, despite the military's declaration that they would bomb the presidential palace if he resisted being deposed. The Socialist Party, along with Cuban advisors, proposed to Allende that he escape to the San Joaquín industrial zone in southern Santiago, to later re-group and lead a counter-coup d'état, but the president rejected the proposition. According to Dr Tanya Harmer, a specialist on the Cold War in Latin America, Allende's refusal to lead an insurgency against the coup was evidence of his unrelenting desire to bring about change through non-violent methods. The military attempted negotiations with Allende, but the President refused to resign, citing his constitutional duty to remain in office. Finally, Allende gave a farewell speech, telling the nation of the *coup d'état* and his refusal to resign his elected office under threat.

The General of the Air Force ordered the presidential palace bombed, but was told that the jet would take forty minutes to arrive. Pinochet then ordered an armored and infantry force under General Sergio Arellano to advance upon the presidential palace. When the troops moved forward, they were forced to retreat after coming under fire from police snipers perched on rooftops. General Arellano called for helicopter gunship support from the commander of the Chilean Army Puma helicopter squadron and the troops were able to advance again. There is no conclusive evidence as to how President Allende died in the attack.

Chile Under Pinochet

The National Stadium in Santiago, with a capacity of fifty thousand, was used as a detention and torture center as the military began to round up the people who had supported Allende. In the first months after the *coup d'état*, the military imprisoned forty thousand people.

Thousands were tortured and raped or were "disappeared" by the death squad Caravan of Death (Caravana de la Muerte). While people sat on benches in the blazing sun for days on end, the junta randomly shot and tortured them or grabbed women to rape. On December 18, 1974, Pinochet was declared Supreme Leader of the nation.[281] Two years later, in 1975, the CIA reported that approximately four thousand prisoners were still being held in the Stadium.[282] The government arrested some one hundred and thirty thousand people in the three-year period after the coup. This was third world capitalist democracy in action.

Surprisingly, General Pinochet lost his "mandate" to rule in a 1988 plebiscite in which people were asked to extend the General's rule for another eight years. The plebiscite was staged for appearances, but the plan backfired because the political parties split on the vote. In response, Pinochet called together the leaders of the junta and demanded "extraordinary powers" to have the military seize the capital. There were dissenting votes within his ranks and the supreme leader was forced to step down. The fact that he respected these developments testify to the fact that big business had tired of him, as well as the fact that he was receiving pressure from the international community and from powerful forces internally.

Nonetheless, the damage to the people's revolution was profound. It was not until thirty years later, in 2019, that the people's reaction against the economic and social injustices to which they were subjected reached such a peak that President Sebastián Piñera's government said it would rewrite Chile's capitalist, dictatorship-era constitution and replace it with one providing a new social pact.[283]

This concession on the part of the government came after months of protests in the streets that had not been quelled by the Chilean government's use of state-sanctioned violence to intimidate civilians, silence dissidents, and retain impunity.

Government forces continue to torture, unlawfully detain, and beat people on a regular basis. Aside from violently dispersing peaceful protests, the military is conducting raids on private residences and schools based on flimsy evidence. Since the beginning of demonstrations in mid-October of 2019, thousands of civilians have been wounded, and over a dozen have been killed by security forces.

Aside from these overt forms of repression, the government has also tried to control the people by censorship, disinformation, withholding evidence, staged looting, staged property damage, and even trying to frame protesters for a military-conducted arson at a bank.[284]

In a style typical of US-supported dictators in South America, the Pinera administration has declared war against an internal "enemy" who the government claimed posed a threat to "order," "nation," and "stability." But as Human Rights Watch tells us:

> This internal "enemy" comprises everyone who is opposing the government's policies and actions, including but not limited to indigenous peoples and ethnic minorities, dissidents, lawyers, students, human rights advocates, independent journalists, feminists, and the poor.[285]

Unfortunately, what continues to happen to Chile is nothing new in Latin America where small countries had long been considered to be the private property of US capital. This capitalist mentality goes back for over a century. For example, in 1920 a United Fruit (now United Brands and Chiquita) manager wrote to a company lawyer describing the company's true intentions in Honduras.

> We must produce a disembowelment of the incipient economy of the country in order to increase and help our aims. We have to prolong its tragic, tormented, and revolutionary life; the wind must blow only on our sails and the water must only wet our keel.[286]

At this writing, in July 2020, people are in the streets fighting US imperialism in Bolivia, Columbia, Venezuela, and Ecuador.

Lessons Learned from Chile

The main lesson from Chile is that the US capitalists will even overthrow democratically elected presidents to ensure that socialism does not succeed.

The US capitalists then install a puppet dictator in power who brutalizes the people with rape, murder, and torture. In the first months of the fascist takeover, the dead and disappeared numbered in the tens of thousands.

Of course, the message of these capitalist thugs to the American people is that the US army is used only to support democracy around the world. In the press, the big lie is that the US army always fights for the freedom of Americans and the freedom of other people. In reality, the army provides the muscle to allow the capitalists to exploit the resources and labor of the world's people.The overthrow of the democractically elected president of Chile was, in fact, the US-based multinational corporations' statement to the poor people of the world as to what would happen to them if they rejected corporate exploitation.

Once in power in Chile, the corporations stole the resources and labor of the people. The Pinochet dictatorship immediately cut tariffs on imports, suppressed union power, and "privatized" hundreds of state-controlled industries. Human services were put under the control of private corporations.

Today Chile has an extractive economy that is driven by exports. Privatization of their infrastructure continues. The government, instead of meeting the needs of the people, has been reduced to mostly working on a regulatory process that serves the foreign-owned corporations.

In November 2019, the people of Chile were back on the streets demanding social reforms and a change in the Constitution that was drafted during Pinochet's reign. That Constitution ignored any responsibility of the government to address the people's need for health care and education. The capitalist government has responded by siccing the military police on the people to murder, torture, and rape.

This lesson is important to learn by all the people of the world, including the revolutionaries who are presently fighting against the Communist governments of Russia and China. Do not look to the US to bring democracy to your country. Establish your own political and economic autonomy and choose allies who are trying to do the same thing.

Nicaragua

Another country that has found itself under the oppressive rule of the United States is Nicaragua.

In 1979, the Nicaraguan people overthrew the long-time dictatorship of Somoza Garcia, who had also cemented his iron rule with the money and political support of the US government. The

Somoza family ruled the country in a cruel and unscrupulous way for more than four decades, but was finally overthrown by the National Sandinista Revolution, when people from all sectors, including workers, businessmen, peasants, students, and guerrillas, joined forces and finally defeated the Somoza dynasty and his National Guard on July 19, 1979.

The symbol for and direct precedent of the revolution was the struggle of General Augusto C. Sandino (1895-1934), a national hero who fought with bravery, supported by an army formed by farmers and workers. He fought against the armed intervention of the United State in Nicaragua, which was perpetrated under the pretext of ensuring peace and democracy in the country.

Sandino and his small army fought US marines in the mountains of the Segovias, the northern part of the country. His tactics where so effective that even though they were outnumbered and even though the Marines were technologically superior, Sandino's army was never eliminated by the foreign invaders.

The Marines finally left the country, leaving behind a Nicaraguan army that was then known as the National Guard ("Guardia Nacional") to continue their dirty work. It was led by a national military man trained in the United States. His name was Anastacio Somoza García.

After the US marines retreated, Sandino, the revolutionary leader, wanted to negotiate peace and allow the lives of his soldiers to return to normal. He was called by the president José María Moncada and by Somoza to negotiate. Unfortunately, Sandino was betrayed, captured, and executed. Even today, the place where the body of this hero is buried remains unknown.

In 1937, Somoza seized political power and established himself as a dictator. His unconditional support by the United States (with whom he negotiated at the country's expense) and his complete control of the National Guard guaranteed his totalitarian "throne." He destroyed his opposition by the use of terror and mass executions. Even so, in September 1956, Rigoberto López Pérez, a young poet, sacrificed his life when he shot the dictator during a social celebration in the city of León. Rigoberto was killed at the same time. Somoza survived the bullets, but died later in a Panamanian hospital.

Somoza García's oldest son Luís Somoza then took over the presidency, while his youngest son, Anastacio Somoza took over the National Guard and immediately started a repressive movement against those who he considered political opponents.

In 1967, Luis Somoza was still occupying the presidency, but suddenly died. His younger brother, Anastacio, "won" the next national election and restored the dictatorial repressive regime.

Even though his father had acquired various real estate properties, businesses, and industries that had become his and his family's patrimony, Anastacio continued to unconscionably appropriate more and more wealth from the Nicaraguan people. On December 23, 1972, a magnitude 6.2 earthquake leveled the capital city, Managua. The earthquake killed ten thousand of the city's four hundred thousand residents and left another fifty thousand homeless. About eighty percent of Managua's commercial buildings were destroyed.

In the face of this suffering by the people of Managua, President Anastasio's National Guard embezzled much of the international aid that flowed into the country to assist reconstruction, and several parts of downtown Managua were never rebuilt. The president gave reconstruction contracts preferentially to family and friends, thereby profiting from the quake and increasing his control of the city's economy. By some estimates, his personal wealth rose to US$400 million in 1974.[287]

Even though social distress was increasing, the National Guard kept any public protest from occurring by torturing and executing political opponents and the general population.

The strong subjugation of the Nicaraguan people ultimately led to the creation of a militant revolutionary movement in 1962, whose intention was to defeat the dictatorship. The movement was named after General Sandino and this is how the "Sandinista National Liberation Front" (Frente Sandinista de Liberación Nacional—FSLN) was born.[288] It was founded by a group of intellectuals and volunteers led by the young Carlos Fonseca, a teacher and librarian who was later killed in the mountains of Nicaragua three years before the FSLN took power.

Fonseca had traveled to Russia and wrote a book entitled *A Nicaraguan in Moscow* in which he praised Russia for its democracy and religious freedom. Above all, Fonseca was primarily influenced by Fidel Castro and the 26th of July Movement that took power in Cuba on January 1, 1959. The Cuban Revolution was a central event in Fonseca's political evolution because it convinced him that a revolution was possible and that organization was necessary. Just as the Cuban revolution was organized outside the framework of the Cuban Communist Party, he came to believe that a Nicaraguan revolutionary movement could be created outside of the PSN. The PSN was the Nicaraguan Socialist Party (*Partido Socialista Nicaragüense*) that operated as the official communist party in the country. At the time of its foundation, the PSN supported the regime of Anastasio Somoza García, which it considered to be an anti-fascist ally.[289]

The Sandanistas had a rough start because they were vastly inferior to the military power of the National Guard. Its first military action ended in a massacre because the group was surrounded by the National Guard and the Honduran army at their shared border in the department of Jinotega. It was in this province where General Sandanista had fought numerous battles with the US marines in the 1920s.

In spite of the initial defeat, the FSLN did not succumb. Peasants, students, and even local landowners joined the organization or became accomplices. As time went by, more people began to adopt the "anti-somocista" ideology and embrace revolutionary views while working to increase their military power and propaganda efforts to build a sense of consciousness in the working sectors, universities, and schools.

Even though the Sandanistas were pushed back to the mountains by the National Guard, the mere existence of this opposition army led to the gradual involvement of more and more people.

Even so, Somoza was still completely in control and guerrilla actions in urban and rural zones were repelled and the remnant Sandinistas were forced to disperse to much more remote areas.

But the movement did not die. In 1969, a small group of Sandanistas was hiding in a house in Managua the capital city. They were discovered by the National Guard who surrounded the house with three hundred soldiers and tanks. Planes were also sent to destroy the house. The battle lasted several hours until there was no more resistance to the attack. When National Guard soldiers entered the house looking for bodies, they discovered that much of the combat was fought by a single young man, Julio Buitrago, who stayed inside the house so that his comrades could escape.

Somoza showed the battle on TV to demonstrate the FSLN's destruction, but the heroic action of a single man turned the entire situation around and the population was moved by the event and identified themselves even more with the revolutionary movement. Other unequal battles took place in several cities.

The Sandanista guerrillas kept up a continuous chain of diversified actions around the country. This provoked strong repressive actions by the National Guard. Through threats, torture, and spying, it was able to apprehend and kill many guerrilla leaders and revolutionaries.

Then, in 1970, the FSLN started what they called a period of "accumulation of forces in silence." During this time, they reduced their attacks to a minimum and focused their concentration on strengthening the organization, its influence in the different sectors, the military preparation of its members, and the compiling of war resources.

Four years later, the movement initiated public combat. They organized attacks all over the country. A breakthrough occurred when an aristocrat, Jose María Castillo, who was a personal friend of Somoza's, held an elite cocktail party at his house. The guerillas entered the house and kidnapped several diplomats and government officials. The National Guard could not attack, so they decided to accept demands, including the publication of a manifesto, the liberty of political prisoners, and a monetary reward. The revolutionaries in the house also negotiated their flight out of the country.

In response, Somoza declared a state of siege and started indiscriminate repression against any suspect. The following years were characterized by strong fighting in cities and rural areas as more and more people got involved with the guerillas.

At the same time, debates within the Sandinistas began to split the organization. Should the struggle be concentrated in the cities or the countryside? Should the political or military aspect of its work be dominant? As the political climate changed, arguments emerged about what relationship should exist between the bourgeois opposition and the FSLN. What demands should be raised within the mass movement? Was the situation already ripe for revolution or did more organizing among the workers need to be done? And most importantly, what should come after the Somosa regime?

As a consequence of the repressive campaign of the National Guard in 1975, a group within the FSLN's urban mobilization arm began to question the viability of the "accumulation of forces in silence" strategy (GPP). In this view, promoted by young orthodox Marxist intellectuals, such as Jaime Wheelock, economic development had turned Nicaragua into a nation of factory workers and wage-earning farm laborers.[290] Wheelock's faction became known as the "Proletarian Tendency." They held that the guerrilla army was acting in isolation from the urban working class. They pointed to the relative decline of the agricultural sector in the 60s, and to the rapid growth and concentration of the urban proletariat. Moreover, the Proletarios emphasized the need for a party of the proletariat to guide the revolutionary struggle. Without such a party, they argued, the workers' movement would stay within the narrow bounds of economism and anti-somocismo. Only political work among the masses, under the party's direction, could lead to the overthrow of Somoza and the transition to socialism. Otherwise, the military efforts of the FSLN would only rebound to the benefit of the bourgeoisie. The military had to be subordinated to the political.[291]

Adherents to the original GPP strategy held firm. GPP stood for "Prolonged Popular War" (*Guerra Popular Prolongada*—GPP). While the urban organization recruited on the university campuses and robbed money from banks, the main cadres were to permanently settle in the north central mountain zone. There they would build a grassroots peasant support base in preparation for renewed rural guerrilla warfare.[292] They regarded the guerrilla units as the essential vanguard of the revolutionary process, while acknowledging the importance of work in the cities as well.

Mountain strongholds and a base in the rural proletariat would prevent the capitalists from usurping the revolutionary impetus. In the final analysis, they argued, it was the rural guerrilla that had undermined the regime and its repressive capacity, thereby advancing the struggle in the cities. With respect to their emphasis on military action, the GPP maintained that the National Guard had to be neutralized militarily before effective political work could be conducted among the masses.

In the heat of Somoza's declared "state of siege" and counter-revolutionary offensive, communication broke down within the National Directorate, which was the leadership body of the FSLN. Positions hardened, turning bitter and polemical. In October 1975, elements of the leadership expelled the Proletarios, who thereafter operated independently, but retained the name of the FSLN.[293]

There is a lesson to be learned here about how internal contradictions can arise among revolutionaries within the course of struggle. This contradiction often revolves around the needs of the rural vs urban populations.

At this point, a group of leaders abroad, acting as a "third force," attempted to reconcile the two positions. In 1976, two members of the ten-person National Directorate, Carlos Fonseca (General Secretary) and Eduardo Contreras Escobar returned to Nicaragua to mediate the dispute. Soon after entering the country they both died in combat. In the same battle, Tomas Borge of the GPP was captured, depriving this faction of any spokesperson in the National Directorate. Thus, the third force, later known as the Terceristas, failing in its reconciliation efforts, gradually evolved an "insurrectionist" strategy of its own.

The Terceristas rejected the shared assumption of the Proletarios and the GPP concerning the protracted nature of the struggle and the need to gradually accumulate forces. Rather, they argued that conditions were already ripe; anti-Somoza sentiment had become generalized and the masses were ready to act. The Terceristas advocated a series of military strikes that would ignite a popular insurrection: "To make war in order to organize, to organize in order to make war until victory."

For the insurrection to succeed, the Terceristas argued that support must be garnered from all anti-Somoza sectors, including elements of the bourgeoisie whose economic pressure would complement armed actions. Only a broad-based coalition, under FSLN hegemony, based on a program of progressive reforms, could isolate the dictatorship and bring about its downfall. Somocismo would be replaced by a provisional popular-democratic government, with FSLN participation.

The Terceristas tempered their revolutionary rhetoric; socialism would not come "overnight," but in stages. Their program demanded the overthrow of Somoza followed by free elections, the disbanding of the National

Guard, and the nationalization of all Somoza's holdings and of all private banks. Until that was achieved, they pledged never to lay down their arms.[294]

At the end of 1977, the FSLN-Insurrectionists organized a strong urban offensive with direct attacks on all National Guard barracks in various cities. At the beginning of the offensive, the majority of the actions were successful but, the superiority of Somoza's army eventually reversed the course of action. Nevertheless, that same year a public manifesto of the "Grupo de los Doce" (Group of The Twelve) encouraged the support of the battle against the dictatorship. The Group of Twelve were all members of the establishment who came to support the Sandinistas after contact with the Terceristas who were in exile in Costa Rica. The Group included a banker, university president, priests, lawyers, and businessmen.[295]

Because the Terceristas, who later became the nucleus of the Insurrectionists, had advocated alliances with the middle and upper class against Somoza, they had approached the members of the Group of Twelve, some of whom were already secretly Sandinistas, to sign a communiqué in support of the FSLN. On October 18, 1977, following the Insurrectionists' offensive, the Group of Twelve issued a manifesto from Costa Rica urging that the Sandinistas must be included in any political process. Their support lent respectability to the Sandinista cause and helped convince many that the rebels were no longer doctrinaire communists.

The following year, in 1978, Pedro Joaquín Chamorro,[296] director of the national newspaper La Prensa, which was the only significant opposition newspaper to the long rule of the Somoza family, was murdered. The people attributed this assassinated to the dictator and there was a large protest in the capital city of Managua.

In light of this development, the Carter administration in the United States pressured Somoza to allow the exiles of the Group of Twelve to return to Nicaragua. Upon their return on July 5, 1978, tens of thousands lined the streets to welcome them back.

In August, an insurrection was started in the local indigenous neighborhood of Monimbó in the city of Masaya. Due to the long history of indigenous people in the area, Masaya is known as "The Cradle of Nicaraguan Folklore" and is the heart of Nicaraguan handicrafts that include hand woven hammocks, embroidered blouses, wood carvings, and hemp weavings. The town also attracts tourists because of its large, diverse market and colorful festivals.

By now the world was following the developments in Nicaragua. Interest was further intensified when a guerilla commando of the FSLN-Insurrectionist branch took over the National Palace in Managua where the Congress met. He kidnapped congressmen and senators. Again, the dictatorship had no other choice than to accept his demands which were to liberate prisoners, publish a revolutionary note, and allow the commando to escape.

A year later, in 1979, several battles were organized all over the country. The Somoza dictatorship reinforced its attacks on the guerrillas, which affected the entire population. The government bombed neighborhoods indiscriminately.

In March, the three branches (tendencies) of the FSLN reached a unification agreement that created a single FSLN. They began an attack against the government at the end of June when all FSLN combatants gathered in Managua. At the same time, a general strike was called in which the workers and even Nicaraguan businessmen participated. The general strike was used by the revolutionaries to mobilize the masses of people in support of the revolution.

Insurrections were soon started in other provinces (departments) and cities. Somoza organized attacks against those cities, killing many civilians. Nevertheless, the FSLN, which was now supported by the entire population, could not be stopped. One by one, each city was liberated from Somoza's control.

These developments caused the Carter administration in the US to look at Somoza as a failure. The people of the US sided with the FSLN when a North American journalist was murdered. Even though the dictator tried to accuse the guerillas of the journalist's murder, another reporter had recorded the event. Eventually, the Carter administration asked Somoza to step down from power before his government was overthrown and thus pave the way for a revolutionary government in Nicaragua.

On July 17, Somoza, his collaborators, and leaders of the National Guard left the country, taking a huge fortune with them. A Somoza supporter, Francisco Urcuyo, became the new president and asked the Sandinistas to stop their attacks. His government lasted only twenty-four hours, after which he abandoned the country. The National Guard finally surrendered.

Two days later, on July 19, thousands of guerillas and civilians entered the Plaza de la Republica, in the old center of Managua, where the whole population celebrated the fall of the Somoza dynasty. It was a day of national celebration and the opportunity to create a new Nicaraguan government lay ahead.

Post Revolution

The Sandinistas inherited a country with a debt of $1.6 billion dollars, an estimated thirty to fifty thousand war dead, six hundred thousand homeless, and a devastated economy. A Government Assembly for the National Reconstruction was organized, formed by five members including three Sandinista militants (Daniel Ortega, Moises Hassan, and Sergio Ramírez (a member of Los Doce "the Twelve") and two opposition members, who were businessman Alfonso Robelo, and Violeta Barrios de Chamorro (the widow of Pedro Joaquín Chamorro). Only three votes were needed to pass a resolution. Reforms immediately began to take the country out of the post-war disaster and to promote social justice at all levels. Banks were nationalized and properties were expropriated from the Somoza family and their allies. Furthermore, as a sign of total victory, a National Crusade of Alphabetization (a literacy campaign) was organized, which reduced the illiteracy rate from fifty-three to twelve percent. This program gained worldwide admiration.

The Sandinistas declared a state of emergency in preparation for a counter-revolution. President Carter sent ninety-nine million dollars in aid. Land and businesses of the Somoza regime were expropriated, the old courts were abolished, and workers were organized into Civil Defense Committees. The new regime also declared that "elections are unnecessary," which led to criticism from the reactionary Catholic Church, among others.

The Revolution ended the Somocista regime, which had seriously deformed the country. It had created a big center, Managua, from which

Somoza's power had emanated to all corners of the territory. Somoza had developed a semi feudalist rural economy with few productive goods that were largely for export, such as cotton, sugar, and other tropical agricultural products. All sectors of the economy of Nicaragua were controlled, in great part, if not entirely, by the Somozas or the officials and others surrounding the regime, whether by directly owning agricultural brands and trusts, or by actively putting them into foreign hands. It is famously stated that Somoza himself owned 1/5 of all profitable land in Nicaragua. While this is not correct, Somoza or his adepts did own, or gave away to cronies, banks, ports, communications, services, and massive amounts of land.

Upon leaving Nicaragua, Somoza first went to the United States, but later settled in Paraguay where he was murdered in 1980 by an Argentinean guerrilla commando. His collaborators, who included many leaders of the National Guard, settled in Miami, Florida in the United States where they became a powerful local political group.

The Sandanista-led government was immediately rejected by the Reagan administration that criticized the preceding Carter administration for its "obsessive concern for human rights." This signaled that the east coast capitalist establishment in the US had tired of the Democrat Jimmy Carter and had decided to throw their money behind the Republican Ronald Reagan. Reagan officials argued that US interests would have best been served by active support of the Somoza dictatorship that the revolution displaced.[297] The new Reagan administration was extremely anti-communist and imposed an economic blockade against the people of Nicaragua that began in 1981. This action further united Nicaragua with the communist bloc of the Soviet Union and created a negative attitude among the Nicaraguan people against the US people.

Despite this, Nicaragua had won great international sympathy and several countries (also from Europe) started to collaborate with the reconstruction of Nicaragua. Organizations and private individuals also helped, which made the existence of various nationalities in the country commonplace.

The Nicaraguan Revolution brought immense restructuring and reforms to all sectors of the economy with the goal of directing it towards a mixed economic system. As a "third-world" country, the Nicaraguan revolutionaries had inherited an agriculture-based economy that was undeveloped and susceptible to the flow of international market prices for its agricultural goods, such as coffee and cotton. The rural economy was well behind in technology and, at the same time, devastated by the guerrilla war and the soon to come civil war against the Contras.

The Nicaraguan Agrarian Reform developed into four phases:

1. First phase (1979): confiscation of property owned by Somocists and its adepts;
2. Second phase (1981): Agrarian Reform Law of July 19, 1981;
3. Third phase (1984–85): massive cession of land individually, responding to demands from peasantry; and
4. Fourth phase (1986): Agrarian Reform Law of 1986, or "reform to the 1981 Law."

Article 1 of the Agrarian Reform Law says that property is guaranteed if it is labored efficiently and that there could be different forms of property. It cites four different kinds of property: (1) *state property* (with the confiscated land from somocists), (2) *cooperative property* (part of confiscated land, but without individual certificates of ownership, to be labored efficiently), (3) *communal property* in response to the needs of indigenous ethnic groups, and (4) *individual property* (as long as it is efficiently used and integrated with national plans of development).

The principles of the Agrarian Reform were based upon the same principles of the revolution: pluralism, national unity, and economic democracy.

In 1984, a presidential election was held. Daniel Ortega, representing the FSLN won the election taking 66.97% of the votes. He ran against representatives from six other parties including the Democratic Conservative Party (PCD), the Independent Liberal Party (PLI), the Popular Social Christian Party (PPSC), the Nicaraguan Communist Party, (PCdeN), the Nicaraguan Socialist Party (PSN) and the Marxist-Leninist Popular Action Movement (MAP-ML).[298]

In 1985, the Agrarian Reform initiative distributed 235,000 acres (950 km2) of land to the peasantry. This represented about seventy-five percent of all land distributed to peasants since 1980. The Reform movement had the twofold purpose of increasing the support for the government among the campesinos, and guaranteeing ample food delivery into the cities.[299]

The Sandanistas also created large scale programs in literacy, health care, education, childcare, unions, and land reform for which they received international recognition.

US Support of Contras

To counter this success on the part of the revolutionaries, the Reagan government secretly provided large sums of money to the remnants of the National Guard who had settled in Honduras, Costa Rica, and Miami. The Reagan administration helped to develop the "Contrarrevolución Nacional" (the National Counterrevolution) whose army became known as "The Contras." This forced the war for independence to start again, which damaged the fragile Nicaraguan economy due to the additional war expenses.

In 1985, the Reagan administration also imposed a commercial embargo on Nicaragua, claiming that the country had adopted a pro-communist attitude. Reagan threatened the Nicaraguan government several times with military intervention.

Nevertheless, the international opinion continued to support the non-communist revolutionary government, especially when the world discovered that the Reagan government had secretly provided the Contras with funds through what became known in the international press as the "Iran-Contra Affair."

The Iran-Contra Affair was a political scandal in which the National Security Council (NSC) of the US government became involved in secret arms transfers to Iran and used the profits to fund the Contras. These activities violated the public policy of the US government. The American public had grown increasingly opposed to funding the Contras and when Congress passed a law banning it, the White House violated the law and resorted to covert and illegal means to continue its support.

The money transfers to the Contras was undertaken by NSC staff member Lieutenant Colonel Oliver North with the approval of head of the NSC, Rear Admiral John M. Poindexter. North and his associates also raised private funds for the Contras. These activities further violated the Boland Amendment, a law passed by Congress in 1984 that banned direct or indirect US military aid to the Contras.

This illegal activity came to light in November 1986 and aroused an immediate public uproar. North claimed that both Reagan and his Vice President, George Bush, had known about the covert operations, though both denied any knowledge of the details and no evidence was ever brought linking them to any wrongdoing. The Tower Commission which was established to investigate the affair accused Reagan of "a lax managerial style and aloofness from policy detail" and the United States suffered a serious, though temporary, loss of credibility as an opponent of terrorism. Moreover, the affair raised serious questions about the power of the executive branch of government and the extent and effectiveness of Congress's oversight of foreign affairs.[300]

The International Court in The Hague ruled that the United States had to stop any indirect hostility and it had to pay a multi-million indemnification to the impoverished Nicaragua. However, this turned out to be merely a symbolic gesture.

Nicaragua was gradually being consumed and destroyed by war. In March 1982, about one year after Reagan set up his initial economic blockade, the Sandinistas declared an official State of Emergency in response to attacks by counter-revolutionary forces. The State of Emergency lasted six years, until January 1988, when it was lifted.

Under the new "Law for the Maintenance of Order and Public Security," the "Tribunales Populares Anti-Somozistas" allowed for the indefinite holding of suspected counter-revolutionaries without trial. The State of Emergency, however, also affected the rights and guarantees contained in the "Statute on Rights and Guarantees of Nicaraguans." Many civil liberties were curtailed or canceled, including the freedom to organize demonstrations, the inviolability of the home, freedom of the press, freedom of speech, and the freedom to strike.

All independent news program broadcasts were suspended. In total, twenty-four programs were cancelled. In addition, Sandinista censor Nelba Cecilia Blandón issued a decree ordering all radio stations to take broadcasts from government radio station La Voz de La Defensa de La Patria every six hours.

The rights affected also included certain procedural guarantees in the case of detention including habeas corpus. The State of Emergency was not lifted during the 1984 elections.

Opponents to the State of Emergency argued its intent was to crush resistance to the FSLN. James Wheelock justified the actions of the Directorate by saying ". . . We are annulling the license of the false prophets and the oligarchs to attack the revolution."

On October 5, 1985, the Sandinistas broadened the 1982 State of Emergency and suspended many more civil rights. A new regulation also forced any organization outside of the government to first submit any statement it wanted to make public to the censorship bureau for prior approval.

While the Sandanistas did not rob the people blind nor oppress them with beatings, rape, imprisonment, and torture, they were beginning to mirror the political behavior of the dictator whom they had fought so hard to depose.

Certainly, there was provocation for their actions, but they lacked the ability to stabilize the country without the use of draconian measures that alienated many of the Nicaraguan people in the process.

The Contras were operating out of camps in the neighboring countries of Honduras to the north and Costa Rica to the south. Backed by the Reagan administration, they were also engaged in a campaign of economic sabotage. With the help of the CIA, they disrupted shipping in and out of the country by planting underwater mines in Nicaragua's Corinto harbour, an action condemned by the International Court of Justice as illegal.

The Contras also carried out a systematic campaign to disrupt the social reform programs of the revolution. This campaign included attacks on schools, health centers, and peasant villages that were sympathetic to the Sandinistas. Widespread murder, rape, and torture were also used as tools to destabilize the government and to "terrorize" the population into collaborating with the Contras. This campaign of terror was fully sanctioned by the Reagan administration.

At the same time, the Reagan administration was engaged in a campaign to turn the sentiment of the American people against the Sandanistas. This campaign came to be called "white propaganda."[301]

White propaganda was a smear campaign coordinated by the National Security Council staff, using CIA experts and Army psychological-warfare techniques against the American people.

The Iran-Contra Committee unearthed hundreds of pages of documents about the domestic propaganda operation and took depositions from many of the principals. House investigators even drafted a chapter about the domestic side of the scandal for the Iran-Contra report. But Senate members of the panel blocked its inclusion in the final report, arguing that the issue fell outside the committee's mandate.

The explosive conclusion of that draft chapter was that the administration's covert propaganda operation had used "one of the CIA's most senior specialists, sent to the NSC by Bill Casey, to create a covert CIA operation aimed at the US media, Congress, and the American people to support the Reagan administration's policies."

The NSC worked to intimidate or discredit those who stood in the way of military aid for the Contras. Ultimately, the campaign came to resemble the sort of covert political operation the CIA runs against hostile forces overseas but is outlawed from conducting at home.[302]

In 1989, Nicaraguan president Daniel Ortega met with the presidents of El Salvador, Costa Rica, Honduras, and Guatemala to hammer out a peace plan for his nation. In exchange for promises from the other nations to close down Contra bases within their borders, Ortega agreed to free elections within a year. These were held on February 26, 1990. Ortega and the Sandinistas suffered a stunning defeat when Violeta Barrios de Chamorro, widow of a newspaper editor assassinated during the Somoza years, polled over fifty-five percent of the presidential vote. The opposition also captured the National Assembly.

Rejection of Sandinistas and Return of Ortega

Chamorro's election came after ten years of Sandinista rule that had been marred by a destructive war with the Contras and a failing economic system in large part foisted upon them by the capitalist government of the United States. The United States promoted Chamorro's victory as validation of its long-time support of the Contras, and many right-wing pundits likened the electoral defeat of the Sandinistas to the crumbling of communist regimes in Eastern Europe during the same period.[303]

This defeat, for many, was the end of the revolution. A major part of the Nicaraguan population now began to reject the Sandinista movement due to its many mistakes and suspension of freedoms. However, the FSLN continues to be one of the strongest Nicaraguan political parties even though many of its historical leaders have left the party as a result of political differences.[304]

Much to the consternation of the George W. Bush presidency in the United States, Sandinista Party leader Daniel Ortega regained the presidency in 2007, and orchestrated a successful strategy of coalition-building with the organized private sector and the Catholic Church. Supported by certain international financial institutions and the Venezuelan Chavista government, Nicaragua achieved strong economic performance with moderate growth.[305]

By 2018, however, the revolution in Nicaragua had greatly disintegrated. The problems for the Sandinistas had begun back in 2006 when President Enrique Bolaños of the Nicaraguan Social Security Institute (INSS) released an actuarial study that forecasted coming deficits in spite of the program's solvency at the time. Subsequent studies by the INSS also suggested the program would face deficits beginning around 2014 and reach insolvency sometime after 2019.[306]

In 2013, senior citizens started protesting in front of the INSS building in Managua. They were fighting on behalf of people who contributed to the social security insurance, but who did not reach seven hundred and fifty weeks of contributions as the law required. The protesters were demanding that at least a reduced pension for such people should be granted. Soon, students and young people joined their protests. After a week of demonstration, the peaceful protesters were violently beaten up by paramilitary mobs associated with the Sandinista Youth within minutes after the police had moved back from the scene. Later, in order to calm down the protests, President Daniel Ortega announced concessions to provide reduced pensions.

He did this despite the fact that the pension fund had already begun to operate in the red. In 2014, another thorny issue developed concerning the building of a Nicaragua Canal. The Canal was a proposed shipping route through Nicaragua to connect the Caribbean Sea with the Pacific Ocean. Scientists and local farmers were concerned about the project's environmental impact because the Canal would be built through Lake Nicaragua which is one of Central America's main freshwater reservoirs.

In June 2013, Nicaragua's National Assembly approved a bill to grant a fifty-year concession to finance and manage the project to the private HK Nicaragua Canal Development Investment (HKND) headed by Wang Jing, a Chinese billionaire. The concession could be extended for another fifty years once the waterway was operational.

As the construction of the Canal was about to begin, several hundred protesters blocked roads and clashed with police during the groundbreaking. Tens of thousands of Nicaraguans then began to protest against President Ortega, not only due to the canal, but also to what they called a corrupt electoral system. Protests began December 24, 2014 with clashes with the police. Farmers were arrested and allegedly beaten by Nicaraguan authorities. It was reported that the government was searching each home in the area to find those who participated in the protests. On December 26, protesters demonstrated outside of El Chipote Prison and were confronted by Sandinista Youth *mobs* on motorcycles, with some protesters being released later that day. On December 30, six of the *campesino* leaders were released from the prison.

In 2015, media reports suggested the project would be delayed and possibly cancelled because Wang's personal wealth declined greatly as a result of the 2015–16 Chinese stock market crash.

By February 2018, the project was widely viewed as defunct although the head of the project insisted work was on-going and HKND retained the legal rights to the concession for the canal as well as side projects. Despite HKND vanishing, the Nicaraguan government indicated that it would go ahead with the 564-mile land expropriations for the canal under Canal Law 840.[307]

In April 2018, the Ortega administration announced reforms to the social security program that imposed by decree a benefit cut of five percent while raising income and payroll taxes. This sparked unrest because of the substance of the reforms, and also to the manner in with they were imposed without a warning or dialogue. Because the Nicaraguan people were left out of any dialogue, their only recourse was to take to the streets. After five days of clashes between protesters and police, Ortega announced the reforms had been canceled.[308]

In the same month, demonstrators marched in Managua and other cities to protest what they regarded as an insufficient government response to forest fires that burned 13,500 acres (5,500 hectares) of the Indio Maíz Biological Reserve, a tropical nature preserve that is home to Rama and Kriol indigenous people, as well as significant biodiversity and endangered species. There were suspicions that the government had an interest in the fire, as it is the largest natural reserve through which the Nicaraguan Canal was planned to be built.[309]

Authorities used live ammunition on protesters while also arming Sandinista Youth members with weapons. At least twenty-six people were killed, including journalist Ángel Gahona an investigative reporter who was streaming live from the protests on Facebook. His death recorded live on air became a symbol of Nicaragua's crisis.[310]

The following day on April 19, the Vice President and first lady Rosario Murillo made a speech mocking the demonstrators and labeling them as "small groups, small souls, toxic, full of hate," bent on the destruction of the country, assaulting peace and development. Murillo also labeled the demonstrators that had been attacked as "aggressors" and the actions by pro-Ortega groups and police as "legitimate defense."

Protests continued to intensify with confrontations occurring in León, Managua, Granada, Boaco, Carazo, Estelí, Rivas, Matagalp a, and Masaya. TELCOR, the state regulator of TV, ordered the suspension of transmissions of four independent TV channels that were reporting the news as well as the Catholic Episcopal Conference's TV channel. The suspension lasted several hours. In the meantime, Murillo accused the protesters of being manipulated and trying to "destabilize" and "destroy" Nicaragua.[311]

Two days after the beginning of protests and the subsequent crackdown by authorities, Ortega made his first public appearance and announced he would hold negotiations for reforms, planned to take effect on July 1, 2018. However, he stated he would meet only with business leaders, alleging that the demonstrators were being manipulated by gangs and other political interests.

Demonstrations increased in response, with protesters objecting to the repression of demonstrators and the exclusion of other sectors from the negotiations, as well as the reforms themselves. Things went downhill from there.

On April 23, marches of citizens, businessmen, and students were held in Managua demanding the end of violence in the country, the release of students arrested by the police, the cessation of censorship of television media, and a response from the government about the students who died during the protests. The protests were the largest seen during the Ortega administration, with hundreds of thousands of demonstrators participating and calling for the president's resignation. The numbers of demonstrators clearly proved that their voice represented the people of Nicaragua and not a few people who were "manipulated by gangs and other political interests."

On April 27, the President of the National Assembly, Gustavo Porras, announced the formation of a commission to examine the deaths and violence during the unrest. The head of the National Police, Aminta Granera, announced her resignation in face of the criticism of her handling of the unrest and police repression of protests.

On April 28, hundreds of thousands participated in marches for "peace and justice" organized by the Catholic churches in Nicaragua in the cities of Managua, Matagalpa, and León. At the events, "bishops, feminists, homosexuals, family members of those killed in the repression . . . and thousands of peasants" gathered in unity to demonstrate. Peasants who lived in rural areas traveled to Managua by a caravan of trucks, arriving to protest against the Nicaragua Canal proposal by Chinese businessmen and the Ortega government.

On May 12, there was heavy fighting in cities across the country between demonstrators, anti-riot police, and youth shock groups of the Sandinista party. The following day, President Ortega called for a cessation of violence, reading a short statement, in which he called for "an end to death and destruction." He proposed a national dialogue.

After weeks of conflict, the National Dialogue began on May 16, 2018. When Ortega and Murillo arrived at the Seminary of Our Lady of Fatima, the site of the national dialogue, the presidential couple was greeted with shouts of "assassins, murderers" by people on the outskirts of the event.

A delegation from the Inter-American Commission on Human Rights (IACHR) arrived in Nicaragua on May 17 to observe the situation of human rights in the country. The IACHR visit occurred as Nicaraguan human rights organizations were reporting between sixty-one and sixty-seven people dead and more than five hundred injured in the repression of protesters.

On May 18, the second day of dialogue, the IACHR, which was now included in the talks, called on the State of Nicaragua to immediately cease the repression of the protesters. The commission also called on the State to guarantee the independence and functioning of the media in the country.

The national dialogue continued on May 21, when the resignation of Ortega and his wife and the Nicaraguan government was requested by participants in order to return the country to normality. The Inter-American Commission on Human Rights (IACHR), issued a preliminary report on the investigations of what happened in the protests in Nicaragua. The IACHR counted at least seventy-six people killed in the protests in Nicaragua and more than eight hundred injured, and denounced serious violations of human rights by the Government of Nicaragua. The report was based on official visits of the IACHR to Man agua, Masaya, León, and Matagalpa.[312]

A week after beginning, the National Dialogue, between the government of Nicaragua and students, the private sector, and civil society was suspended indefinitely. The leader of the Nicaraguan Catholic Church, Bishop Leopoldo Brenes, who had acted as a mediator of this dialogue, explained that the lack of agreement on an agenda of issues to be discussed prevented negotiations from continuing.

On May 30, a march was led by the Mothers of April Movement, the Student April 19 Movement, and representatives of civil society and private enterprise to honor the victims killed during the protests. The march was repressed by the national police in the company of paramilitary groups and pro-government mobs, leaving approximately fifteen dead. Most of the victims died from shots to the head, neck, and chest.

After more such incidents during June and the first week of July, on July 11, a Nicaraguan opposition leader and academic Felix Maradiaga was attacked in the city of León (northwest Nicaragua) by a group of Sandinista sympathizers. Protests and clashes continued.

The international community intensified pressure on the Government of Nicaragua. On July 16, the United States, thirteen Latin American countries, and the Secretary-General of the UN, Antonio Guterres, demanded that Ortega end the repression of the demonstrators. The Office of the United Nations High Commissioner for Human Rights also denounced the *Law on Terrorism* that was recently approved by the pro-Ortega Parliament of Nicaragua, which criminalized peaceful protests.

The government responded by defending the Law on Terrorism and claiming that the UN was creating "biased declarations." President Ortega said that he would not resign from the presidency of Nicaragua before finishing his term in 2021.

In early-September a general strike was organized to demand the release of political prisoners. Ninety percent of the Nicaraguan people participated costing the country twenty to twenty-five million dollars a day.

On September 29, 2018, President Ortega declared that political protests were "illegal." The United Nations condemned the actions as being a violation of human rights regarding freedom of assembly.[313]

The struggle of the people of Nicaragua against the Ortega government continued into 2019. In July, many government supporters filled the square in Managua to celebrate the fortieth anniversary of the overthrow of Somoza. President Ortega called for a divided nation to come together. The sceptics, however, including many former Sandanista leaders, felt that Ortega had betrayed the revolution.[314]

In the meantime, the capitalist government of the United States had kept up its assault on President Ortega's government. Trump's National Security Advisor, John Bolton announced restrictions and sanctions against Cuba, Nicaragua, and Venezuela.[315]

But to focus on this overt attack on the economy of Nicaragua is to lose sight of the capitalist propaganda campaign that continues to dupe the people. As an example of this, we only have to review Nicaragua's online newspaper, *Today Nicaragua*. As of this writing in June, 2020, the paper is filled with criticism of the Ortega government for its slow response to the Covid 19 pandemic. The government is charged with covering up statistics in order to confuse the people. The paper complains that because of the government's response to Covid 19, American's now had to flee the country. What is interesting is that between the paragraphs of such articles are colorful ads that promote Donald Trump. Nicaraguans can fill out a form and vote on the job that Trump is doing. They can sign a card wishing him a happy birthday; they can read about what a high IQ Barron Trump has. They can even vote whether immigrants should be granted social security benefits when they go to the United States. While these highly polished pop-up ads come and go. There were no articles that portrayed the suffering and the breakup of families that the Trump administration caused at the US border.[316]

The capitalists have proven beyond a doubt that they will stoop to whatever means necessary to destroy any opposition to their fascist dictatorships in Latin America.

Lessons Learned from Nicaragua

Regardless of Ortega's good intentions when the Sandanistas created their revolution against capitalist exploitation, their lack of a universal ideology, caused them to fall victim to the same fearful, materialistic response from which they were trying to escape. This failure resulted in their not being able to take the people along with them when the counter-revolutionary forces began their propaganda campaign against the Sandanista government.

The main lesson from Nicaragua is that if revolutionaries lose their service mentality due to pressures from the counter-revolutionary forces and imperialist powers, they may resort to killing and brutalizing the same people whom they fought to liberate from the previous dictatorship. They do this as a matter of course when they see the people echoing the counter-revolutionaries' criticism of their government. When this occurs, they begin to see the people themselves as counter-revolutionaries.

The Sandinistas had inherited a country with a debt of 1.6 billion US dollars, an estimated thirty to fifty thousand war dead, six hundred thousand homeless, and a devastated economic infrastructure. They organized a Government Assembly for National Reconstruction. Their reforms immediately began to take the country out of the post-war disaster and to promote social justice at all levels. Banks were nationalized and properties were reclaimed from the Somoza family and their allies. An Agrarian Reform movement was set up on the same principles of the revolution: pluralism, national unity, and economic democracy.

However, when faced with pressure from the counter revolutionary movement, the revolutionary leaders stumbled. They created a six-year State of Emergency that allowed for the indefinite holding of suspected counter-revolutionaries without trial. This effectively punished the Nicaraguan people as well. Many civil liberties were curtailed or canceled, including the freedom to organize demonstrations, the inviolability of the home, freedom of the press, freedom of speech, and the freedom to strike.

Seeing their enemies everywhere, the Sandanista government under Daniel Ortega turned on their own people whom they had fought so hard to liberate.

By 2018, the Sandanista revolution in Nicaragua had greatly disintegrated.

In the face of such relentless oppression and repression, nuclear revolutionaries around the world must have a strong moral purpose so that greed, oppression, and exploitation cannot prevail against them. Such an all-pervading moral conviction can only be sustained by subscribing to a universal ideology that is nurtured by a personal quest for spiritual unity with an absolute loving consciouness and the desire to serve the well-being of all.

Chapter Five: Contemporary Revolutions Under Communism

Revolutions in the Soviet Bloc

IMMEDIATELY AFTER WWII, THE Soviet Union, which was created by the Communist Party in Russia, installed communist-leaning governments in the Eastern European countries that the Russians had liberated from Nazi control during the war. The Americans and British feared that communism would spread into Western Europe and worldwide. In response, the US, Canada, and their European allies formed the North Atlantic Treaty Organization (NATO) in 1949. This alliance was a political show of force against Russia and its allies.

In 1955, in response to NATO, the Soviet Union consolidated power among the Eastern bloc countries under a rival alliance called the Warsaw Pact. This set off the Cold War.

The Cold War was waged on political, economic, and propaganda fronts between the Eastern and Western blocs and persisted in various forms until the fall of the Soviet Union in 1991 when Mikhail Gorbachev, the eighth and last leader of the Soviet Union, resigned his post. The Soviet Union ceased to exist on December 31, 1991.[72]

When the Soviet government fell, the Russian mafia, composed of capitalist oligarchs, who had struggled to survive during the height of communism, stepped in to fill the power void. The state's infrastructure, ranging from public utilities to police services, mostly evaporated during the collapse. Additionally, government payrolls almost completely disappeared, which resulted in ex-KGB officers, police officers, and Soviet Army soldiers flooding the mafia's ranks in search of steady employment. Mafia oligarchs seized state-owned assets and enterprises throughout Russia, including the mines, the oil and gas industry, telecommunications, energy networks, and forests. The mafia also extorted the public in exchange for providing security and enforcing laws wherever the Russian government was unable to do so. The Putin administration initially tried to curb the oligarchs, but soon cut a deal with them that allowed them to continue their exploits as long as they supported his government.

Today, under Putin's autocracy, anyone who speaks in opposition to government corruption will be arrested, exiled, or even murdered under mysterious circumstances. This oppression stymies Russia's chances of establishing a true democracy and allows government corruption to continue to expand.[73] In this way, the Russian revolution eventually died and Russia became just another capitalist country.

The Ukraine

In looking at the revolutions in the Soviet Bloc, we see that the Russian ruling class plays the same imperialist role as the US by setting up puppet dictators and seizing the resources and labor of people in other countries.

A decade after the fall of the Soviet Union, the people of the Ukraine were still fighting for independence from Russian control. They organized the Orange Revolution which was a series of public protests and political events that took place in Ukraine from November 2004 to January 2005, in the immediate aftermath of the run-off vote of the Ukrainian presidential election. The election was marred by massive corruption, voter intimidation, and electoral fraud. Kiev, the Ukrainian capital, was the focal point of civil resistance, with thousands of protesters demonstrating daily. Nationwide, there were also acts of civil disobedience, sit-ins, and general strikes.

The protests were prompted by reports from domestic and foreign election monitors, as well as widespread public perception, that the results of the run-off vote of the 2004 between leading candidates Viktor Yushchenko and Viktor Yanukovych were rigged by the local authorities linked to Russia in favor of the latter. The protests succeeded when the results of the original run-off were annulled and a revote was ordered by Ukraine's Supreme Court. The second run-off was declared to be "fair and free." The final results showed a clear victory for Yushchenko, who received about fifty-two percent of the vote, compared to Yanukovych's forty-four percent. Yushchenko was declared the official winner and with his inauguration in January 2005 in Kiev, the Orange Revolution succeeded in its goal.

Putin, who had controlled Yanukovych, was not happy with this development and continued to manipulate politics behind the scenes. In the 2010 presidential election, Yanukovych "won the election" for Ukrainian President against Yulia Tymoshenko who had co-led the Orange Revolution and was the first woman appointed Prime Minister of Ukraine.[74] Tymoshenko said she had proof of cheating by the Yanukovich camp in the run-off and vowed to contest his victory in court. Even so, Yanukovych, who was supported by Putin and the oligarchs, was not concerned. In his support, the Central Election Commission declared that the presidential election was conducted fairly.[75]

Yanukovych was a political strong man who rose to power in Donetsk, an industrial city in Eastern Ukraine and the capital city of the unrecognized Donetsk People's Republic, which was led by Russian patriots. Donetsk is the fifth-largest city in Ukraine.

Once in power, Yanukovych centralized power in his family by taking control of state assets, including real estate and cash. To appease the people, he told them that he would sign a trade agreement with Europe, but after a meeting with President Putin of Russia, he cancelled the agreement. People saw this betrayal as a continuation of the authoritarian mafia-state controlled by Putin and his inner circle.

The last-minute decision to pull out of the trade agreement sparked mass protests by students that lasted seven days and nights in Kiev during February, 2014. The riot police were ordered to clear the square on the eighth night. Videos were circulated of students being beaten by police. This let to mass demonstrations of the Ukrainian people against the Yanukovych government. These demonstrations soon went beyond the trade deal to the question of justice and accountability in Ukrainian society.

A tent city was set up in Maidan Square in Kiev where thousands of people camped. They exchanged goods and services and set up barricades and defense systems. The tents were warm and complete with kitchens, electricity, and TVs. Maidan Square became a city within a city.

The people were ready for a long struggle and economic hardship. They knew that if they wanted independence and freedom it might take ten years or more to achieve it. They understood that the new system would be difficult to set up and succeed, but it was better than living under the old system.

The people looked to their youth as their leaders and vowed to stand behind them. In Russia, Putin characterized Maidan Square as full of alcoholics, drug addicts, and homeless, but the Ukrainian youth were not dissuaded. They said "my soul calls out to me to help the people."

In Maidan Square, the people prepared for a revolutionary war. Yet, the protests remained peaceful even after the police attacked the people's advocates, beat them, tortured them, and left them to die. Many of those attacked said their attackers spoke with a Russian accent.

On December 8, the crowd toppled a Lenin statue nearby. Protests and clashes increased in January, after the Ukrainian parliament passed a group of anti-protest laws.

Protesters occupied government buildings in many regions of Ukraine. The protests climaxed in mid-February when riot police advanced towards Maidan Square and clashed with protesters, but did not fully occupy it.

Police and activists fired live and rubber ammunition at each other. There was fierce fighting in Kiev from February 18 through 20, during which seventy-seven protesters were shot dead. As a result of these events, an agreement was signed on February 21 by Yanukovych and the leaders of the parliamentary opposition under the mediation of the European Union and the Russian Federation. The signing was witnessed by the Foreign Ministers of Germany and Poland, and the Director of the Continental Europe Department of the French Foreign Ministry. The Russian representative refused to sign the agreement.

Shortly after the agreement was made, Yanukovych and other high government officials fled the country. Protesters gained control of the presidential administration and Yanukovych's private estate where, in remembering the old women begging for bread crusts on the streets of

Kiev, they were shocked by the opulent display of wealth that he had stolen from the people.

Afterwards, the parliament removed Yanukovych from office, replaced the government with Oleksandr Turchynov, and ordered that former Prime Minister Yulia Tymoshenko be released from prison.

Unfortunately, despite the ousting of Yanukovych, the installation of a new government, and the adoption of a Ukraine–European Union Association Agreement, this was only the beginning of the people's problems. Soon afterward, the Russian army occupied and annexed the Crimean Peninsula in the south and moved an army into the Donetsk region in the east.

Since April 2014, Donetsk and its surrounding areas have been a war zone in which pro-Russian separatist forces are battling against Ukrainian military forces for control of the city and surrounding areas. Donetsk International Airport became the epicenter of the war with almost a year-long battle, which led to massive casualties among civilians and a total ruination of the northeastern neighborhoods of the city. As of May 2019, the Russia separatists had full control of the city, with Ukrainian forces still engaged in combat outside of the city.

Lessons Learned from Ukraine

In looking at the revolution in the Ukraine, we see that Russia plays the same imperialist role as the US by setting up puppet dictators and seizing the resources and labor of people in other countries.

However, even if the Ukrainian people are able to throw off the yoke of Russian domination, they would still face Western capitalists who would seek to control their resources and labor and impose a capitalist puppet government upon them. The problem that all the people of the world face is the capitalist class and the capitalist psychology, regardless of label. As such, the Ukranian people should not be too quick to place themselves at the feet of the Western capitalists, but should make plans for economic self-reliance at least in so far as food, shelter, clothing, health care, and education are concerned. So also, they should ensure that they remain in control of their physical infrastructure, which includes water, energy, the transportation system, etc.

Perhaps the effort by President Trump in 2019 to corrupt the Ukrainian government in exchange for weapons will help the people understand why this recommendation is so important.

In the Ukraine, the revolution was led by the youth. We also see that revolutionaries, who occupy city squares from which to stage their protests, must be prepared to meet their own basic necessities as well as face a running battle with the police and hired goons. It is fruitless for the victims of Russian imperialist to turn to the West for any long-term support. The West will only become the new imperialists if they get a chance.

As of this writing in the spring of 2020, the brave, freedom-loving people of the Ukraine are still fighting against the Russian invaders. When they turned to the US for help, President Trump promised them weapons in exchange for their government digging up dirt on Joe Biden, his Democratic opponent in the presidential race.

This reinforces our contention that people can trust neither the communist nor the capitalist powers to align themselves with. Rather, people's best course of action is to build a local economy that can ensure autonomy in meeting basic needs and align with people and governments who are doing the same thing. Only in this way can local people overcome the exploitation and oppression of the global capitalists whether in the guise of capitalists or communists. Only in this way can we create a world that serves the interest of all the people.

Belarus

Belarus was another country that purportedly gained independence from Russia with the fall of the USSR. In this country, however, the people did not even get a whiff of freedom. By not acting quickly enough, the window of freedom was slammed shut almost immediately.

The capitalist oligarchs in the form of the dictatorship of Alexander Lukashenko have been in power since 1994. The political opposition has not held a seat in parliament since 1996. They are virtually ignored by the state-affiliated media and the government arrests them on sight if they exercise their right to protest.

The Belarusians proved more cautious than the people of the Ukraine and have suffered their oppression in virtual silence. The risk of a state collapse, civil strife, and Russian interference seemed too high to them. As a consequence, the secret police, still referred to as the KGB, are everywhere watching and arresting and torturing anyone they want to. A culture of fear and isolation blankets the country.

In March, 2015, Lukashenko told police there would be no "Maidan" style protests in Belarus, referring to the protests in Ukraine that unseated the Yanukovych government. Two days after this warning, the opposition Belarusian People's Front Party proposed to abandon plans for street protests scheduled to take place after the presidential election in November. The leaders of the opposition argued that a better future was "not worth people's blood."

But it was not just this lack of revolutionary fervor that hobbled the people of Belarus. They also did not have the political infrastructure upon which to build a revolution. Unlike the people of the Ukraine, they did not have the organizational capacity to effectively participate in an election and be able to challenge the results. Secondly, Lukashenko and the KGB threatened, imprisoned, tortured, and even killed the editors and journalists of the private press who, as a result, did not address the issue of election fraud, inform the people about it, or cover ensuing protests. Thirdly, the people were sufficiently intimidated by the state that they did not have the capacity to plan and organize. Fourthly, the country was closed off to political involvement by foreign and international donors and potential partnerships from democracy-oriented NGOs.

There were also international politics at play that worked against the people of Belarus. The country received no attention from the US government because the capitol city of Minsk was actively engaged with radical governments, including Saddam Hussein's regime in Iraq. Thus, Belarus possessed little strategic value to the US.

As of 2015, the West actually looked favorably upon Lukashenko's dictatorship. They told the dictator that independence from Russian control was more important than democracy. The West was clearly not interested in a Belarus democratic revolution.

In 2015, the Belarusian left-wing website, Prasvet, said that election fraud has led the opposition to lose interest in campaigning with the public. Their logic was that if votes are not counted there is no point in campaigning. They concluded that it was better to find groups prepared to come to the square and confront the police.

By 2017, the government was continuing its crackdown on the people but the West turned its face from these violations of human rights.

The government made maximum use of the death penalty to silence the leaders of the people. Those condemned to death were executed by a shot to the head. Authorities did not inform families of the execution date or the burial place.

This oppression eventually led the people to organize protests in thirteen cities during February and March of 2017.

Police arbitrarily detained at least seven hundred people in connection with the protests, including one hundred journalists and sixty human rights activists and charged them with fabricated offenses, such as hooliganism or participating in unsanctioned protests, and sentenced them to twenty-five days' detention. The protesters were denied access to lawyers and could not call defense witnesses.

Laws and regulations governing public association continue to prevent human rights groups or political opposition movements from registering and operating freely. Involvement in a non-registered organization is a criminal offense. Most registration rejections are based on arbitrary bureaucratic pretenses.[76]

In 2018, civil society activists, lawyers, rights groups, and independent media continued to face government harassment and pressure.

In June, the government amended the Law on Mass Media, adding excessive restrictions requiring that all online media outlets keep public records of the names of people who submit comments online and disclose that information to authorities. The amendments also provided for holding owners of registered online media criminally liable for any content posted on their website.

Freedom of assembly and even freedom to associate remained severely restricted and penalties increased.[77]

The people refer to the constant surveillance by the KGB as being "Under the Hood." One woman, the wife of a journalist, explains what they mean by that. Being Under the Hood is when you cannot leave the city limits, and you have a 10:00 pm curfew. And if you violate these restrictions you get a visit from the KGB. You know you are under the hood when you realize there is an invisible man in your bedroom and you can feel his breathe while on the phone. Or when your Skype account is hacked and you hear ugly things said about you and your family. Or when you get a threatening email sent to an address known only to you and another friend. Or when you walk down the street and not know if a person is a normal pedestrian or one whose job it is to watch you. It is when you learn to always take different routes to places you frequent. It is when you are driving and keep looking in the rear-view mirror to figure out which car is following you. It is when you learn to recognize KGB people after a glance at their faces.

If a person is arrested by the KGB, they are often tortured. To stop the pain and protect their families, they are forced to join the KGB and become spies and informers on their friends and acquaintances.

At minimum, to go against the state means losing your job, your livelihood, and putting your family at risk. Such is the condition that results when the people are not united and prepared for revolution when the opportunity presents itself.

In the summer of 2020, the people of Belarus faced another corrupt "election" and the continuation of Lukashenko's dictatorship. Without having independent observers at the vote count, the people across the country took to the streets in protest, but were met by police using truncheons and rubber bullets. Demonstrators were rounded up and tortured. People were killed. Metro stations were closed and the internet was shut down.

Lukashenko, who claims to have received over eighty percent of the vote, blamed the nation-wide protests on foreign intervention and asked for assistance from Russia to quell the people's protests.

Lessons Learned from Belarus

Belarus is a clear example of what happens to a people when they ignore the opportunity to take power and establish freedom, and, by not acting at the appropriate time, lose any chance of conducting a revolution.

When the Soviet Union collapsed, the government was taken over by the puppet dictator Lukashenko. He controlled the press and arrested any journalists who reported on any protests or statements by his political opponents.

The people of Belarus proved more cautiously than the people of the Ukraine and suffered their oppression in silence. The risk of state collapse, civil strife, and Russian interference seemed too high of a price for them.

Further, the people of Belarus lacked the political infrastructure upon which to build a revolution. Feeling isolated, they were easily intimidated by the government and lacked the resources to plan and organize.

Finally, the people were not able to create allies or potential partners abroad because the government closed off any possible communication with foreigners.

The US government, for its part, was not interested in a revolution in Belarus. They told Lukashenko that the independence of Belarus from Russia was more important than democracy.

Under these conditions, the oppression of the people grew exceedingly worse. The secret police turned every one against their neighbor by convincing them that their neighbors were spying on them for the secret police.

Bulgaria

Bulgaria represents another case of a failed revolution. Like Belarus, the revolution for freedom did not even get off the ground. But what happened in Bulgaria is somewhat more deceptive to the outside observer.

Beginning in 1989, after forty-five years of rule by the communist party, which took its orders from Russia, the Bulgarian people created a political democracy. But this was far from adequate to bring the people the freedom they desired.

The former communist party was restructured and succeeded by the Bulgarian Socialist Party (BSP), which won the first post-communist elections for the Constitutional Assembly in 1990 with a small majority. Meanwhile, Zhelyu Zhelev, a communist-era dissident helped to form the Union of Democratic Forces (SDS). He was elected President by the Assembly in 1990.

This party aimed for Bulgaria to join the European Union and NATO. The contest to move the country toward Western capitalism or toward Russian mafia-style capitalism was set in play.

In the first parliamentary elections held under the new constitution of Bulgaria, in October 1991, the Union of Democratic Forces (SDS) party won a plurality of the seats, having won one hundred and ten out of the two hundred and forty seats, and created a cabinet along with the support of the Movement for Rights and Freedoms that was a liberal party (DPS) widely perceived as a party of the ethnic Turks minority in Bulgaria. Yet, their government collapsed only one year later in 1992. It was succeeded by a team put forward by the Bulgarian Socialist Party (BSP), which served until 1997 when it collapsed due to a severe economic crisis. The President dissolved the government and appointed a provisional one to serve until early parliamentary elections could be held in April, 1997.

The April 1997 elections resulted in a landslide victory for the SDS, winning a majority of one hundred and thirty-seven seats in parliament, and allowing them to form the next government. This proved to be the first post-communist government that did not collapse and served its full four-year term until 2001.

In 2001, the former monarch of Bulgaria, Simeon Saxe-Coburg-Gotha, returned to power, this time as Prime Minister with his nationalist movement (NDSV), having won half (one hundred and twenty) of the seats. His party entered a coalition with the DPS and invited two functionaries of the BSP (who sat as independents). In opposition were the two previously governing parties, the Russian-leaning BSP and the Western-leaning SDS. After four year, the SDS suffered the defection of numerous splinter groups while the ruling party NDSV itself ruptured into a pro-right core and a pro-liberal fringe group. Bulgaria entered NATO in 2004.

In the aftermath, the BSP won the parliamentary elections in 2005 with eighty-two out of the two hundred and forty seats, but as it did not get the majority of the seats, a coalition government was formed by the three biggest parties—BSP, NDSV and DPS. The elections also put in parliament some right-wing parties, as well as the extreme-right nationalist coalition led by the Ataka party as an answer to the former coalition government of NDSV and DPS. Bulgarian entered the European Union in 2007.

In the parliamentary elections of 2009, the center-right party of the mayor of Sofia, GERB, won with one hundred and seventeen seats. The party formed a minority government with the support of the right-wing parties. Once becoming the governing party—the National Movement Simeon II—did not amass enough votes to enter the parliament. The austerity measures required by the stagnation of the Global Financial Crisis led to massive protests and the resignation of the cabinet in early 2013, months before the end of GERB's term.

In the early elections the former leading party GERB received the most votes from the people. This was the first time since 1989 that a ruling party was re-elected; in the past painful and unpopular reforms had to be implemented. However, as GERB received only ninety-seven of the two hundred and forty seats, and failed to make a coalition, they refused the mandate, handing it down to the next party BSP. The socialist party chose the non-party former Minister of Finance Plamen Oresharski to form a cabinet. His cabinet was supported by the BSP and the DPS, opposed by GERB, while Ataka was absent.

Only two weeks after its initial formation, the Oresharski government came under criticism and had to deal with large-scale protests some with more than eleven thousand participants. One of the main reasons for these protests was the controversial appointment of media mogul Delyan Peevski as a chief of the National Security State Agency.[78]

Now we come to the heart of the problem of why the people of Bulgaria never experienced freedom, despite political democracy.

Peevski was part of the Bulgarian mafia that served the same role as the Lukashenko's dictatorship in Belarus. The Bulgarian mafia controls the streets of Bulgaria and the people live in fear of these organized thugs. These corrupted warriors, controlled by a capitalist psychology, are involved in drug trafficking, cigarette smuggling, human trafficking, prostitution, illegal import-export dealings, extortion (often under the cover of ostensible security and insurance companies), racketeering, car theft, arms trade, and financial crimes like falsification of IDs, organized bank fraud, mortgage fraud, and credit card fraud.

They control all branches of the government through fear of beatings and execution and by donations to all major political parties. Not a single major mafia figure has been punished by the Bulgarian law, though there are on-going mafia-related assassinations. Because they "own" municipalities and individual members of Parliament, the Bulgarian governments have done little to reduce the presence of organized crime.

The mafia also operates behind a guise of legitimacy by running their operations through large insurance companies such as SIC and VIS whose activities range from local street extortionists to international drug dealers and money launderers.

The mafia organizations got their start after the collapse of the USSR. Their initial members were athletes, particularly wrestlers. Some referees became criminals, who then favored the fighters whom they could hire to work for them.

The wrestlers, who were the darlings of the Communist regime, soon discovered that the new government had neither the desire nor the funds to support them. Politicians had cut the huge subsidies for sports clubs, leaving thousands of athletes out of work. Many wrestlers, karate students, weightlifters and boxers then turned to petty crime.[79]

> The image of the Bulgarian "mug", including a sturdy muscular build, a black suit, sunglasses, a shaved head, and golden jewelry, became synonymous with the so-called Bulgarian "transition" (to the market economy). These wrestlers were also known to own expensive cars with license plates with double numbers so that strangers would recognize their status as elite criminals.[80]

The wrestlers began their criminal activities by stealing cars, shaking down tourists, and offering "protection" to small businesses through their newly created "insurance" companies.

Soon, however, they began to partner with members of the state security apparatus composed of Special Forces operatives and spies. Retired colonels and generals of the old older also came on board. Moreover, you could not become a mafia boss without the support of the police. This is how the warrior elite and leaders of the old communist order transitioned to capitalism.

In short time, the mafia took over the fashion industry. Here they made lots of money and used the models as "beautiful and refined whores."[81]

The intellectuals were silenced and the artists and writers spied upon. The media never challenged the mafia and their accounts of murders and street killings only served to keep the people in a psychosis of fear.

Because of the appearance of democratic process, however, the Western press praises Bulgaria for its freedom of speech and human rights records and gives Bulgaria high marks for political rights.[82]

As of July, 2020, we can see the results of that corrupt position.

In an effort to defuse weeks of anti-government protests against government corruption, long time Prime Minister, Boyko Borissov, called for an overhaul of the constitution.

The protesters quickly dismissed his plan as an empty gesture and promised to continue their campaign. They wanted Borissov out of office for corrupting state institutions, especially the judiciary, to serve the oligarchs.

During a moment of opportunity for greater freedom, it is always easier for a strong and organized small group of self-interested people to seize power than for a large group of unorganized people. If the service-minded revolutionaries are not prepared for eventualities as those presented by Ukraine, Belarus and Bulgaria, this will be the people's fate.

So far, we have discussed attempted revolutions in countries seeking to gain freedom from the old communist order.

Lessons Learned from Bulgaria

Bulgaria is another example of a revolution that never got off the ground. In this case, it was coopted by political party divisiveness. It is a classic example of the limitations of political democracy when the people do not also create the infrastructure for economic autonomy insofar as meeting their basic needs is concerned.

After forty-five years of being controlled by the communist party that took its orders from Russia, when the Soviet Union fell, the people created a political democracy, but without creating an economic democracy.

While the intellectual political "leaders" wasted their energy on political infighting and party combats, the Bulgarian mafia, basically capitalist thugs, controlled the streets and kept people in fear for their lives. The mafia came to control all three branches of government by making donations to the myriad of political parties.

Consequently, while the people may protest the government and change their votes between political parties, this exercise is fruitless when it comes to improving their living standards and gaining them the freedoms they desire.

All these things destroy the collective consciousness of the people and demonstrate why spiritual revolutionaries must make plans for a nuclear revolution and be willing to lead the people in a struggle for the benefit of all and the benefit of the natural world.

The lesson learned from Bulgaria is that political democracy without economic and cultural democracy can easily be turned into a farce. Warriors of the old order will move quickly to regain power, in one guise on another, and they can easily succeed when the people are not organized and able to move quickly to establish their own socio-economy.

In the Far East

Hong Kong

The Political Situation

On July 1, 1997, Britain returned Hong Kong to China's control after it had been a British colony since 1941. Part of the agreement was that Hong Kong would be part of China but retain its capitalist economy. It was defined as "one country, two systems."

The divisiveness between Hong Kong and mainland China was set from the very first days following the 1984 Sino-British Joint Declaration, in which it was stated that Hong Kong was to be returned to China and become its first Special Administrative Region on July 1, 1997, under the principle of "one country, two systems."

As a consequence of this agreement, Hong Kong has a different political and economic system from mainland China. The Hong Kong Basic Law, the constitutional document drafted by the Chinese side before the handover, is founded on the terms enshrined in the Joint Declaration. It governs its political system, and stipulates that Hong Kong shall have a high degree of autonomy in all matters except foreign relations and military defense. The declaration stipulates that the region may maintain its capitalist economic system and guarantees the rights and freedoms of its people for at least fifty years after the 1997 handover.

The guarantees of the territory's autonomy and individual rights and freedoms, while enshrined in the Hong Kong Basic Law, which outlines the system of governance of the Hong Kong Special Administrative Region, is still subject to the interpretation of the Standing Committee of the National People's Congress (NPCSC) in Beijing.

According to the Basic Law, the leader of Hong Kong, the Chief Executive, is currently elected by a twelve hundred-member Election Committee, while "the ultimate aim is the selection of the Chief Executive by universal suffrage upon nomination by a broadly representative nominating committee in accordance with democratic procedures."

Since that agreement, China has continuously tried to exert more control over Hong Kong, leading to a series of protests and anti-government actions.

Although the revolution in Hong Kong was inspired by the worldwide Occupy movement that began with the Occupy Wall Street (OWS) protest movement in New York City on September 17, 2011, which was organized to bring attention to economic inequality and the pervasive greed, corruption, and influence of corporations on governments, the Hong Kong Occupy movement is directed at the inability of the people of Hong Kong to choose their own elected officials by universal suffrage.[83]

A series of sit-ins and street protests, often called the Umbrella Revolution or Occupy Movement, occurred in Hong Kong from September 26 to December 15, 2014.

The protests began after the communist government in Beijing issued a decision proposing reforms to the Hong Kong electoral system. The decision was widely seen to be highly restrictive, and tantamount to the Chinese Communist Party (CCP)'s pre-screening the candidates for the Chief Executive of Hong Kong.

Students led a strike against the decision and started protesting outside the government headquarters during the 26th of September.

Two days later, on September 28, the Occupy Central with Love and Peace movement announced the beginning of its civil disobedience campaign. Students and other members of the public demonstrated outside government headquarters, and some began to occupy several major city intersections.

Police retaliated with tear gas while Triad thugs (mafia capitalists) attacked protesters. This led more citizens to join the protests and to occupy two of the main business and shopping districts. The number of protesters, which peaked at more than one hundred thousand at any given time, overwhelmed the police and gave rise to "containment errors."

Government officials in Hong Kong and in Beijing immediately denounced the occupation of intersections as "illegal" and a "violation of the rule of law," while Chinese state media claimed repeatedly that the West had played an "instigating" role in the protests, and warned of "deaths and injuries and other grave consequences."

The protests precipitated a rift in Hong Kong society, and galvanized the youth—a previously apolitical section of society—into political activism and heightened awareness of their civil rights and responsibilities. Not only were there fist fights at occupied sites and flame wars on social media, family members also found themselves on different sides of the conflict.

Demonstrations lasted for almost seventy-eight days despite numerous incidents of intimidation and violence by triad thugs, and several attempts at clearance by the police, who used tear gas and pepper spray, and the rape of women protesters. By December 15, the protesters were completely routed.

The violent police action during the occupation was widely perceived to have damaged the reputation of what was once recognized as one of the most efficient, honest, and impartial police forces in the Asia Pacific region.

The protests ended without any political concessions from the government. Rhetoric from the Chief Executive of Hong Kong, CY Leung, and Beijing officials about rule of law and patriotism increased and there was an assault on academic freedoms and the civil liberties of activists.

As early as March 2013, the Occupy Central with Love and Peace (OCLP) had begun holding rounds of deliberations on electoral reform proposals and strategies. In June 2014, the OCLP conducted a "civic referendum" on its own electoral reform proposal in which 792,808 residents, equivalent to over one fifth of the registered electorate, participated.

In response to these popular deliberations, in early 2014, Hong Kong's Chief Executive C Y Leung submitted a report to the National People's Congress in Beijing inviting them to deliberate whether it was necessary to amend the method of selection of the Chief Executive.

In June 2014, the State Council of the National People's Congress in Beijing issued a white paper called *The Practice of the 'One Country, Two Systems' Policy in the Hong Kong Special Administrative Region* in which it claimed for China, "comprehensive jurisdiction" over Hong Kong. "The high degree of autonomy of the Hong Kong Special Administrative Region is not full autonomy, nor a decentralized power," it said. "It is the power to run local affairs as authorized by the central leadership."

In an atmosphere of growing discontent, the annual July 1 protest march attracted the biggest numbers in a decade and ended in an overnight sit-in in Central with five thousand police conducting over five hundred arrests.

Tensions mounted again on August 31, 2014 when the Standing Committee of the National People's Congress in Beijing set limits for the 2016 Legislative Council election and 2017 Chief Executive election. While notionally allowing for universal suffrage, the decision imposed the standard that "the Chief Executive shall be a person who loves the country and loves Hong Kong," and stipulates "the method for selecting the Chief Executive by universal suffrage must provide corresponding institutional safeguards for this purpose."

The decision further stated that for the 2017 Chief Executive election, the Chief Executive-elect "will have to be appointed by the Central People's Government." The process of forming the 2016 Legislative Council would be unchanged, but following the new process for the election of the Chief Executive, a new system to elect the Legislative Council via universal suffrage would be developed with the approval of Beijing.

In response to the State Council's decision, the minority pro-democracy legislators promised to veto the framework for both elections as being inherently undemocratic. In addition, Occupy Central with Love and Peace (OCLP) announced that it would organize civil disobedience protests.

On September 13, 2014, representatives of Scholarism, a pro-democracy student activist group, including seventeen year-old Agnes Chow Ting, staged a small protest against the State Council decision outside the Central Government Offices and announced a class boycott for university students for the week commencing September 22. Agnes Chow encouraged students unable to join the protest to wear a yellow ribbon to signify their support. The Hong Kong Federation of Students (HKFS) and Scholarism mobilized students for the class boycott, beginning with a rally that attracted thirteen thousand students on the Chinese University of Hong Kong campus on the afternoon of September 22. From that point onward, public protests and demonstrations lasted through the end of September to December.

The protests caused strong differences of opinion in Hong Kong society, with a "yellow (pro-occupy) vs. blue (anti-occupy)" war being fought. The media reported conflicts within peer groups over values or what positions may be orthodox, and rifts formed between mentor–mentees over the extent to which the movement should go.

Parents fought with their children over their attending protests. Those who opposed the Occupy protests did so for different reasons. A significant part of the population were refugees from Communist China in the 1950s and 1960s and had lived through the turmoil of the Hong Kong 1967 Leftist riots which were large-scale riots between pro-Communists and their sympathizers and the Hong Kong government. Others felt that the protesters were too idealistic, and feared upsetting the Beijing leadership and the possibility of another repeat of the crackdown that ended the Tiananmen Square protests of 1989.

Being realistic, however, the overwhelming reason that people opposed the demonstrations was that they disrupted the lives of ordinary citizens by causing road blocks, traffic jams, school closures, and financial loss to businesses (in particular, those run by the Triads in Mong Kok).

The BBC showed video footage from a Hong Kong TV network, which appeared to show 'anti-Occupy protesters' being hired and transported to an Occupy protest site. The 'protesters', many of whom were initially unaware of what they were being paid to do, were secretly filmed on the bus being handed money by the organizer. Anonymous police sources informed the BBC Newsnight investigation that "back-up was strangely unforthcoming" to scenes of violence. *The South China Morning Post* also reported claims that people from poor districts were being offered up to eight hundred HK dollars per day, via WhatsApp messaging, to participate in anti-Occupy riots.

The Hong Kong police stated that up to two hundred gangsters from two major Triad gangs may have infiltrated the camps of Occupy Central supporters. A police officer explained that the police could not arrest the Triad gangsters there "if they do nothing more than sing songs for democracy."

In a televised interview on October 19, Chief Executive CY Leung echoed Chinese claims about foreign responsibility for the protests, but declined to give details until an "appropriate time." Six months later, on April 22, 2015, a reporter asked Leung, "has that time come yet?" Leung simply responded, "Well, I stand by what I say."[84] Three years later, Leung had yet to provide the promised substantiation.

The US State Department categorically rejected accusations of interference, calling the charges "an attempt to distract from . . . the people expressing their desire for universal suffrage."

On October 1, *China News Service* criticized the protesters for "bringing shame to the rule of law in Hong Kong." *The People's Daily* said that the Beijing stance on Hong Kong's elections is "unshakeable" and legally valid. Stating that the illegal occupation was hurting Hong Kong, it warned of "unimaginable consequences." Some observers remarked that the editorial was similar to the April 26 editorial that foreshadowed the suppression of the Tiananmen Square protests of 1989.

By October 6, official Chinese media outlets called for "all the people to create an anti-Occupy Central atmosphere in the society." The protesters were described as "going against the principle of democracy." But the protesters were damned if they did and damned if they did not. A commentary in the *China Review News* claimed that "the US is now hesitant in its support for the Occupy Central. If those campaign organizers suddenly soften their approach, it will show that their American masters are giving out a different order."[85]

Jumping ahead to the next election of the Hong Kong Chief Executive on March 26, 2017, CY Leung announced that he would not run for a second term.

Despite leading in the polls, candidate John Tsang struggled to receive nominations from the pro-Beijing electors and had to rely heavily on pro-democrats. Carrie Lam, who had Beijing's Liaison Office actively lobbying for her, attracted five hundred and eighty nominations, almost half the votes in the Election Committee and only twenty-one votes short of winning the final election, while Tsang and Woo received one hundred and sixty-five and one hundred and eighty nominations respectively, most of which came from the pro-democracy camp which consisted of about three hundred and twenty members.

As a result, Lam became the first female Chief Executive of Hong Kong.[86] Lam turned out to be a puppet of Beijing just as the protesters had always feared she would be. After doing nothing to advance the democratic cause of the Hong Kong people, the protests began again in 2019.

Proposed Extradition Legislation

Beginning in January, 2019, black-clad protesters wearing hard hats and face masks broke into the Hong Kong legislative chamber and began to spray slogans on the walls and tear down portraits of legislative leaders.

At the same time, tens of thousands of protesters marched through the city to demand expanded democracy, the full withdrawal of newly proposed extradition legislation, and the resignation of Carrie Lam.[87] The extradition legislation had become a major issue with the protesters. While this extradition legislation was justified by the Hong Kong government as a way to prevent Hong Kong from becoming a shelter for criminals, it caused widespread criticism from the people of Hong Kong as well as the legal profession, journalist organizations, business groups, and foreign governments fearing the erosion of Hong Kong's legal system and its built-in safeguards. Critics said people would be subject to arbitrary detention, unfair trial, and torture under China's judicial system where the Communist Party-controlled court system has a conviction rate as high as ninety-nine percent.[88]

Human Rights Watch's Sophie Richardson said in a statement, "The proposed changes to the extradition laws will put anyone in Hong Kong doing work related to the mainland at risk. No one will be safe, including activists, human rights lawyers, journalists, and social workers."[89]

Lam Wing Kee, a Hong Kong bookseller, said he was abducted, detained, and charged with "operating a bookstore illegally" in China in 2015 for selling books critical of Chinese leaders. In late April, Mr. Lam fled Hong Kong and moved to Taiwan where he was granted a temporary residency visa. Now living in Hong Kong, he said, "If I don't go, I will be extradited. I don't trust the government to guarantee my safety, or the safety of any Hong Kong resident."[90]

On June 9, an estimated one million people marched to the government headquarters to show they were against the proposed extradition bill. It was largely a peaceful rally, though some small skirmishes broke out.

Three days later, on June 12, another large demonstration took place where police fired tear gas and rubber bullets. The people did not back down and the stand-off developed into the worst violence Hong Kong had seen in decades.

The protesters, who were led by young Millennial and Generation Z activists were now prepared for handling police surveillance and cyber-monitors. They had gotten rid of all their Chinese apps like WeChat, Alipay, and Taobao and had installed a private network on their smartphones to use with a secure messaging app called Telegram. They hid their identities referring to each other by first names only.

During demonstrations and other activities, they bought only single-ride subway tickets instead of prepaid stored-value cards. They left their credit cards at home and avoided mobile payments in favor of cash. They took no selfies or photos of the demonstrations.

In addition, they wore face masks to obscure themselves from CCTV, fearing facial-recognition software, and bought fresh pay-as-you-go SIM cards.

"The Chinese government will do a lot of things to try to monitor their own people," said Bonnie Leung, a leader of the Hong Kong-based Civil Human Rights Front.

Leung cited media coverage of China's use of artificial intelligence to track individuals and its social-credit-score system. "We believe that could happen to Hong Kong, too," she said.[91]

Antony Dapiran, who wrote a book on protest culture in Hong Kong said, "They (the young activists) don't trust Beijing, they don't trust their authorities and the legal system, and they don't like the blurring of lines between Beijing and Hong Kong."[92]

Unlike the pro-democracy movement in 2014, the latest demonstrations have remained intentionally leaderless in another attempt to foil the police. The protesters that gathered around government buildings were without clear leaders.

Demonstrators shared protest tips and security measures with people they had met just hours before to avoid a similar fate. Meetups were primarily planned on Telegram, which became the top trending app at the iPhone app store in Hong Kong in the days leading up to the protest.

Protesters received a barrage of on-line advice from each other that included changing their usernames on Telegram to sound nothing like their actual names, changing the phone number associated with the app and using SIM cards without a contract.

Even these measures were not always enough to eliminate cyberspying by the police. For example, Hong Kong police arrested Ivan Ip in his home. Ivan was a coordinator of a Telegram group that had thousands of people. Telegram also reported a massive cyberattack, which the company said probably originated from China and was timed with the July 12 protest.

The fears of the encroachment of mainland China on their liberties is fed by the knowledge that the Hong Kong police force sends about one hundred and fifty officials every year for "ideological and practical" training at elite mainland police academies. A larger number also receive regular training in "hand-to-hand combat, interrogation skills, criminal investigation, and gun use," according to news releases from the Chinese government.

When a high-speed rail link connecting Beijing to Hong Kong opened in 2018, Chinese police were allowed to enforce mainland laws in the rail terminal. This was the first time that mainland police were allowed to patrol Hong Kong as part of joint immigration checks.

Despite such developments, activists told *The Washington Post* that they remain undeterred and will continue to show up at demonstrations. They have masks and goggles prepared, they say, both as a shield against police tactics such as the use of pepper spray and to avoid potential facial-recognition and other surveillance software.[93]

Another three days later, on June 15, Hong Kong leader Carrie Lam issued a dramatic reversal, saying she would indefinitely delay the extradition bill. Throughout her press conference she adopted a resolute, unapologetic, and defensive tone that her critics labelled arrogant and insincere. She stuck to her justification of the extradition bill, insisting it was to seek justice for a young Hong Kong woman murdered by her boyfriend in Taiwan—a rationale that many doubted because Taiwan had already said it would not receive the murder suspect. She said she was acting out of "empathy" towards the victim's family. Lam defended the bill as "a good thing" that would facilitate international cooperation in bringing fugitives to justice and said she was "perplexed by the extreme reaction" of people who opposed such a "laudable objective."[94]

Despite her appeals for calm, the following day, on June 16, an estimated two million people took to the streets demanding that the bill be withdrawn completely. There were calls for Lam's resignation.

To understand the reason for this great outpouring of public protest, we must understand that the police had fired rubber bullets and one hundred and fifty rounds of teargas to break up a huge rally on June 12, just days after a million people took peacefully to the streets to protest against the extradition legislation.

The people believed that by permitting the extradition of anyone on Hong Kong soil to face trial in China would effectively remove the firewall between Hong Kong's common law system and the mainland's party-dominated legal system. Anyone in Hong Kong could then be whisked away to Beijing in the dead of night without any legal protections.

Ever since the Occupy Movement began in 2014, Hong Kongers had seen irrevocable changes to their way of life. Four popularly elected pro-democracy lawmakers had been disqualified by the courts for saying their oaths too slowly or with the wrong intonation. Politicians had been forbidden to stand for election. The pro-democracy Hong Kong National Party had been banned in the interests of 'national safety," and activists had been sent to prison on public-order offences. And now the police had used violence against their own people.

The unseemly rush to pass the unpopular extradition law has also weakened each of the territory's institutions. The legislature descended into unseemly brawls, with fist fights breaking out as committees dueled. The civil service and judiciary were no longer seen as politically neutral. The police force was now an object of popular hatred, and its relationship with the public was irretrievably damaged.

The chief executive, Carrie Lam, was so unpopular that protesters carried pictures of her face stamped with the word "Liar" and six thousand mothers turned out to accuse her of not being fit for office. Even though the bill had been put on hold, the process had already permanently devalued the institutions that Hong Kong people held dear.

The ability to demonstrate had long become an important expression of local identity that distinguished Hong Kong from mainland China, and over the years, Hong Kongers had enthusiastically marched with performative flair, including shopping actions, carol singing rallies, and artistic protests against censorship with blank placards.

Yet the police designation of the June 12 protest as a riot, combined with court verdicts a month earlier in which nine pro-democracy campaigners have been convicted over their leadership of the Occupy Movement rallies in Hong Kong in 2014, prompted renewed alarm about the city's political freedom.

The protest leaders found guilty included the sociology professor Chan Kin-man, sixty, the law professor Benny Tai, fifty-four, and the Baptist minister Reverend Chu Yiu-ming, seventy-five. The men were charged with conspiracy to cause public nuisance and of inciting others to cause public nuisance. The sentences carried a maximum of seven years in prison.[95]

The core values that Hong Kongers cherish include religious freedom, press freedom, judicial independence, and civil rights, including the ability to protest. These values are seen by Beijing as among the "seven unmentionables," putting Hong Kongers on the frontline of the clash between "universal" values and the Communist party's need for authoritarian control.

Lack of a Grand Strategy

Faced with these existential threats, Hong Kong's default position has in recent years been a defensive crouch. "We don't have a grand strategy," the political scientist Ray Yep from City University told a reporter before this round of protests had broken out. "In every situation, you just defend what you can the most. This is how you defend Hong Kong values."[96]

Hand-scrawled messages posted around town read "Stop shooting students." "Is protesting a crime?" "Is speaking a crime?" One message simply said: "Keep going till the end."[97]

Over the following days, anger continued to mount towards the police and on June 21, protesters blockaded police headquarters for fifteen hours. They demanded that the protesters who were arrested during previous protests be exonerated.

On June 22, Yonden Lhatoo, the chief news editor at the *South China Morning Post*, a newspaper that strives for neutrality in its publication, wrote an article entitled, "There will be blood if Hong Kong protesters continue to blur the line between people power and mob rule." While admitting that the young people of Hong Kong had good reason to feel "alienated, downtrodden, disrespected, and left behind," by the "deaf government officials who remained nestled in their ivory towers," he was seriously concerned that the young people had besieged police headquarters for fifteen hours straight and what the consequences would be for the rule of law in Hong Kong. He complained that they "barricaded the doors and pelted the building front with eggs, covered CCTV cameras with adhesive tape, defaced the walls with obscene and hateful messages, and pretty much held top brass, frontline officers, and support staff hostage for hours on end."

He referred to the incident as an act of "anarchy" and wanted to know how this could be acceptable, by any measure, for any cause? He made the distinction between the two million people who had marched peacefully on the streets on June 16 and the takeover of roads and government buildings on June 21. He declared that the former event was an exercise in people power, but the latter was an example of mob rule.

He said that it was understandable why no city leaders appeared to help quell the violence because "many young protesters seem unafraid of batons or tear gas and a police crackdown would only provoke them further."[98] We have to ask the meaning of these words. Is he saying that the politicians should only appear when the people are afraid of repression?

It was at this point in time when the pro-establishment political parties began to speak up.

Starry Lee Wai-king, chairwoman of the Democratic Alliance for the Betterment and Progress of Hong Kong announced that a complete withdrawal of the controversial extradition bill and holding talks with protesters could help solve the problems facing Carrie Lam, but Lee also warned that demands for an independent inquiry into police use of force could "easily come up with biased conclusions." Lee said Lam's party would understand if the government announced a complete withdrawal of the extradition bill, in an effort to heal society.

Lee also said the government could consider holding talks with protesters, even though they had no leaders. "Dialogue must be better than confrontation. But . . . it's not easy to handle this new mode of processions, because there is no leader," she said.[99]

The leader of the pro-business Liberal Party, Felix Chung Kwok-pan also spoke up. He called for the appointment of new advisers to Chief Executive Carrie Lam's cabinet to provide a wider spectrum of views. He also said the government should consider a broad and comprehensive inquiry into the recent controversy, such as whether foreign intervention was involved.

Frederick Ma Si-hang, former Secretary of Financial Services and the Treasury, urged the government to examine why young people were so disgruntled. "Young people are the pillar of society. If they have great grievances about society, we will have a hard time ahead," he said. "Hong Kong has a very serious sickness, and the government must think of ways to heal our home."[100] Of course, even though he was retired from public office, Frederick Ma understood the background economics of the situation.

Meanwhile, signatories of an urgent appeal for Lam to withdraw the bill, signed by retired lawmakers and ex-officials, including former security chief Peter Lai Hing-ling, increased from twenty to thirty-two. The officials also proposed that an independent inquiry be set to look into all aspects of the government's handling of the bill.

Tam Yiu-chung, Hong Kong's only delegate to the National People's Congress Standing Committee, China's top legislative body, said it would not support an inquiry that targeted police.

"Police need to keep their morale, fight crime, and protect the people. It won't do society any good to target them," Tam said, adding Beijing remained firmly supportive of Lam.[101]

At the same time this was going on, student unions at nine higher education institutions issued a joint statement to reiterate that protests would continue until the government met their four demands: (1) withdrawal of the extradition bill; (2) retraction of all references to the clashes on June 12 as a riot; (3) all charges dropped against protesters; and (4) an independent inquiry on the use of force by police. Street booths were set up to promote their cause.

Two days later, on June 24, a reporter for the *Guardian* newspaper living in Hong Kong wrote an article on the rise in numbers of people in Hong Kong seeking to emigrate. The article pointed to the political crisis as a cause, but also to the exorbitant housing prices and high cost of living.

Although the government did not keep official statistics on the number of Hong Kongers moving abroad, emigration consultants said that the desire to leave the district jumped after the 2014 Occupy Central Movement, but failed to force Beijing to grant Hong Kong people an unfettered vote to choose their leader. The desire to move abroad spiked again in 2019 amid the recent protests.

Elisa Wong, who was seeking to emigrate to Australia, spoke to the hard decision she had to make: I love Hong Kong but we have no other choice . . . who would want to leave the place where you grew up and leave behind your elderly parents and friends?"[102]

The One Country-Two Systems policy, put in place after the handover of sovereignty from Britain to China in 1997 promised to maintain Hong Kong's status quo for fifty years, but it would run out in 2047. It is still unclear what Hong Kong's status will become once that deadline passes, but the political crises in recent years added uncertainty to its political future.

For Angel Chan, a thirty-year old mother of two preschool children, the looming extradition bill, even temporarily suspended, still filled her with trepidation. The "mainlandisation" trend in Hong Kong, including the indoctrination of political ideology and the shift to use Mandarin rather than Cantonese for Chinese classes in schools, cast a shadow over her children.

"You don't know when they will bring the law back and it will constantly hang over us," she said. "And what will happen in 2047, when our children are in their thirties? I don't want my children to be in a place where there is no freedom of thought."[103]

Willy Lam, an adjunct professor at the Chinese University of Hong Kong, said that even though the extradition bill has been suspended, "it won't remove the anxiety because Beijing is determined to promote political integration between Hong Kong and China." This includes infrastructure projects such as a bridge linking Hong Kong to China, a daily quota of one hundred and fifty mainland Chinese migrants to Hong Kong, the indoctrination of ideology at schools, and the silencing of critical media.

Lam said both waves of emigration, in 2014 and 2019, demonstrated "a serious degree of distrust of the Chinese Communist Party" and are linked with the fear that the Chinese system may encroach upon Hong Kong's core values like the rule of law, civil rights, and freedoms.[104]

Young people, active in the protests, expressed helplessness and said they too would leave if their efforts to uphold Hong Kong's current freedoms failed. "If there is nothing we can do to save Hong Kong, I would focus on working hard in the next few years and seek to move abroad," said a twenty-four-year-old athlete who identified himself as Tom.[105]

On June 28, Chery Kang on CNBC's "Squawk Box" interviewed Joshua Wong the Poster Boy of the pro-democracy movement. Wong wanted the G-20 leaders, who were scheduled to meet within the week in Osaka, Japan, to help protect Hong Kong's freedoms by raising their concerns with Chinese President Xi Jinping.

"It's a must for world leaders to share their concern," Wong said, urged US President Donald Trump and others to follow the lead of Japanese Prime Minister Shinzo Abe, who appeared to obliquely reference the situation in Hong Kong in a meeting with the Chinese leader the day before.

Abe told Xi of Japan's support for an "open and free" Hong Kong under the "one country, two systems" agreement. A senior Chinese foreign ministry official, however, had said on Monday the 24th that China would not allow Hong Kong to be brought up at the G-20.

Joshua Wong, the student leader who had risen to prominence in the 2014 protests of the Occupy Central movement, said that, "It's not only related to political freedom but also related to economic freedom that's being strongly eroded by Beijing."[106] He said that the territory's status as an international financial center must be maintained to ensure a stable business environment.[107]

These remarks provided fodder for the pro-Beijing camp and for socialists of the world who said that Joshua Wong and the Occupy Movement protesters were puppets of the international capitalists who are bent on destroying socialist China. Granted that Wong's remarks sound pro-capitalist at worst and extremely naïve at best, the Occupy protests are not so much about capitalism vs communism as they are about a local people wanting to live in a democracy in which they can choose they own leaders and have their human and civil rights realized. The protesters' fear of and aversion to Beijing's iron rule are legitimate even though they might not be fully aware at the moment of their exploitation by the capitalist multinationals. While they might not understand the dynamics that are exploiting them economically, it is very clear to them that they risk losing their political freedoms with each passing day that Beijing continues its squeeze on them.

On Sunday June 30, there was a show of support for the police as thousands gathered in heavy rain and sweltering heat, some waving the Chinese flag, to observe a moment of silence. The police estimated that fifty-three thousand attended the rally.

Former police chief Tang King-shing said he felt hurt when protesters besieged police headquarters during the week, while pro-Beijing spokesperson and member of China's National People's Congress, Maria Tam, praised what she described as this rally attendees' civilized ways of expressing their opinions. "We can't tolerate those who . . . attack the rule of law, attack the Legislative Council, attack the government," Tam said. "Some have been taking directions from foreign countries."[108]

Before dawn, on Monday morning, July 1, which was the twenty-second anniversary of Hong Kong's transfer of sovereignty from the UK to China, riot police and hundreds of protesters had gathered on roads leading to a square where the Hong Kong and Chinese flags were set to be raised. The ceremony, which was attended by Hong Kong leaders and dignitaries, including Carrie Lam, was moved indoors as protesters gathered. Officials said the event, which had never been held inside before, was moved because of "inclement weather."

As helicopters carrying the flags of China flew over, protesters waved middle fingers at them. Earlier in the morning, demonstrators had replaced a Chinese flag with a black flag featuring a withered Bauhinia flower, a riff on the Hong Kong flag. That flag was still flying on Monday night.

Across the city, more than five hundred thousand demonstrators were on the march. They forced major thoroughfares to shut down.

The scene of defiance was the latest indication that anger in Hong Kong was not to be easily quelled. Despite warnings that the police considered the marchers to be a "serious safety threat," demonstrators turned up in the hundreds of thousands once again.

The protests were well organized. Participants set up first-aid, water and food stations, and used hand signals to indicate police mobilization or use of pepper spray. Demonstrators urged one another and members of the public not to take photos of the crowd to ensure their anonymity.

A splinter group of pro-democracy protesters stormed the Legislative Council building again. They ripped down metal fencing around the building and broke through shutters, windows, and glass doors. Once inside, they wrote graffiti on walls, tore down portraits of pro-Beijing officials, and emptied rooms of chairs and desks, pushing weeks of tensions and mass demonstrations to a new level.

The young people were particularly angered by the Legislature because it was going to pass the China extradition bill. They also understood that Hong Kong had a very twisted and convoluted kind of election system. Pro-Beijing candidates are guaranteed at least half of the seats. As such, Beijing has made sure they will have enough votes to pass anything, any policy, any bill, any funding application before the Hong Kong government. The pro-democrats are always outnumbered, not because they are unpopular, but because they are not allowed the seats. They have more votes, but fewer seats. That sums up the Hong Kong legislative election process. And the young people knew it.

The protesters produced a declaration that included a call for overthrowing the "puppet Legislative Council and the Government," and they vowed not to leave the building. Outside the building, protesters cheered as more windows and doors were broken.

But just after midnight Tuesday, police equipped with riot shields, tear gas, and projectiles began removing protesters from the streets surrounding the complex, sending them fleeing. Officers then retook the complex, stopping and frisking those who remained nearby.

At a 4:00 a.m. news conference, Carrie Lam criticized protesters for vandalizing the Legislative Council building. "I hope the community at large will agree with us that with these violent acts that we have seen, it is right for us to condemn them and hope society will return to normal as soon as possible," she said.[109]

Protesters, however, vowed to press on. "Unless universal suffrage and a just election system are in place, we shall never stand down," they said in a statement.[110]

After weeks of spiraling tensions, Monday's protests took on a different flavor. In the face of an increasingly assertive Beijing, protesters viewed the occasion as their final chance for a massive stand against a government they believed was not working in their interests.

"We are exhausted," said a twenty-two-year-old protester who did not want to give his name for fear of retribution from authorities. "But today's march is special. We think it will be the last one that people will come out [to] on a large scale. We have to show our disappointment and anger."[111]

Five years ago, when Hong Kong residents flooded the streets to demand true democracy, the movement was named *Occupy Central with Love and Peace*. On Monday, however, when the protesters overran the Legislature Council building, they hung a banner with a harsher, more pessimistic message borrowed from the Hunger Games films: "If we burn, you burn with us."

While the protests raged across Hong Kong and the Legislative Council building was under attack, the city's leader Carrie Lam was giving a pre-fab speech inside the city's exhibition center, protected by massive barricades and a large security force, telling participants that "I will actively reach out to young people of different backgrounds through various channels to listen to their thoughts."[112]

Internal Dissent

In mainland China, there was no mention of Monday's protests on social media. Instead, state media played up news of Sunday's pro-police rally and highlighted official celebrations of the anniversary of Hong Kong's return.[113]

The attack on the Legislative Building caused blowback, not only from the public, but also from within the ranks of the protesters themselves. Internal divisions and public criticism emerged as the people became fearful of a brutal backlash from Beijing. Many saw the attack on the Legislative Building as a step too far that ultimately played into the hands of Beijing.

The blowback against the occupiers of the Legislative Building was met head-on by one of the main occupiers, in an interview with the *South China Morning Express* on July 5.

Brian Leung Kai-ping, twenty-five, who was the only one who had openly revealed his identity that night, explained his actions in an interview and told why he had no regrets.

> SCME: After an hour and a half, reporters observed you removing your mask and asking everyone to stay. Why did you do that?
>
> Brian Leung Kai-ping: At the time, more and more people, wary of police countermoves, started to leave the Legco chamber. I made a risky move to step on the desk of one lawmaker, removed my face mask, and shouted at the top of my voice: The more people here, the safer we are. Let's stay and occupy the chamber, we can't lose no more.
>
> Some protesters warned me not to remove my mask, but I felt it was the defining moment of the night. I felt we ought to appeal to the crowds to join in and form a

barrier and support those inside the Legco complex. No one could tell when we would step foot in Legco again.

As police were drawing closer and closer, after some deliberation, most decided to end the siege. I volunteered to be in front of the camera to read out the key demands of protesters in the chamber.

The last thing I wished to see, after all the action taken, was to have no clear demands put on the table.

If we didn't do that, the public might only remember the vandalism and point fingers at us as a mob. That would also hand the government a convenient reason to prosecute each and every one of us, which would mark yet another setback to civil society like in the 2014 Occupy movement.

SCME: So what was the young protesters' state of mind in being part of the July 1 protest and other sieges?

Brian Leung Kai-ping: The pursuit of freedom and democracy is what fundamentally drove hundreds of protesters on Monday into Legco, the same goal shared by hundreds of thousands who took to the street earlier. The government has thus far turned a blind eye to our demands, and there was no real change nor real actions tabled. If Carrie Lam claimed herself ready to be more humble, why did she not make clear the suspended bill was completely withdrawn, a move that could easily settle the controversy?

Or, the government could choose not to charge protesters arrested earlier, which we saw happened to those in Taiwan's Sunflower movement. Or, it could task an independent inquiry into police's excessive use of force.

Any of these would be welcomed by the civil society, but the government refused to take these calls on board.

At the outset of the protests against this bill, people called for Carrie Lam to step down. But gradually, we

> came to realize the root of the problem stems from the undemocratic system in selecting the chief executive. Whether Lam remained in office became a secondary issue.
>
> If there was universal suffrage, when 2 million people—almost one-third of the population—took to the streets, any leader would have no choice but to step down.
>
> Also, the government needs to realize that even by improving livelihoods or setting up more consultation platforms to absorb the opposition from our generation, it would do little to help.
>
> We do not have the luxury of our parents to settle down in another place. Nor do we have the burden of a 30-year mortgage to worry about. Young people have nothing to lose, their only hope is to stay safe to see the sun rise, and hope to join protest another day. We want democracy, now.[114]

On July 7, two hundred and thirty thousand protesters marched in Kowloon, an area popular with mainland tourists, in a bid to explain their concerns.

The Chinese State media coverage of the protests that have roiled Hong Kong for months, which ended in police spraying tear gas and firing rubber bullets, largely focused on condemning demonstrators for engaging in violent clashes with the police.

Beijing authorities said little except to scold the UK and other countries for meddling in Chinese affairs when they urged the Chinese government to uphold its end of the Joint Declaration.

The waves of protesters out on the street actually confounded the Chinese tourists, some of whom were visiting Hong Kong for the first time and had never seen a demonstration—ever. "I don't really understand the issue," said Miao Yiwen, twenty, a university student, who arrived in Hong Kong two days ago and was reading a local newspaper on a street corner half a block from the protests.[115] "When I arrived in Hong Kong, I learned that there would be an event here today, but I don't get what's going on. Why is everyone so easily excitable?" she said, as the sound of protesters chanting in unison rose into the air.

Another visitor, Summer, twenty, said it was "stupid" of Hong Kong people to organize such a demonstration. "If you do this—have a lot of people to come out to demonstrate—then for sure there will be some unforeseen impacts on the economy, on tourism," he said, declining to give a surname.[116] He thought the demonstrations would leave a bad impression on foreigners, and dissuade others from visiting.

Other tourists were upset that their travel plans had been upended because trains going between Hong Kong and mainland China were canceled Sunday and some tour groups had to reschedule their outings.[117]

On July 9, Carrie Lam reiterated that the extradition bill was "dead," urging protesters to stop their actions. However, she still refrained from fully withdrawing the bill. The *Hong Kong Free Press* saw through the ruse. It reported,

> On Tuesday morning, Chief Executive Carrie Lam appeared at a much-anticipated conference. Not only has she merely hinted at the potential withdrawal of the Extradition Bill ("the bill is dead")—thus falling short of agreeing to any of the Anti-Extradition protesters' many demands—her statement only reiterated the same platitudes the SAR government has been spouting for the past two decades.
>
> Like an automaton speaking from a skipping record, Lam pledged to sincerely listen to the public and to broaden the basis for communication.
>
> But we've heard this before, and we all know that the government's version of "listening to the people" usually runs like this: "The government is now consulting the public on political reform. Our proposal is to increase the share of voters from 0.00015 per cent to 0.00017 per cent of the population. Together, let's roll forward democracy! Send us your views!" They would be advised to add this disclaimer: "options other than those laid out in this leaflet will *not* be considered due to [insert dubious legalistic or procedural constraint or fallacious argument here]."

> There is only one answer to Hong Kong's political woes and for its government's legitimacy crisis: democracy. And to get there, Hong Kong's civil society cannot simply oscillate between relative compliance and sudden bursts of rage every half-decade or so. It needs to put constant and persistent pressure on the government, and avoid cooptation traps that are meant to bribe the opposition and strengthen the system's legitimacy."[118]

On July 21, protesters defaced China's Liaison Office in Hong Kong. That same night, mobs of men wearing white-shirts attacked commuters in Yuen Long underground station near mainland China in a new escalation of violence. The protesters suspected these were members of Triad gangs and pointed out that police arrived very late to stop them.

A day later, an article in the *South China Morning Post* reflected on the day's developments. Comparing the situation to a game of chess, the reporter concluded that the opportunity for checkmate was no longer possible because the window for reconciliation had closed. He asked how reconciliation could any longer be possible when the chief executive spent her time in hiding and when she did emerge yesterday it was only to denounce the defacing of the national emblem at Beijing's liaison office, while only, as an afterthought, did she condemn the Triad-linked mobs that attacked the Yuen Long subway commuters and sent many of them to the hospital.

As for Wang Zhimin, the director of the liaison office, he chose only to condemn the defacing of the national emblem, without even mentioning the savage Yuen Long attacks.

The reporter, who was also a TV show host, added, "by treating the defaced emblem as far more serious than the mob attacks that left over forty injured, Lam and Wang have shown how divorced they were from public opinion. Wang proudly proclaimed that the vandalized emblem was replaced overnight so a clean one would see the sunrise. The reporter concluded, "It is impossible to see the sunrise in the dismally overcrowded wards of our public hospitals."[119]

On Friday, thousands of demonstrators staged a sit-in at the Hong Kong International Airport with the aim of informing visitors of the ongoing political unrest. The sit-in was organized by aviation workers, including airport staff and flight attendants.

Beijing Weighs In

Protests continued on Saturday, despite the threats from mainland China to send in the People's Liberation Army to quell the protests.[120] It was a prospect dreaded by many in Hong Kong, and debate was growing in mainland China about whether the central government should end weeks of upheaval in Hong Kong by employing the People's Liberation Army (PLA).

The PLA has had a presence in Hong Kong since the city's return to Chinese sovereignty but, and unlike in mainland China, memories of the military's bloody suppression of pro-democracy students and activists in Beijing in 1989 are still strong in Hong Kong three decades later. Therefore, the threat to activate the army against protesters filled many of them with an overwhelming a sense of helplessness.

But images of protesters vandalizing Beijing's liaison office in downtown Hong Kong last Sunday fanned nationalist anger across the mainland, prompting calls for PLA intervention.

On Wednesday, the Chinese Defense Minister Wu Qian publicly stated: "We are closely following the developments in Hong Kong, especially the violent attack against the central government's liaison office by radicals on July 21. Some behavior of the radical protesters is challenging the authority of the central government and the bottom line of 'one country, two systems.' This is intolerable."[121]

On July 27, thousands again demonstrated in Yuen Long condemning the subway station attack that occurred on July 21. Police responded by firing tear gas at the unauthorized protesters. This led to more violent clashes between police and protesters.

Yang Guang of the Hong Kong and Macau Affairs Office under Beijing's State Council, which is Beijing's top policy office on Hong Kong, held a press conference on Monday, July 29, which suggested that China's top leaders had arrived at a view on Hong Kong and formulated a response to the protests. The response was tempered, but completely missed the point of the protesters.

Yang praised the Lam government and condemned the violence of the protesters. He praised the work of the Hong Kong police department. Surprisingly, he even praised the peaceful protesters which was somewhat unusual because protests are not allowed in China. This indicated to pundits that Yang's remarks were addressed to an international audience and that China was now looking at the protests in Hong Kong as strictly a domestic Hong Kong problem.

Xu Luying, of the Affairs Office, addressed economic problems in Hong Kong like youth opportunity, social mobility, and housing affordability which had merit and which could correctly be laid at the feet of the international corporations and local "tycoons," but it also showed that China remained unwilling to address the core issue of the protests, which could be determined from the slogans chanted by the marchers in every demonstration in the past year: " "Democracy now!" and "Reclaim Hong Kong! Revolution of our times!"

The protesters are addressing Hong Kong's undemocratic government that is not operating in the interests of the people, but rather on behalf of Beijing's communist party and their partners among the international capitalist class.[122]

Unaware of the economic plans for their city transpiring in the background, on August 2 and 3, even the civil servants in Hong Kong, supposed to be politically neutral, joined demonstrations by the thousands. Police again fired tear gas, rubber bullets, and bean bag rounds at protesters, but by this time the protesters were wearing masks and protective gear at every demonstration.

During the same weekend as this protest, James Dorn, a senior fellow and China specialist at the Cato Institute in Washington, DC published an article in the *South China Morning Post* proffering that "If protesters want to protect Hong Kong's way of life, they must win the war of ideas."

He suggested that if Hong Kong could export its system of limited government and individual freedom to the mainland, it would show that a free market in ideas is far superior to "socialism with Chinese characteristics."

He suggested that the protesters needed a long-term strategy, instead of just a tactical approach to reform that could backfire if the violence continued to grow worse.[123]

This was good advice on its surface, but it masked the fact that western style capitalism was also responsible for exploiting the people economically under the guise of the free market and limited government. Therefore, his conjecture that, "The moral force of the voices of freedom and limited government should not be underestimated" continues to ring hollow for those who are looking at the struggle in Hong Kong from the perspective of a value system that seeks the welfare of humanity as a whole. From this perspective neither the western capitalists nor the Chinese capitalists have a solution to human problems. Neither do the ideologies of capitalism or communism per se.

On August 5, another city-wide strike was held, bringing services across Hong Kong to a standstill. Carrie Lam gave her first media address in two weeks, saying Hong Kong was "on the verge of a very dangerous situation."

She warned the protesters not to "play with fire," not to "underestimate the firm resolve of the central government" and not to "mistake restraint for weakness." It was one of the strongest warnings Beijing had issued about the protests. Clearly, the velvet glove had come off and the iron fist was again on full display.

The protesters were undeterred. On August 11, police stormed enclosed railway stations, firing tear gas at protesters, leading yet again to head-on confrontation.

One week after the airport clashes, a new rally took place on August 18 amid reports of Chinese police and army gathering across the border in Shenzhen. Tensions were high.

Nonetheless 1.7 million people showed up for the Sunday pro-democracy rally which remained peaceful. For the first time in weeks, there were no major clashes between police and activists.

Even so, violence resumed the following weekend, with police using tear gas and water cannons against protesters, who returned fire with projectiles, including bricks and petrol bombs. For the first time in the unrest, police fired a live round of ammunition to disperse the protesters.[124]

Several days later, on Thursday, a number of prominent pro-democracy activists and lawmakers were arrested by police in Hong Kong. The activists included Joshua Wong and Agnes Chow who had been active in the pro-democracy movement since 2014. The lawmakers were Cheng Chung-tai, Au Nok-hin, and Jeremy Tam Man-ho, who were arrested for offences related to past protests.

Joshua Wong, who had become the poster boy for the movement, was arrested by suddenly being pushed into a private car on the street as he was walking toward a train station.

Amnesty International called the arrests "an outrageous assault" on free expression and the European Union foreign policy chief Federica Mogherini labelled the latest developments "extremely worrying." "We expect the authorities in Hong Kong to respect the freedom of assembly, expression, and association as well as the right of people to demonstrate peacefully," she said.[125]

The following Saturday, large protests again occurred with protesters lighting fires and attacking the parliament building with petrol bombs. In response, police used tear gas, rubber bullets, and water cannons to disperse crowds.

Police also entered subway cars and indiscriminately attacked people with batons and pepper spray.[126]

On September 4, Hong Kong's leader Carrie Lam finally announced she would withdraw the highly controversial extradition bill that had started the protests. She claimed she had wanted to do it sooner, but Beijing would not let her. She also unveiled other measures that were designed to soothe unrest, but her opponents said it was too little, too late.

The movement has been largely leaderless but everyone now had the same end goals. They are:

- The withdrawal of calling the June 12 protests a riot, a term that can bring a heavier jail sentence;
- An amnesty for all arrested protesters;
- An independent inquiry into alleged police brutality;
- Universal suffrage in elections for Hong Kong's chief executive (the territory's leader), and Legislative Council; and

- The resignation of Carrie Lam, whom they view as a puppet of Beijing.[127]

Role of Social Media

With protests being staged almost daily for an entire year, how was it possible that this large popular movement, which contains millions of people, could operate without strong leadership?

The answer lies in the role of social media. Millions of protesters communicate through social media. For example, on August 11 when the police stormed enclosed railway stations and fired tear gas at protesters that led to the head-on confrontation in which a woman protester was injured in her eye, videos of the incident, circulating on message boards and apps the entire day, showed police firing tear gas and projectiles at short range inside a subway station, as well as officers who impersonated demonstrators to infiltrate their ranks. Images also showed the young woman with blood streaming out of her right eye after police shot bean-bag rounds into a crowd.

Within minutes from when the picture of the young woman began to circulate on-line, more photos of protesters expressing sympathy also began to circulate. In this way, the young woman became an instant symbol of the people's struggle.

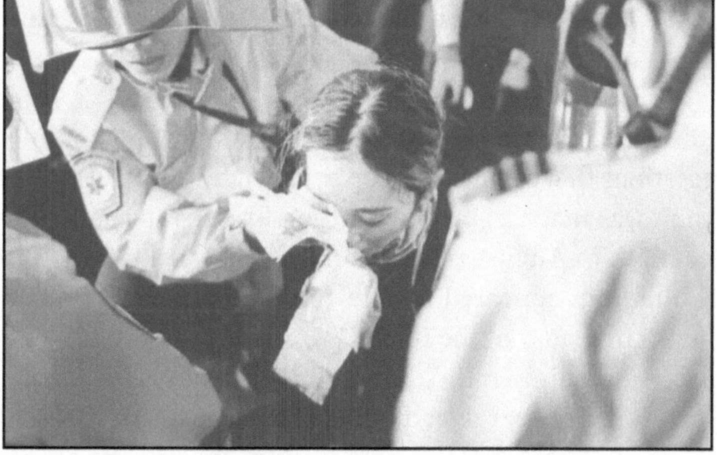

Fig. 1: Woman Sustains Severe Eye Injury During Hong Kong Protest

The tech-savvy protesters did not need to wait for a leader to tell them how to respond.

Fig. 2: Where's My Right Eye

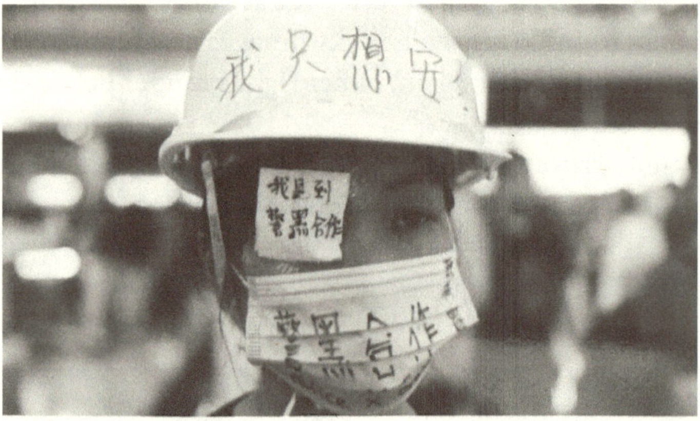

Fig. 3: One Woman's Eye Patch Became Symbol of Hong Kong Protests

All decisions are reached in a similar manner. They flood online forums with suggestions that can be voted up or down by their peers. This is how the decision was made to take over Hong Kong's international airport.

On the night of August 11, one of the first suggestions to swarm the terminal buildings appeared in a Telegram group at 8:50 p.m., shortly after photos circulated of the injured woman.

"If people marching in the different districts today all came here to the airport, it can be paralyzed," posted a user called "Shiba Taro" in a group used to organize an earlier sit-in at the arrivals hall. "They're going to fire shots and aim at the head anyway—let's take them down together with us."[128]

A debate about the pros and cons followed in various online forums. The airport had some practical advantages, such as air conditioning, Wi-Fi, nice toilets, and decent food options. Most importantly, proponents said, it would rattle the economy and the government would have trouble dispersing the crowds, upping the stakes dramatically.

Not everyone was convinced. Some users called the airport a "very last resort," and suggested targeting shopping malls or the subway system instead. Others like "Strawberry Milkshake" called for calm, while someone named "Moni K" noted an airport shutdown would hurt their image among foreigners who generally supported the movement.

After more than an hour of discussion, the Telegram channel administrator called a vote at 10:02 p.m. It was a landslide: seventy-nine percent of more than four thousand users supported escalating the struggle with an airport protest. The rest is now history.

These digital town hall meetings serve as the backbone of the pro-democracy movement in Hong Kong. Denied full democracy by the Communist Party in Beijing, the young people have been able to create their own form of democracy.

"When there's no single individual decision maker, it's hard for the government to predict what will happen and launch any kind of suppression," said Joshua Wong, a main leader of the 2014 Occupy protests in Hong Kong, who spent time in jail and now considers himself part of the rank and file. "We don't have a leader this time, but we have a platform."

The app Telegram plays a large role in this platform. It allows discussion groups in which anyone can have a say. In this way, the people can mobilize large crowds quickly. One of the largest groups, known as "Scorched Earth Team," in August, 2019 raised more than one million dollars in less than an hour to place ads about police brutality in global newspapers. The "supplies team" helps store helmets, gas masks, and other protest gear, while graphic designers in the "marketing team" create artwork to highlight key moments and advertise events.

The protesters use forums like LIHKG, which is a multi-category forum website based in Hong Kong that is similar to Reddit. LIHKG is well-known for being the ultimate platform for discussing the strategies for the leaderless protests.

Sometimes this method of communicating degenerates into a mob mentality as can be seen in the incident where a group of protesters beat up a mainland Chinese man they suspected was a police officer seeking to infiltrate the group.

The crowd then refused to let paramedics evacuate the man for the next few hours. This led to a standoff on live television in which it was unclear whether the man would survive or not. In a poll that asked if the group should let the man go, about seventy-five percent said he should either be detained or severely beaten.

Though the man survived, the incident shocked the conscience of Hong Kongers and westerners alike who supported the broader push for greater democracy, but also saw firsthand the pitfalls of a leaderless movement.

"It was purely a mob mentality," said Antony Dapiran, a Hong Kong-based lawyer who authored the book "City of Protest" about the financial hub. "People in a crowd lose their individual thinking, and without some sort of moral authority or leader figure there was no one to talk the crowd back and calm things down."[129]

The soul-searching began shortly afterward. By 1:00 a.m. Wednesday morning, a Telegram poll showed that seventy-four percent of nearly eleven thousand respondents voted to apologize for the violence and the disruption to travelers.

In the days that followed, the chastened protesters took a backseat to the more traditional Hong Kong pro-democracy forces. The Civil Human Rights Front, which organized some of the largest demonstrations over the past few months, led a peaceful mass rally through the rain on Sunday that helped reset the movement.

While the Civil Human Rights Front sticks to legal activities, it is pursuing the same ultimate goal as protesters who engage in civil disobedience like the siege of the Legislative Council on June 12. Bonnie Leung, the group's vice convener, credits that move for prompting the government to shelve the bill that would have allowed extraditions to China. This was the issue that sparked the unrest in early June.

"Being peaceful is the backbone of the movement," Leung said. "But there are still limitations: Peaceful protesters can't stop the hearing from happening. The clashes can." That understanding between moderates and more radical groups stands in contrast to 2014, when infighting undermined their ability to occupy strategic areas of the city. This time all of them have adopted the motto "no pointing fingers, no selling out, no severing ties" to sustain the movement.

The technology underpinning the movement is a big reason the protests have no end in sight: Authorities cannot simply lock up the masterminds and send everyone home. By the same token, no single leader has enough power to call off the protests, cut a deal with Chief Executive Carrie Lam, or even tell fellow demonstrators what to do next.

Wong himself is a case in point. Lionized in a 2017 documentary on Netflix titled "Joshua: Teenager vs. Superpower," he is now just one of thousands of voices discussing ideas on LIHKG and Telegram.

In the mainland, China can simply use its Great Firewall to shut down the online discussion. But in Hong Kong, doing so may damage the city's autonomy as much as putting Chinese troops on every street corner.

Despite Beijing's efforts to curtail free speech in Hong Kong, the ability to freely surf the web and exchange ideas remains a key attraction for foreign investors. As long as protesters can use online tools, including those banned in China like Google Drive, they will remain a threat to both Hong Kong's government and Chinese President Xi Jinping.[130]

The people of Hong Kong were back on the streets during 2020 as the Chinese government, instead of honoring their rights, imposed a new national security law that would unravel more of their civil liberties.[131]

Police used water cannons, tear gas, and pepper spray to disperse protesters who retaliated by throwing plastic bottles and umbrellas at the police.

The Economic Situation

At this writing, the extradition order is no longer the ultimate ground for all the disturbance. While there is certainly a fear of being swallowed up by a shadowy criminal justice system that is known for its authoritarian and punitive methods and while being sent miles from home without support is grounds for alarm, most of the demonstrators would not even be suspected by the mainland government if it were not for their public protests. As such, there must be something more at stake than fear of political oppression.

Today Hong Kong is one of the largest population centers in the world. It has a population of more than 7.4 million people who live in an area of just four hundred and twenty-six square miles, making Hong Kong one of the most densely populated places in the world. The largest city in Hong Kong is Hong Kong, with a population of 7,012,738 people.[132]

Hong Kong is also in the top five financial centers in the entire world. Located in China's Pearl River Delta, it provides access for multinational banks and corporations to the economy of the entire Far East.

As such, Hong Kong is one of the world's leading tax havens and also a principal offshore jurisdiction. It offers incorporation of offshore companies, the registration of vessels and boats, offshore insurance, and investment management among other services.

Like other tax havens of the world, it provides zero or low taxation schemes to primarily foreign corporations, but not to residents of the respective jurisdictions. In addition, it provides high levels of privacy and confidentiality for multinational corporations so no one can actually determine how much wealth they are hiding from the world's people.

The taxes paid by a multinational corporations to Hong Kong are only derived from profits gained from conducting business or trade within the jurisdiction itself. Income gained abroad is not taxed. Neither are corporate interest, capital gains, or dividends subjected to any form of taxation. Local taxes for such companies are fixed at 17.5% which is less than the income tax rate most Americans pay in the United States. Needless to say, the fact that they do not have to pay taxes has led to an escalation in the number of international companies setting up offices in Hong Kong.

Hong Kong is also very popular with offshore bankers because they can easily set up accounts and interest is not taxed. Again, the tax haven of Hong Kong is strict about providing privacy for offshore bank account holders.

Companies in Hong Kong are not obligated to prepare financial statements for any authority in Hong Kong. They can register using nominee corporate directors and shareholders. This hides the identity of the capitalists who actually own the companies registered in Hong Kong. Their names are not filed with the Registrar of Companies. This means that their information does not become public record nor can they be held accountable for any malfeasance.

Tax haven Hong Kong also has no exchange controls in place. This allows corporations based in Hong Kong to engage in trade without restrictions. The movement of money is made easier where there is no exchange controls in place.

Regarding wealthy individuals, Hong Kong provides privacy and confidentiality for bank account holders and, at present, Hong Kong has no estate tax.[133]

What this means, essentially, is that multinational corporations can come in and pay virtually no money to the people of Hong Kong. Moreover, the use of tax havens like Hong Kong results in a loss of tax revenues to all the countries that are not tax havens.

Estimates of the scale of taxes avoided vary, but the most credible estimates have a range of one hundred to two hundred and fifty billion US dollars per year. In addition, capital held in tax havens can permanently leave the tax base (base erosion). Estimates of capital held in tax havens also vary, but the most credible estimates are between seven to ten trillion US dollars (up to ten percent of global assets).

The negative impact of corporate tax havens on the world's people can be particularly witnessed by looking at the lives of people in developing nations, where tax revenues are not available to build the infrastructure to meet the people's basic needs for food, shelter, health care, and education.[134]

We find that the capitalist economy of Hong Kong is as much to blame for the people's aggravation as is the socialist political oppression by mainland China. "One country, two systems" has in fact given the people of Hong Kong, particularly the young, the worst of both systems—economic exploitation by the capitalists and political oppression by the socialist government using the military to enforce its will on an unwilling population.

Let us take a look at the class structure of Hong Kong as a means to start our investigation into the economic exploitation of the people of Hong Kong.

The word *class* is understandably a sensitive word in China. Marxist-Leninist rhetoric which includes concepts like "class enemies," "class conflict," and "class struggle" are rarely seen in the country's media these days, but since China began its push toward integrating itself into the global capitalist system in 1979, social stratification has emerged in a society that had hitherto tried to eradicate the very concept.

Now, after thirty-five years of breakneck-speed development, social class has become increasingly entrenched while opportunities for upward mobility have become increasingly limited.

While the western capitalist countries can be held responsible for initiating this dilemma, the communist party, many of whose members have since become capitalists themselves, have ensured that the exploitation of the people of Hong Kong continues.

Karl Marx had predicted that the working class would eventually find their economic positions intolerable, but now the new working class in Hong Kong is literate, has smart phones, and are finding ways to resist. And China is no longer anti-capitalist, but in fact has become the new face of global financial capitalism. The capitalists in China have become the new ruling class, even if many of them go to business meetings wearing Communist Party pins on their lapels or wearing their military uniforms.

Increasingly, the people of Hong Kong have been unable to escape the knowledge of their exploitation by the rich, either in the guise of western corporations or the Chinese "tycoons" from the mainland. One quarter of the population of Hong Kong lives below the poverty line while another quarter is made up of millionaires.[135]

As the working people are forced to live in costly "shoebox" apartments in order to survive in the urban jungle created by the international capitalists, the capitalists themselves have been enjoying Hong Kong as their gateway by which to exploit the labor and resources of the entire Far East.

Currently, in Hong Kong, the richest households earn forty-four times what the poorest families earn. As such, Hong Kong is ranked second in the world in terms of inequality of income. New York City ranks first.[136]

Hong Kong has a population of seven million people. In a comprehensive study of class structure in Hong Kong, made in 2003, about 13,000 households enjoyed a minimum per capita income of 101,000 US dollars. Another 53,000 households enjoyed a minimum per capita income between $35,000 and $100,000. There was another 135,000 households that have a per capital income of $24,000. This means that less than ten percent of the population actually has a per capital income of twenty-four thousand dollars or more. Ninety percent of the people of Hong Kong earned less than $24,000 a year. Of these, five hundred thousand households earn between $12,000 and $23,999. This gave family members in this group an average per capital income of about $7,575. Finally, the rest of the people, approximately 4.7 million people have a per capita income of less than $7,575 per year.[137]

While the western and Chinese capitalists like to use the measure of per capita gross domestic product, which measures an average income across all classes, to claim that Hong Kong is one of the richest cities in the world, their intentionally-deceptive calculation hides the gross income inequality just indicated above. Their measure of the city's wealth is inflated by hundreds of billions of dollars that bring little or no benefits to the population. As James Tien observed during a discussion, "'the majority of the population does not benefit from such economic activities that are concentrated in the banking and financial sector."[138]

To make matters worse, the number of households living below the poverty line in Hong Kong has been rising since 1996. The numbers demonstrate that there are more people living in poverty in Hong Kong than there are in Indonesia, even though Indonesia is considered to be a poor third-world country and Hong Kong a wealthy economic tiger.[139] So much for the benefits of capitalism to the people living in the glittering capitalist paradise of Hong Kong.

Understanding this underlying economic reality of Hong Kong, we may better understand the virulent anger that characterizes the young people's revolution there. Let us look at what these economic numbers actually mean in regard to the people's lifestyle. A look at the housing crisis in Hong Kong can help us paint the picture.

In Hong Kong, housing is extremely expensive. Real estate prices are between one hundred and fifty to two hundred thousand Hong Kong dollars per square foot for apartments. That comes out to between twenty thousand to twenty-five thousand US dollars for an apartment per year. While the majority of the middle class are crammed into towering apartment buildings, the majority of the population, particularly the young, cannot afford these prices. Around two hundred thousand people, who are in the lower-income, bracket live in miniscule, illegally subdivided housing units called "coffins." A coffin contains only fifteen square feet. Other low income residents live in "squatter huts" and "cage homes" with subdivided units and wire mesh-enclosed rooms. There is another group of people who cannot even afford these subhuman spaces and are known as "McRefugees" because they spend their nights sleeping in twenty-four hour McDonald fast food joints.[140]

Fig. 4: Residents Sleeping at McDonald

The rich, by comparison, live in comfortable and opulent homes that sit atop the city's many lush mountains. Hong Kong ranks as the number two city in the world with the most billionaires. Most derive their wealth either directly or indirectly from the real estate and financial services sectors that gouge the people.

The business-friendly policies of the Chinese President Xi Jinping has only led to their wealth increasing. As of the beginning of 2018, Hong

Kong's richest fifty people saw their wealth increase by sixty billion dollars to now total three hundred and seven billion dollars. So much for Chinese socialism.Only a systemic change implemented by the Hong Kong government can resolve this issue directly. For its part, however, the government seems more concerned with increasing the prosperity of the country's upper echelon than taking care of the disenfranchised.[141]

Given the structure of the Hong Kong capitalist economy, most young people in Hong Kong are locked into their lower-class position. This means they have no access to meaningful careers, affordable housing, a better quality of life, or increased discretionary spending. At the same time, their high level of education encourages expectations that the economy is unable to fulfill. As hope fades, anger builds.[142]

Older people also responded to the protests in an article published in the *South China Morning Post* entitled, "Hong Kong people, young or old, have it rough: no wonder they have taken to the streets." Ying Sau Lo told how, after retiring, she discovered that her life-long payments into her mandatory pension fund disappeared and how she tried to make an appointment with a hospital regarding acute knee pain and was told the earliest appointment she could get was over four years away.

Philip Leung reminded everyone that the issue was not about politics or the extradition bill; it was about people's livelihoods. He said that couples, after twenty years of hard work, could still not afford a three hundred square foot apartment despite neither going to restaurants nor taking vacations. He pointed out that college graduates in 2017 made less in entry-level jobs than their predecessors twenty-five years earlier.

He said that "When young people don't see hope, they will do anything to vent their anger, they will be destructive, and any excuse will do." He reminded the paper's readers that," Not one civilization has advanced culturally, politically, or economically without bloodshed—without students and idealists sacrificing their own lives and sometimes, unfortunately, the lives of others. This is the reality."[143]

Ringo Yee offered that there is a popular perception in Hong Kong that the government has a "cozy relationship" with the "tycoons" and they support each other. He said that "The storming of the Legislative Council reflects what the young "vandals" are thinking. They are against the rich, the '1 per cent' who control most of the city's wealth and a government that staunchly protects the rich."[144]

As the protests continued, the western capitalists and the Beijing capitalists were both becoming alarmed about the protests for democracy in Hong Kong because the security of their money was being called into question. The negative impact of the freewheeling bastion of finance capital on the local working people, especially the younger generation was beginning to be exposed. A dystopian vision of late-stage capitalism, characterized by severe economic inequalities, social unrest, and environ mental degradation, was now on full display in the streets of Hong Kong.

The capitalists were beginning to feel the heat. The youth of Hong Kong have nothing to lose. Capitalist or Communist, what does it matter to them when they are exploited by both.

On July 31, the *South China Morning Post* reported that because of the protests against the whittling away of the people' democratic freedoms, Hong Kong's economic expansion had ground to a halt, showing a paltry 0.6 increase in annual gross domestic product, well below market expectations. It was the lowest growth since 2009 when the city was trying to recover from the crash of the global economy due to the housing mortgage crisis in the US.

A government spokesman said in a statement that economic sentiment had "deteriorated visibly" in the face of worries about the global economy, with private spending growing only modestly and investment falling further.

Now Hong Kong's economic future was again plagued by uncertainty. Local business groups, including the Triad mafia, complained that the protests had also hurt the local retail and tourist trade.

Edward Yau Tang-wah, the city's commerce minister stated, "The updated situation . . . showed that the local economy had rapidly and negatively declined. This is more or less due to the social sentiment, and the overall poor environment."[145]

Options to Hong Kong

In other developments, the Beijing capitalists, in a trade war with the US capitalists, began to look to Shenzhen instead of Hong Kong to restart their economic future. Shenzhen is a city on the mainland just north of Hong Kong that is a leading global technology hub similar to

Silicon Valley in the United States. Shenzhen also hosts the Shenzhen Stock Exchange as well as the headquarters of numerous China-based multinational companies such as JXD, Vanke, Hytera, CIMC, SF Express, Shenzhen Airlines, Nepstar, Hasee, Ping An Bank, Ping An Insurance, China Merchants Bank, Tencent, ZTE, Huawei, DJI, and BYD. It has one of the busiest container ports in the world and is ranked fourteenth in the 2019 Global Financial Centers Index.

To spearhead this decision, President Xi Jinping endorsed a document to give Shenzhen special status to carry out bold reforms as a model for other cities. Guidelines were endorsed by the Central Committee for Deepening Overall Reform, the top body overseeing economic and administrative reforms that were designed to promote Shenzhen as a "pilot demonstration area of socialism with Chinese characteristics."

Zhang Yansheng, chief research fellow with the China Center for International Economic Exchanges, said Beijing wanted Shenzhen to experiment with blending innovation and socialism. "In the past forty years, Shenzhen has been very successful," he said. "In the next forty years, can Shenzhen become a trailblazer in terms of regulations and modernization so that it will become a model for the whole country?"[146]

Li Xiaobing, an associate law professor at Nankai University in Tianjin, said China needed new ways to boost its economy, particularly given its rivalry with the United States. Li added that Shenzhen's role would go beyond industry to also test changes to the legal process.

"The guidelines also mentions legal institutions. The advantage of Hong Kong is its rule of law and we don't want Shenzhen just to be a successful commercial area . . . We would want its legal institutions to become compatible with its size and influence," Li said.[147]

With China now locked in a trade war with the US, Beijing is counting on the hi-tech sector to spur development and cut reliance on imports of key technologies. Amid threats by the US to stop selling microchips to Chinese companies such as ZTE and impose sanctions on Chinese telecom giant Huawei, Xi has repeatedly called for industry to innovate and become more self-reliant.

In February, Beijing unveiled the ambitious Greater Area Bay blueprint aimed at transforming Hong Kong and ten southern cities on the mainland into a combined economic powerhouse, with Hong Kong, Macau, Shenzhen, and Guangzhou identified as the four "pillars" of development.

The blueprint highlighted the advantages of each of the four pillars and reassured Hong Kong that it would continue its role as the leading financial center.

Li said Hong Kong still had an edge, but the central government would not wait if the city's turmoil continued.

Tian Feilong, an associate law professor at Beihang University in Beijing, said Beijing had been struggling unsuccessfully to implement its policies in Hong Kong and so had a strong incentive to shift the focus of the Greater Bay Area to the mainland.

"In the future, the reform and opening up of the Greater Bay Area will focus on the mainland instead of Hong Kong," he said. "The traditional advantages of Hong Kong are fading and it is getting more difficult for central government's policies to be implemented there."[148]

In 2018, Shenzhen's economy surpassed Hong Kong's for the first time. While economic growth in Hong Kong rose by just 3 per cent to HK$2.85 trillion (US$363 billion) Shenzhen's gross domestic product grew by 7.6 per cent to 2.42 trillion yuan, or HK$2.87 trillion based on the 2018 official exchange rate.[149]

Regarding the response of the Western capitalists, it was in mid-June, 2019, that the US government began to protest against China's behavior in regard to Hong Kong.

The English newspaper, *The Telegraph*, reported that: "The more that Beijing chips away at Hong Kong's civil liberties and autonomy, the more certain it is that Washington will strip the enclave of its special status."

The article went on to explain that once this occurs, the US will cease to recognize the territory as an independent member of the World Trade Organization and will cut off access to sensitive technologies. Hong Kong will be subject to the same tariffs faced by exporters from mainland China.

"Many of the fourteen hundred US firms operating there will decamp, setting off an exodus towards Singapore. The Hong Kong model as we know it will no longer be viable."

The Telegraph article referenced the "Hong Kong Human Rights and Democracy Act" introduced in the US House and Senate the week of June 9, which was backed by both parties. It promised sanctions covering trade, finance, and technology and proposed sanctions against those "complicit in suppressing basic freedoms in Hong Kong"—a catch-all phrase that targets the top political leadership of Hong Kong.

"If all this rolls out, Hong Kong is going to be just another Chinese City," said one of the drafters.[150]

The rating agencies had begun to issue warnings. Fitch said its AA+ grade "rests on the assumption that the territory's governance standards, rule of law, policy framework, and business and regulatory environments remain distinct from that of mainland China".

Of course, the US is not concerned about the extradition bill because it would violate the democratic rights of the people of Hong Kong. Rather, the business elites are alarmed by other clauses in the bill. It seems the bill calls for "freezing and confiscating the assets of persons wanted for crimes in other jurisdictions." In this way, the bill would bring the vast store of foreign capital within reach of the Communist Party's control for the first time.

Not only are the private stashes of the fourteen hundred US-based multinational corporations at risk, but Credit Suisse says there are over eight hundred and fifty individuals in Hong Kong who each hold more than one hundred million dollars. Reuters reported that one tycoon with political "exposure" in China was already shifting sums of this value to accounts in Singapore.

Paul Chan Mo-po, Hong Kong's financial secretary, issued a warning in May about capital flight, but blamed the stress on large outflows from China through the Shanghai Connect link, mostly by foreign investors alarmed over the deepening trade conflict. But that is unlikely to be the whole story.

President Trump's trade war with China had consequences not fully grasped by the American people and the world's people.

The Hong Kong Monetary Authority (HKMA) has had to intervene repeatedly over recent months to defend the long-standing dollar peg. It forced interbank Libor rates to go up to levels not seen since 2008, triggering a "short squeeze" to burn the fingers of speculators and shore up its currency.

Such action drains liquidity and is untenable over time. Monetary tightening by the Hong Kong Monetary Authority, forced to comply with the US Federal Reserve, has slowed economic growth to a ten-year low. Retail sales suffered over the months from February to December, 2019.

A long squeeze risks popping Asia's most overheated property boom. The Bank for International Settlements says Hong Kong's "credit gap," an early warning indicator for banking crises, was running at over thirty percentage points of GDP above its long-term trend as recently as early 2018. This was the most extreme reading anywhere in the world. It has left an overhang of leverage.

Hong Kong is the financial gateway in and out of China, a conduit for Chinese companies that hold eight hundred and forty billion of US dollar debt. Its banking system is 8.3 times the entire GDP of Hong Kong. A sudden loss of confidence in Hong Kong would have global systemic consequences.

The HKMA has four hundred and twenty-nine billion dollars in reserves to defend its currency peg. This is double the monetary base and more than enough in normal circumstances. However, investors and rich locals might view matters differently if the exchange regime loses the blessing of Washington. A political crisis of confidence might overwhelm the institutional architecture.

Ultimately it is the prerogative of the US president to decide whether to suspend Hong Kong's special status. In the meantime, Hong Kong's democracy protests offer the US another irresistible weapon against Beijing. One bad tweet from the Oval Office, however, and East Asia's super-rich may take matters into their own hands.[151]

That being said, the average person in Hong Kong seem oblivious to this economic reality playing out in his or her background.

Up to this point, the people of the United States and England had not been informed by the media in their countries about the economic exploitation of the people of Hong Kong by the international capitalist elites.

During the month of September, 2019, most news about Hong Kong was no longer about the protests or about Beijing's anti-democracy stance. Rather, it was about the survival of capitalism in Hong Kong and around the world.

The discussion began in the socialist publications that critiqued the western press for trying to dismantle communist China by painting the Hong Kong protesters as heroes. They pointed to how the press glorified the protestor's march to the US Consulate on Sunday, September 8, where they waved US flags and sang the Star-Spangled Banner.

Those who participated in the march to the US Consulate called for the US Congress to pass the bipartisan Hong Kong Human Rights and Democracy Act. By increasing pressure on China in this manner, they believed it would bring genuine democracy to Hong Kong. The signs calling on President Trump to "Liberate Hong Kong" were merely a way to catch the attention of US media. Few, if any, would believe that Trump would have had the power and the will to "liberate" Hong Kong.

But recent US-based reports on the Hong Kong protests had sowed doubt among American leftists as to the legitimacy of the Hong Kong protests. Writing in *The Grayzone*, Dan Cohen claimed Hong Kong's protesters were backed by the US and marked by "nativism and mob violence." He criticized the fact that pro-democracy advocates had met with John Bolton, Nancy Pelosi, and Marco Rubio[152] In another article written by John V. Walsh and published in *Global Research*, the author argued that Lam's *suspension* of the extradition bill meant the same as withdrawal. He also criticized the fact that protest leaders had met with an official from the US Consulate claiming that this "contradicted the contention that there is no US "black hand," as China calls it, in the Hong Kong riots. In fact, here the "black hand" was caught red-handed."[153]

Reuters also published an article on September 9, revealing how ex-Pentagon chief Mattis said that the US should support the protesters, if not militarily, at least morally.[154]

While accusations that pro-democracy leaders had met with US officials is true, this does not mean that the US is responsible for orchestrating the months-long protests that have brought two million people to the streets every day. It would be as if ninety-three million Americans were to march against the government for months on end. To get this great number of people on the streets would entail people from a wide political spectrum.

While the Hong Kong people's movement might not be far left, neither is it far-right. It is wrong to try to force the protests in Hong Kong into an American lens of "progressive versus conservative," or even capitalist vs communism.

Let us not forget that whereas the western capitalist press may try to manipulate people to support US and British capitalism, so also the Chinese press is trying to manipulate people to support Chinese capitalistic communism.

China had also gone to great lengths to control information about the protests, especially on social media platforms. Twitter discovered nearly one thousand fake accounts that originated in China that, the tech company said in a statement "were deliberately and specifically attempting to sow political discord in Hong Kong, including undermining the legitimacy and political positions of the protest movement on the ground."[155]

Facebook found a smaller number of accounts it says were linked to the Chinese government focusing on the Hong Kong protests. And Google said it "disabled two hundred and ten channels on YouTube" that were behaving in "a coordinated manner while uploading negative videos related to the ongoing protests in Hong Kong."[156]

So also the western socialist press tended to avoid any mention of police brutality in Hong Kong, which had been directed not only against protesters but also against reporters, pro-democracy politicians, and the public at large.

Certainly, the people of Hong Kong turned to the US for help because the US is the enemy of their enemy even though they may not have fully realized at the time that their suffering is also due to the US and European-based capitalist corporations. These corporations not only exploit the people of Hong Kong by not paying their fair share of taxes, but they also use the people of Hong Kong to hide the money that they exploit from the people of other countries. By seeking help from Trump, who is one of the most divisive human beings on the planet, the people of Hong Kong were making a strategic mistake, but under the circumstances the protesters found it tactically useful.

In July 2020, China passed a new national security law in Hong Kong that has led to the arrest of student organizers and protesters, the banning of pro-democracy politicians from office, the cancelation of local elections, and the arrest of journalists and publishers. To impose its iron rule on the people of Hong Kong, China opened a National Security

Office in Hong Kong to ensure that the local police would enforce the national security law.

To counter this initiative on the part of China, Google, Facebook, WhatsApp, Twitter, Telegram, and LinkedIn announced that they would temporarily stop processing law enforcement requests for user data.

More dramatically, Britain recently revealed details of a possible path to citizenship for 2.9 million Hong Kong residents who hold a British National (Overseas) passport status.

Separately, US President Donald Trump signed an executive order ending Hong Kong's preferential trade status, and enacted the Hong Kong Autonomy Act, which requires sanctions against foreign individuals and banks seen as contributing to the erosion of the city's autonomy.

In retaliation, Beijing's foreign ministry suspended Hong Kong's extradition agreements with Canada, Australia, and Britain. It has also threatened not to recognize the BN(O) passport.

Regardless of the battle between the overlords for control of the people of Hong Kong, the people's revolution is a long way from over. In the course of their revolution for true democracy, they will eventually have to battle both the US and China for their political and economic freedom.

Lessons Learned from Hong Kong

The chief lesson concerning Hong Kong is that the upheaval, over the past year, has been the result of the people's frustration, especially the young, who lack any political representation and who live in dire poverty so pervasive that it strips them of their ability to meet their basic needs.

Yet, while the people risk their lives in battles with the police and the Chinese army, the Western capitalists in their gleaming skyscrapers are content to let them die for the sake of profits. It is interesting to note that in September of 2019, the same year of mass protests, the Hong Kong Stock Exchange offered to buy the London Stock Exchange for thirty-seven billion dollars.[157]

The British capitalists rejected the deal, citing technical difficulties, but one wonders what the real motivations behind the offer and its refusal were.

In any case, it proves that the Chinese communist party has adopted the ideology of neoliberalism which contends that transferring ownership of the collective wealth to the 1% is beneficial to society. China has joined in the quest to create global monopolies that prey on the world's people and, in so doing, equate any popular resistance with terrorism.

Yet beneath the political restrictions, the people of Hong Kong are still being exploited by the Chinese and western capitalists. We have also learned that both the western capitalists and the so-called communist countries are willing to abandon any local people's democratic struggles if they sense they can no longer control the local people and profit off them.

As far as concrete lessons for revolutionaries are concerned, we have learned from the revolutionaries in Hong Kong how to protect ourselves from police surveillance and arrest, and stay organized by using social media. We have learned how a revolutionary movement can appear to be leaderless by using social media as a tool to make tactical decisions.

We have learned that political freedom is important to those who have experienced a taste of it, but it is not understood by people who never had it.

We have learned how the revolutionaries in Hong Kong were able to maintain a good working relationship between the mainstream peace movement and the more radical revolutionaries. Activists from both camps appreciated the specific roles each played in revolution. In other words, the intellectual and the warrior revolutionaries joined forces for the good of the people as a whole.

Finally, we have learned the high cost that local revolutionaries must pay when they challenge an overlord empire. It behooves the revolutionary people of the world to keep Hong Kong on their radar and provide assistance to the revolutionaries in Hong Kong by any means necessary.

Chapter Six: Contemporary Revolutions in Africa and Middle East

In Africa

Sudan

Political Situation

THE HISTORY OF THE revolution in Sudan is much more complex than the battles against capitalist and communist overlords that have characterized the revolutions we have investigated so far. The revolution in South Sudan is also mixed with religious feuds, tribal warfare, and the earlier interference of French and English colonialists.

The history of Sudan also includes portions of a larger region known as the Sahel. The Sahel is the ecoclimatic and biogeographic zone of transition in Africa between the Sahara desert to the north and the Sudanian Savanna to the south.[158]

The Sahel today includes parts of northern Senegal, southern Mauritania, central Mali, northern Burkina Faso, the extreme south of Algeria, Niger, the extreme north of Nigeria, parts of Cameroon and Central African Republic, central Chad, central and southern Sudan, the extreme north of South Sudan, Eritrea, and the extreme north of Ethiopia.

Historically, the western part of the Sahel was sometimes known as the Sudan region.

In 1896, a Belgian expedition claimed portions of southern Sudan that became known as the Lado Enclave. The Lado Enclave was officially part of the Belgian Congo. An 1896 agreement between the United Kingdom and Belgium saw the enclave turned over to the British after the death of King Leopold II in December 1909.

At the same time, the French also claimed several areas: Bahr el Ghazal and the Western Upper Nile up to Fashoda. By 1896, they had a firm administrative hold on these areas and they planned on annexing them to French West Africa. An international conflict known as the Fashoda Incident developed between France and the United Kingdom over these territories. In 1899, France agreed to cede the area to the Anglo-Egyptian Sudan.

To cement their hold over Africa, the colonialists divided up territory into administrative districts they called nation-states regardless of the local people's previous relations to the land or to each other.

From 1898, the United Kingdom and Egypt administered all of present-day Sudan as the Anglo-Egyptian Sudan, but northern and southern Sudan were administered as separate provinces of the joint administration. In the very early 1920s, the British passed the Closed Districts Ordinances that stipulated that passports were required for travel between the two zones, and permits were required to conduct business from one zone into the other, as totally separate administrations prevailed.

In the south, English, Dinka, Bari, Nuer, Latuko, Shilluk, Azande, and Pari (Lafon) were official languages, while in the north, Arabic and English were used as official languages. Islam was discouraged by the British in the south where Christian missionaries were permitted to work. The UK/Egyptian governors of south Sudan attended colonial conferences in East Africa, not in Khartoum, the capitol of the North, and the British hoped to add south Sudan to their East African colonies.

Mostly, the British focus was on exploiting the economy of the north. Southern political arrangements were left largely as they had been prior to the arrival of the British. Until the 1920s, the British had limited authority in the south.

In order to establish their authority in the north, the British promoted the power of Sayyid Ali al-Mirghani, head of the Islamic Khatmiyya sect, and Sayyid Abd al-Rahman al-Mahdi, head of the Islamic Ansar sect.

The Khatmiyya is a Sufi order founded by Sayyid Mohammed Uthman al-Mirghani al-Khatim, the grandson of the Imam of the Al-Haram mosque in Mecca and the Mufti. It is the largest Sufi order in Sudan, Eritrea, and Ethiopia. It also has followers in Egypt, Chad, Saudi Arabia, Somalia, Uganda, Yemen, and India.

Al-Sayyid al-Khatim's lineage went back to the Islamic prophet Muhammad through his grandson Al-Sayyid al-Hussain.[159]

The Ansars are a religious movement in the Sudan whose followers are disciples of Muhammad Ahmad (August 12, 1844 – June 22, 1885), the self-proclaimed Mahdi. After taking power in Sudan between 1883 and 1885, he established the Mahdist regime, which was ruled by a modified version of the sharia.[160]

Under British rule, the Khatmiyya became the Democratic Unionist Party, while the Ansar sect essentially became the Umma party.

In 1943, the British began preparing the north for self-government, establishing a North Sudan Advisory Council to advise on the governance of the six North Sudanese provinces: Khartoum, Kordofan, Darfur, and Eastern, Northern, and Blue Nile provinces.

Then, in 1946, the British administration reversed its policy and decided to integrate north and south Sudan under one government dominated by the north. The South Sudanese authorities were informed at the Juba Conference of 1947 that they would, in the future, be governed by a common administrative authority with the north. From 1948, thirteen delegates, appointed by the British authorities, represented the south on the Sudan Legislative Assembly.

Many people in South Sudan felt betrayed by the British and the Egyptians, because they were largely excluded from the new government. They interpreted the British deal solely as a method to make South Sudan a colony of North Sudan. The language of the new government was Arabic, but the bureaucrats and politicians from southern Sudan had, for the most part, been trained in English. Of the eight hundred new governmental positions vacated by the British in 1953, only four were given to people from South Sudan.

Because the political structure in the south was not as organized as that in the north, political groups and parties from the south were not represented at the various conferences and talks that established the modern state of Sudan. As a result, many southerners do not consider Sudan to be a legitimate state. As Dr. Riek Machar Chairman and Commander-in-Chief of the Southern Sudan Independence Movement/Army, (SSIM/A) states:

> . . . the people of South Sudan regard themselves as 'internally colonized people'. The deliberate hand over of South Sudan to North Sudan by the British was one of the greatest blunders ever made in the diplomacy of the British colonial history. If the British had felt that South Sudan was not yet ready to become an independent state by itself then, they should have either handed over its administration to an international body like the UN instead of North Sudan or should have left North Sudan get independence separately as they did with North Rhodesia.[161]

In time the SSIM/A became known as the Sudan People's Liberation Army United (SPLA United), and was one of the guerilla factions fighting for independence for South Sudan. Its leader, Reik Machar, eventually became the Vice President of the autonomous Southern Sudan.

In February 1953, the United Kingdom and Egypt concluded an agreement providing for Sudanese self-government and self-determination. The transitional period toward independence began with the inauguration of the first parliament in 1954.

Sudanese Civil War

On August 18, 1955 a revolt in the army in Torit, Southern Sudan, a forerunner of the SSIM/A, broke out, which although quickly suppressed, led to a low-level guerrilla insurgency by former Southern rebels, and marked the beginning of the First Sudanese Civil War.

On December 19, 1955 the Sudanese parliament, unilaterally and unanimously, declared Sudan's independence. The British and Egyptian Governments recognized the independence of Sudan on January 1, 1956.

The United States was among the first foreign powers to recognize the new state.

However, the Arab-led Khartoum government in the north reneged on promises to southerners to create a federal system, which led to a mutiny by southern army officers that sparked seventeen years of civil war (1955–1972).

The National Unionist Party (NUP), under Prime Minister Ismail al-Azhari, dominated the first cabinet, which was soon replaced by a coalition of conservative political forces. In 1958, following a period of economic difficulties and political maneuvering that paralyzed the public administration, Chief of Staff Major General Ibrahim Abboud overthrew the parliamentary regime in a bloodless coup d'état.

General Abboud did not carry out his promises to return Sudan to civilian government, however, and popular resentment against army rule led to a wave of riots and strikes in late October 1964 that forced the military to relinquish power.

The Abboud regime was followed by a provisional government until parliamentary elections in April 1965 led to a coalition government of the Umma and National Unionist Parties under Prime Minister Muhammad Ahmad Mahjoub.

Between 1966 and 1969, Sudan had a series of governments that proved unable either to agree on a permanent constitution or to cope with problems of factionalism, economic stagnation, and ethnic dissidence. The succession of early post-independence governments was dominated by Arab Muslims who viewed Sudan as a Muslim Arab state. The constitution proposed by the Umma/NUP coalition in 1968 was arguably Sudan's first Islamic-oriented constitution.[162]

Dissatisfaction culminated in a second *coup d'état* on May 25, 1969. The coup leader, Colonel Gaafar Nimeiry, became prime minister, and the new regime abolished parliament and outlawed all political parties.

Disputes between Marxist and non-Marxist elements within the ruling military coalition resulted in a briefly successful coup in July 1971, led by the Sudanese Communist Party. Several days later, anti-communist military elements restored Nimeiry to power.

In 1972, the Addis Ababa Agreement led to a cessation of the north-south civil war and a degree of self-rule for the south. This led to a ten years hiatus in the civil war.

The goal of the Agreement was to address and appease concerns of the southern Sudan liberation and secession movement, because the First Sudanese Civil War had proven costly in lives and resources to the northern Sudanese government and the southern population.

The Addis Ababa Agreement established the Southern Sudan Autonomous Region. It meant South Sudan would no longer be divided into the three separate regions of Al-Istiwā'iyyah (Equatoria), Bar al-Ghazāl, and A ālī al-Nīl (Upper Nile). The Southern Sudan region would run itself through a separate legislative and executive body. The separatist South Sudan soldiers of the Anya Nya would be integrated into the Sudanese army and police force. The Addis Ababa Agreement gave Nimeiry popularity and prestige both within Sudan and outside of the country.

A decade of relative peace followed even though the Addis Ababa Agreement failed to dispel the tensions that had originally caused the civil war. Continued resource infringements and marginalization of the south by the north led to increased unrest in the south starting in the late 1970s.

The US and Russia inserted themselves in the struggle between the Sudanese people as arms suppliers. Since independence in 1956, the Sudanese army had been trained and supplied by the British, but relations were cut off after the Arab-Israel Six-Day War in 1967. At this time, relations with the USA and West Germany were also cut off with Sudan. From 1968 to 1971, the Soviet Union and eastern bloc nations sold large numbers of weapons and provided technical assistance and training to Sudan. At this time, the army grew from a strength of eighteen thousand to roughly fifty thousand men. Large numbers of tanks, aircraft, and artillery were acquired at this time, and the Soviets dominated the army until the late 1980s.

Relations cooled between Sudan and the Soviet Union after the coup in 1971, and the government sought to diversify its suppliers. Egypt became the most important military partner in the 1970s, providing missiles, personnel carriers, and other military hardware.

The United States began selling Sudan a great deal of equipment around 1976. Military sales peaked in 1982 at one hundred and one million dollars. The alliance with the United States was strengthened under the administration of Ronald Reagan. American assistance for military programs increased from five million dollars in 1979 to two hundred million dollars in 1983 and then to two hundred and fifty-four million dollars in 1985. Sudan thus became the second largest recipient of US aid to Africa (after Egypt). The construction of four air bases to house Rapid Deployment Force units and a powerful listening station for the CIA near Port Sudan were created.

In 1983, the Second Sudanese Civil War erupted. It occurred because even though the Addis Ababa Accords were incorporated into the Constitution of Sudan there were significant violations on the part of Nimeiry's government in the North.

The first violations occurred when Nimeiry attempted to take control of oil fields straddling the north-south border. Oil had been discovered in Bentiu in 1978, in southern Kurdufan and Upper Blue Nile in 1979, the Unity oilfields in 1980, Adar oilfields in 1981, and in Heglig in 1982. Access to the oil fields meant significant economic benefit to whoever controlled them.

Furthermore, the Islamic fundamentalists in the north were discontented with the Addis Ababa Agreement, which gave relative autonomy to the non-Islamic majority in the Southern Sudan Autonomous Region. As this faction continued to grow in power, in 1983, President Nimeiry declared all Sudan an Islamic state, terminating the Southern Sudan Autonomous Region.

Immediately, the Sudan People's Liberation Army (SPLA) was founded as a rebel army to reestablish an autonomous Southern Sudan by fighting against the central government. In 1984 and 1985, after a period of drought, several million people were threatened by famine, particularly in western Sudan. The regime tried to hide the situation internationally.

In March 1985, the IMF demanded that the Sudanese government increase the prices of local basic necessities as part of its Structural Adjustment Program debt collection process (for details see the section on The Economic Situation). This triggered popular demonstrations against the IMF. On April 2, eight unions called for mobilization and a "general political strike until the abolition of the current regime." On April 3, massive demonstrations shook Khartoum, but also the country's main cities. The strike paralyzed institutions and the economy. On April 6, 1985, a group of military officers, led by Lieutenant General Abd ar Rahman Siwar adh Dhahab, overthrew Nimeiry, who took refuge in Egypt. Three days later, Dhahab authorized the creation of a fifteen-man Transitional Military Council (TMC) to rule Sudan.[163]

In June 1986, Sadiq al Mahdi, head of the National Umma Party and Imam of the Ansar, formed a coalition government with the Democratic Unionist Party (DUP), the National Islamic Front (NIF), and four southern parties. Unfortunately, however, Sadiq proved to be a weak leader and incapable of governing Sudan. Party factionalism, corruption, personal rivalries, scandals, and political instability characterized the Sadiq regime.

After less than a year in office, Sadiq al Mahdi dismissed the government because it had failed to draft a new penal code to replace the sharia, reach an agreement with the IMF, end the civil war in the south, or devise a scheme to attract remittances from Sudanese expatriates. To retain the support of the DUP and the southern political parties, Sadiq formed another ineffective coalition government.

Rise to Power of al-Bashir

In 1989, the government and southern rebels began to negotiate an end to the war, but a coup d'état brought a military junta into power which was not interested in compromise. The leader of the junta, Omar al-Bashir, consolidated his power over the next two decades, declaring himself president.

Since the coup, he has been "elected" three times as President in elections that have been under scrutiny for electoral fraud. In 2003, al-Bashir was accused by northern rebel groups, the Sudan Liberation Movement (SLM), and the Justice and Equality Movement (JEM) [a powerful anti-government faction] of oppressing Darfur's non-Arab population. The government responded by carrying out a campaign of ethnic cleansing against Darfur's non-Arabs. This resulted in the death of hundreds of thousands of civilians and the indictment of Sudan's president, Omar al-Bashir, for genocide, war crimes, and crimes against humanity including rape and pillage, by the International Criminal Court in 2009.[164]

Because of al-Bashir's assault on Darfur, the area has been in a state of humanitarian emergency since 2003. By 2010, about three hundred thousand people had died and almost three million were forced into refugee camps. Many of the displaced were children.

At the same time al-Bashir was waging war against the people of Darfur in the western region of Sudan, he initiated peace with southern rebels, although skirmishes in parts of the south continued. The two sides agreed to a peace treaty that was signed on January 9, 2005 in Nairobi. The terms of the peace treaty were:

- The south will have autonomy for six years, followed by a referendum on secession.
- Both sides of the conflict will merge their armed forces into a thirty-nine thousand-strong force after six years, if the secession referendum should turn out negative.
- Income from oilfields is to be shared evenly between north and south.
- Jobs are to be split according to varying ratios (central administration: seventy to thirty, Abyei/Blue Nile State/Nuba mountains: fifty-five to forty-five, both in favor of the government).
- Islamic law is to remain in the north, while continued use of the sharia in the south is to be decided by the elected assembly.

Islamization of Sudan

The decade of the 1990s had seen a "top down" Islamization of Sudan under the National Islamic Front (NIF), founded by Hasan al-Turabi. It was one of only two Islamic revival movements to secure political power in the twentieth century (the other being the followers of Ayatollah Ruhollah Khomeini in the Islamic Republic of Iran). Al-Turabi became one of the ideological leaders of the Pan-Islamic movement that spread throughout the Middle East following the assault on oil supplies by Western corporations and the wars for oil that followed.

The NIF emerged from Muslim student groups that first began organizing in the universities during the 1940s, and its main support base has remained the college educated. It supported the maintenance of an Islamic state run on sharia and rejected the concept of a secular state and has proven to be both politically adept and ruthless in its use of violence, in particular in the internal conflicts of the Second Sudanese Civil War with the south and the Darfur conflict, as well in the provisioning of proxy forces such as the Lord's Resistance Army, West Nile Bank Front, and Uganda National Rescue Front II against Uganda.

In the late 1990s, the Front changed its name to National Congress and the "gross human rights violations" of the regime's early years gave way to "more subtle methods of social control such as restrictions on the right to freedom of expression, opinion, religion, association, and movement.[165] Education was overhauled to focus on the glory of Arab and Islamic culture, and memorizing the Quran. School uniforms were replaced with combat fatigues and students engaged in paramilitary drills.

Religious police in the capital ensured that women were veiled, especially in government offices and universities. The relaxed political culture became much harsher, with human rights groups alleging a proliferation of torture chambers known as "ghost houses" used by security agencies. The war against the non-Muslim south was declared a jihad. On state television, actors simulated "weddings" between jihad martyrs and heavenly virgins (houris). Turabi also gave asylum and assistance to non-Sudanese jihadi, including bin Laden and other Al Qaeda members.

In 1999 al-Turabi and his supporters were expelled from the Congress by Sudan's ruler Omar Hassan al-Bashir, and subsequently founded the rival Popular Congress Party which has remained in opposition.

On August 31, 2006, the United Nations Security Council approved Resolution 1706 to send a new peacekeeping force of seventeen thousand three hundred to Darfur. In the following months, however, the United Nations' Mission in Sudan (UNMIS) was not able to deploy to Darfur due to the Government of the Sudan's steadfast opposition to a peacekeeping operation undertaken solely by the United Nations. In response, the UN transferred its authority to a joint African Union/United Nations peacekeeping operation. Following prolonged and intensive negotiations with the Government of the Sudan and significant international pressure, al-Bashir finally accepted the peacekeeping operation in Darfur.

In 2009, the International Criminal Court issued an arrest warrant for al-Bashir, accusing him of crimes against humanity and war crimes. The Sudanese government and the Justice and Equality Movement signed a ceasefire agreement ending the Darfur conflict in February, 2010.[166]

South Sudan

In January 2011, the referendum on independence for Southern Sudan was held, and 98.83% of the population voted for independence. The new country became the Republic of South Sudan on July 9, 2011, with its capital at Juba. The new constitution established a presidential system of government headed by a president who is head of state, head of government, and commander-in-chief of the armed forces. It also established the National Legislature comprising two houses: a directly elected assembly, the National Legislative Assembly, and a second chamber of representatives of the states, the Council of States. The constitution also provided for an independent judiciary, the highest organ being the Supreme Court.

John Garang, the founder of the Sudan People's Liberation Movement (SPLM), was the first president of the autonomous government until his death on July 30, 2005. Salva Kiir Mayardit, his deputy, was sworn in as First Vice President of Sudan and President of the Government of Southern Sudan on August 11, 2005. Riek Machar replaced him as Vice-President of the Government.

Al-Bashir announced that he accepted the result, but relations between the North and the South still remained tentative. A month earlier, on June 6, 2011, armed conflict had broken out in South Kordofan between the forces of Northern and Southern Sudan. On June, 20, the parties agreed to demilitarize the contested area of Abyei where Ethiopian peacekeepers were then deployed. The area is now administered by the United Nations Interim Security Force for Abyei.

The division of oil revenues was at the heart of the conflict as seventy-five percent of all the former Sudan's oil reserves are in South

Sudan. The region of Abyei still remains disputed and a separate referendum will be held in Abyei on whether they want to join Sudan or South Sudan.

Within South Sudan, things were not much better. The new country was immersed in tribal wars with at least seven armed groups in nine of its ten states. The wars had already displaced tens of thousands of people. The fighters accuse the government of plotting to stay in power indefinitely, and of not fairly representing and supporting all tribal groups while neglecting development in rural areas. The infamous Joseph Kony's Lord's Resistance Army (LRA) also operates in a wide area that includes South Sudan. This is the organization noted for its widespread enslavement of children and turning them into armed killers.

In December 2013, a political power struggle broke out between President Kiir and his former deputy Riek Machar, as the president accused Machar and ten others of attempting a coup d'état. This struggle ignited the South Sudanese Civil War. Ugandan troops were deployed to fight alongside South Sudanese government forces against the rebels. The United Nations stationed peacekeepers in the country as part of the United Nations Mission in South Sudan (UNMISS).

Numerous ceasefires were mediated by the Intergovernmental Authority on Development (IGAD) between the Sudan People's Liberation Movement (SPLM) led by President Kiir and the SPLM-in-opposition party led by Riek Machar. They were all subsequently broken.

IGAD is an eight-country trade bloc in Africa that includes governments from the Horn of Africa, Nile Valley, and the African Great Lakes. Its headquarters is in Djibouti City. A peace agreement was signed in Ethiopia under threat of United Nations sanctions for both sides in August 2015. Machar returned to Juba in 2016 and was appointed vice president. Following a second breakout of violence in Juba, Machar was replaced as vice-president and he fled the country as the conflict erupted again. Rebel in-fighting has become a major part of the conflict. Rivalry among Dinka factions—led by the President Kiir Mayardit and Malong Awan Malong, who had been appointed chief of general staff of the Sudan People's Liberation Army by President Kiir Mayardit—also led to fighting. In August 2018, another power sharing agreement came into effect.

About four hundred thousand people are estimated to have been killed in the civil war, including notable atrocities such as the 2014 Bentiu massacre committed by the SPLM-in-opposition faction.

Although both men have supporters from across South Sudan's ethnic divides, subsequent fighting has been communal, with rebels targeting members of Kiir's Dinka ethnic group and government soldiers attacking Nuers. More than 4 million people have been displaced, with about 1.8 million of those internally displaced, and about 2.5 million having fled to neighboring countries, especially Uganda and Sudan.

The war, as we have seen began as a feud between forces loyal to President Salva Kiir and to then-Vice President Riek Machar. It soon spiraled into fighting among several factions, engulfing the country in ethnic violence and eventually producing a devastating humanitarian crisis.

On September 12, 2018, a peace treaty was signed by all parties, but it did not get at the real issues dividing the combatants and many do not expect it to hold.

In the meantime, over seven million people, two-thirds of the population of South Sudan, are in need of aid, almost every one of them going hungry. Food security is expected to deteriorate more, with another seven hundred thousand people estimated to face crisis levels of hunger with the onset of the July to August lean season, the period of time between harvests when food stores are low.

In 2017, famine was declared in two counties in South Sudan, and famine has remained a persistent threat since. Without peace and consistent humanitarian access, another famine occurred in 2019.[167]

While hunger and disease rack the country, millions flee to neighboring countries. Their hunger is aggravated by human rights abuses, mass rape, and war crimes committed by both sides of the conflict.[168]

Oppression in Northern Sudan

In northern Sudan (Sudan) the situation is not much better. After months of mass rallies, on April 11, 2019, the Sudanese people finally ousted the dictator Omar al-Bashir after two decades of his rule. Even so, the revolution continues because al-Bashir's regime was quickly replaced by violent military aggression carried out by the Rapid Support

Forces (RSF), which are Sudanese paramilitary forces that were originally created by al-Bashir and placed under the administration of the National Intelligence and Security Service.

The RSF had been used by the government of al-Bashir to put down rebellions. It is also believed that the European Union funded al-Bashir's government to control migration of Sudanese people across the Sahara to Libya. Although the EU consistently denies it, many Sudanese believe that this gave license to the RSF to police the border, extract bribes and ransoms, and also participate in human trafficking.[169]

During the mass rallies in 2019, the military junta, who took control of the country, employed the RSF to violently crackdown on pro-democracy demonstrators.[170] Along with other government security forces, the RSF carried out the Khartoum massacre on June 3, 2019 in which more than one hundred people were killed and seven hundred were injured during an attack on a peaceful sit-in. In this massacre, at least seventy demonstrators were also raped, both women and men.[171]

This violent suppression of the Sudanese people is occurring at a time when the World Food Programme (WFP) reported that seven million people face food shortages, with more than twenty thousand dangerously near famine. This situation is predicted to worsen.[172]

On July 24, 2019, the Sudan's military council that seized power after al-Bashir's ouster, reported that it had thwarted a coup attempt and arrested an unspecified number of senior officers in connection with a plot to restore the party of ousted President Omar al-Bashir to power.

The military said in its statement: "The failed attempt aims to abort your glorious revolution and to return the former National Congress regime to power, and to disrupt the path before the expected political solution that aims to establish a civilian state."

The statement went on to claim that the coup attempt involved General Hashim Abdel Mottalib Ahmed, head of the Joint Chiefs of Staff; a number of high-ranking officers from the armed forces and the National Intelligence and Security Service; and leaders of the National Congress Party and the Islamic Movement party that al-Bashir also headed.

Hamid Eltgani Ali, a Sudanese professor at the American University in Cairo, cast doubt on the coup attempt, however, saying the ruling military council was trying to consolidate power. "These are fabricated attempts by the (Transitional Military Council) [TMC] to strengthen their grip on power," he said.[173]

Sudan's military junta and pro-democracy factions have yet to sign an agreement regarding power sharing going forward. The parties signed a political declaration that outlines an agreement during the week of July 14, 2019. The agreement stipulates that the parties would form a joint sovereign council that will rule for a little over three years while elections are organized. The eleven-member Sovereign Council, will be composed of five military personnel chosen by the military junta and five civilians selected by the pro-democracy forces. The remaining member will be a civilian picked by consensus between the two sides.

The body will be headed by a general during the first twenty-one months of the transition, followed by a civilian for the remaining eighteen months.

The umbrella protest movement will nominate the prime minister, who will name a cabinet of twenty ministers, excluding the interior and defense ministers who will be appointed by the Council's military personnel.[174]

However, the second, more contentious part of the power-sharing deal, the so-called constitutional agreement, which is meant to specify the division of powers during the transitional period, has been stalled.

The most heated bone of contention is the military junta's demand for "absolute immunity" from prosecution for the military figures who will be part of the new Sovereign Council.

The pro-democracy forces have strongly rejected blanket immunity amid concerns over the effectiveness of an independent investigation with the military still in charge. Protest leaders have suggested a "temporary immunity" be offered that would be valid as long as the member is in service.

The sticking point for the pro-democracy forces is that the commander of the Rapid Support Forces (RSF), who carried out the June 3 massacre against peaceful demonstrators, General Mohamed Hamdan Dagalo, is also the deputy head of the Transition Military Council with whom they are negotiating. The General has callously dismissed the accusations of the pro-democracy forces against the RSF as an attempt to distort the image of his troops.[175]

The pro-democracy forces, who are known as the Forces for Declaration of Freedom and Change (FDFC), have been meeting in Ethiopia with leaders of the Revolutionary Front, an alliance of Sudanese rebel groups who are also part of the movement.

For decades, Sudan had been the scene of rebellions in the provinces by ethnic and religious minorities who felt marginalized or oppressed by the Khartoum government, which, under al-Bashir was dominated by Arab Muslims. The Revolutionary Front includes rebel groups from Darfur as well as Blue Nile and South Kordofan provinces.[176] The Revolutionary Front considered the first agreement of the pro-democracy forces with the TMC to be a "betrayal of the revolution." The leaders of the FDFC, however, say they had to make concessions to avoid further bloodshed. "We had a civilian revolution, and the very idea of power sharing with the military was already rejected by the Sudanese people, but this is what the balance of power dictated."[177]

Unfortunately, the Saudi and UAE monarchies have also sided with the leader of the paramilitary RSF, General Mohamed Hamdan Dagalo, to destroy the local people's revolution.

These Middle East dictators want Sudan to provide foot-soldiers for their plans for regional domination. In 2015, then-President Omar al-Bashir agreed to join the Saudi-led coalition that invaded Yemen to fight the Houthi rebellion, hoping the Saudis would help his regime by injecting cash into the collapsing Sudanese economy.

But that did not happen. Neither the Saudis in Riyadh, nor the leaders of the United Arab Emirates (UAE) in Abu Dhabi, ever rewarded al-Bashir according to expectations, and instead focused on courting Mohamed Hamdan Dagalo, the young and ambitious commander of the Rapid Support Forces (RSF) militia, who was tasked with supplying Sudanese soldiers for the war in Yemen.

For Mohammed bin Zayed, the de facto ruler of the United Arab Emirates (UAE), the RSF was a godsend. To realize his regional ambitions, bin Zayed cannot use his own national army, as well-equipped and trained as it is. He needs proxies and militias that are not associated directly with him to do his dirty work. In Sudan, he is dependent upon the rule of the RSF. The RSF has supplied Sudanese fighters to secure his bases in Yemen.

With an impoverished population of over seven million in Sudan, the RSF, if not stopped, could expand its militia for hire operation into a massive war enterprise, with Mohammed bin Zayed as its main client and patron.

This would mean the continuous victimization of Sudan's children and young men. While they are officially being paid for their service, their fate is no different from that of their enslaved ancestors who were forced into the military service of a foreign master.

And while their commanders in Khartoum proclaim that their mission is to protect the two holy cities of Islam, they are, in fact, being used as tinder in a regional conflict in which they have no stake.

To ensure Sudan does not break it off and withdraw its fighters from Yemen, the Gulf countries of UAE and Saudi Arabia have made it their priority to crush the Sudanese revolution in order to secure RSF's position of power.[178]

But even as the RSF brutalizes the people of the Sudan, nothing seems to deter the people nor diminish their resolve.

In the meantime, the United Nations and the World Bank have ignored the economic roots of the conflicts in Sudan and South Sudan claiming that human rights violations are at the heart of violent conflict and its prevention. Since 2016, for example, the World Bank, together with the United Nations Department for Peacekeeping Operations (DPKO), has supported efforts spearheaded by the African Union to strengthen the integration of human rights into the African Union's conflict response instruments. The World Bank's Global Program for Reintegration Support (GPRS), with support from the World Bank's Nordic Trust Fund (NTF) on human rights, and in collaboration with the UN Office of the High Commissioner for Human Rights (OHCHR), has provided support for strengthening the African Union's peace keeping initiatives.[179]

We do not intend to minimize the valuable contribution of these international agencies in supporting the Sudanese people, many of whom are children, women, persons with disabilities, ethnic minorities, etc. We only want to say that for these international agencies to deny and cover up the root cause of the environmental and human destruction in Sudan and South Sudan does a great disservice to the people of the world. It blames the local people for the crisis when they were only reacting to the exploitation of their environment and social structures by the international capitalist elite in the form of the World Bank and multinational corporations. By hiding the root cause of the people's oppression and suppression, it leads the world community to believe that the people of Sudan and South Sudan are just a bunch of savages who do not know how to act in a civilized manner and therefore need the capitalist institutions to look after them.

The Economic Situation

While the war between Sudan and South Sudan certainly has strong elements of religious, ethnic, and racial conflict, the underlying conflict is based on economics and its impact on the natural environment.

As Mohamed Suliman tells us:

> Conflicts are processes, not static events, and over the last three decades, developments in the Sudan have gradually but consistently changed the nature of the conflict from being a classic ethno-religious conflict to one mainly over resources, with the economic and resource crisis in the North emerging as a driving force in the Sudanese civil war.[180]

Before describing how the economic crisis in Sudan and South Sudan originated, it is important to understand the democraphics of both countries. More than eighty percent of its fifty-three million people live in rural areas. The population is composed of one hundred and thirty-two tribes and sub-tribal groups in an area of nine hundred and sixty-five thousand square miles. Black Africans dominate in the high rainfall savannah of the South, with peoples of Arab origin in the desert scrublands of the North, and mixed tribes in the central low rainfall savannah.[181] The last full census (1955/6) estimated that forty per cent of the Sudanese population were 'Arab' in the sense of cultural rather than racial identity. Sudanese Arabs are a mixture of Arab, Nuba, and black African stock.

The Arab tribes in central Sudan, such as the Shaygiyya and Ja'aliyyinare are mainly riverain farmers, while farther from the Nile live the Arab pastoralists and rain-dependent farmers. In the poor savannah of the north and west, the camel breeders such as the Shukriyya and Kababish roam, and south of them the Baggara Arabs herd cattle.

Northerners of non-Arab descent comprise about thirty per cent of the population. These include the Nubians in the far North bordering Egypt, the Beja in the Red Sea Hills, the Nuba in the Nuba Mountains of southern Kordofan, the Ingessana and other peoples in southern Blue Nile Province, and the Zaghawa, Fur, Masalit, and others in Darfur. In addition, there are many long-term immigrants from West Africa collectively known as 'Fellata,' who spread six hundred and twenty-one miles across the country from West to East. The great majority of the non-Arab Sudanese are Muslims.

The Southern Sudanese, who amount to thirty per cent of the population of Sudan and South Sudan, consist of two main groups. One is the Nilotic group of primarily cattle herding Dinka, Nuer, and Shilluk who inhabit the central grasslands of the South. The second group, including the Azande, Bari, Mural, Topes, and many others, cultivate the wooded lands along the southern borders. Islam and Christianity have had a greater impact on these people than on their Nilotic neighbors.

The river Nile runs 1,403 miles from south to north, and the country's capital, Khartoum, is sited where the White Nile is joined by the Blue Nile from Ethiopia.

All rivers in Sudan are part of the Nile waters system. In addition to the White Nile and Blue Nile, the Bahr al-Arab, Dander, Rashad, and Atbara rivers flow into the main Nile. With the exception of the Bahr al-Arab, all the other perennial tributaries of those rivers originate either from outside the Sudan or from Southern Sudan.

This had an important bearing on the civil war because rainfall, the only other water source, is characterized by wide variability of distribution within the same area in any one season, and this is reflected in both run-off and seepage variability. Crop production and the welfare of livestock fluctuate greatly in accordance with the rainfall patterns.

Despite the variability of the culture and different socio-economic needs of the people, when the British and French introduced their capitalist economy in Sudan towards the end of the nineteenth century, they restricted participation by the Sudanese in order to maintain political and economic control. Consequently, foreign invasions and domestic uprisings did little to change the basic structure of the economy. It remained essentially a subsistence economy, with some commercial agriculture based on the use of serfs and slaves, as well as some long-distance trade ties with Egypt and with the East.

After independence, a new Sudanese capitalist class evolved from the indigenous, urbanized merchant class known as Jellaba, who have spread all over the Sudan and into some neighboring countries. The Jellaba constitute the largest body of the indigenous capitalist class and have exerted considerable economic and political influence on Sudan's modern history.

The Jellaba descended from Arab traders who arrived in Sudan during the time of the great expansion of Islamic civilization a thousand years ago.

Instead of diversifying into production in the manner of the European middle classes, the would-be Arab bourgeoisie became entrenched in the role of 'Jellaba,' (literally bringers of goods) rather than manufacturers. For a thousand years their ancestors specialized in short and long distance trade, thus the Jellaba prefer the intricacies of commercial dealings to long-term investment in industrial enterprise, which for them remained a little-known activity.

By virtue of their trading connections and geographic distribution, the Jellaba are well organized and adaptable. Their political talent has been underestimated on many occasions, and when challenged, they have regained their hold on the state either by the power of the vote or the power of the gun.[182]

In the early nineteenth century, when the Turks invaded Sudan, the Jellaba and their private armies collaborated with the Turkish rulers in the penetration and plunder of the South. The memory of the brutal slave trade conducted mainly by mercenaries of the northern Jellaba has lived on in the culture of the South. The experience of such aggression by Arab Muslims against black Africans gave rise to southern resistance to Islam and the embrace of Christianity, which Southerners perceived as being on their side against oppression.

To this day, the majority of Northerners have chosen to ignore rather than admit the shameful history of the Jellaba, although a tendency to refer to Southerners as "slaves" still persists. Slave trading magnified and distorted cultural and ethnic differences and left a lasting sense of grievance and mistrust.

When independence came to Sudan in 1956, it created the political conditions for the Jellaba to break away from the constraints of direct colonialism. By the 1960s, their focus had shifted from the pump-irrigated cotton schemes of the 1950s (such as the White Nile schemes) to large-scale mechanized farming of sorghum and sesame in rain-fed land areas. Because this new capitalist class lacked the prerequisites for industrialization, namely capital, technical and scientific knowledge, and markets, they shifted from manufacturing production to extraction of natural resources.

Introduction of Mechanized Farming and Its Impact

Until the early 1970s, Sudan's agricultural output was mostly dedicated to internal consumption. In 1972, the Sudanese government, under Nimeiry, became more pro-Western, and made plans to export food and cash crops. This spread from eastern Sudan southwards into Blue Nile Province and then west into southern Kordofan and Darfur.

Today there are almost 10 million acres under licensed mechanized cultivation, while there are only 9.4 million acres under traditional rain-fed agriculture. The former 'supports' eight thousand largely absentee farmer-landlord capitalists, while the latter is the livelihood of two to three million "peasant" farmers.

The tractorization and intensification of agriculture dealt a severe blow to traditional peasant farming, with great social and environmental changes inflicted on peasant and pastoralist societies.

Traditional agro-pastoralism began to collapse across the central clay plains of northern Sudan, and a new and burgeoning category of impoverished people emerged who were dependent on selling their labor to survive. Many migrated to the towns, considerably swelling the numbers of the urban poor.

Unlike the migration of the European peasantry during the Industrial Revolution, however, this move was not towards centers of higher economic production, but to areas of greater food availability and food aid. Nonetheless, a historical pattern is being echoed in tragic fashion.

Another consequence of the rapid impoverishment of the northern Sudanese traditional peasants and pastoralists is the abandonment of relatively benign methods of exploitation of nature and their replacement with aggressive hi-tech methods which assume that natural resources are limitless.[183]

In creating a class of local resource-extractors, using capitalist technologies and funded in large part by the IMF so as to incorporate Sudan into the global market economy, the consequence has been to directly impoverish both the environment and the rural people who depend on it for survival.

Their exploitation of these resources was so "thoughtless and unscrupulous," however, that they endangered the farmers and pastoralist tribes in Northern Sudan.

The use of agricultural land, upon which eighty percent of the population depended, once produced food for local consumption, but now the main crops are sorghum, millet, cotton, groundnuts, sesame, and gum Arabic.

An additional two hundred and fifty million acres are used for grazing land, but range resources are being reduced by the expansion of cultivation, by deforestation, and by grass fires, which burn up to thirty per cent of the total forage production.

One hundred and fifty-two million acres are in natural forest, but the forests are also being decimated in the North by the expansion of mechanized farming and increasing demand for fuelwood. At current rates of consumption versus regeneration and afforestation, all forest areas in Northern Sudan will soon be denuded.

As the large-scale mechanized farms expanded, they took over great stretches of traditional farm land, water supplies, grazing lands, and herding routes. They displaced millions of small farmers. They also cleared large areas of forest (including almost ninety-five per cent of the forest in eastern Sudan), which destroyed local access to fuel wood and gum Arabic that the people used in their meals as a source of fiber.

The Mechanized Farming Corporation (MFC) was established in 1968, upon the insistence of the World Bank to provide development capital for the creation of large mechanized farms. This corporation supervised the expansion of mechanized agriculture into southern Kordofan, White Nile, and Upper Nile Provinces.

This led to the usual problems faced by third world countries who adopted the capitalist plan for their "development." Global commodity prices declined throughout the 1970s causing economic problems for Sudan. At the same time, debt servicing costs, from the money spent mechanizing agriculture, rose. In 1978, the International Monetary Fund (IMF) negotiated a Structural Adjustment Program with the government. This further promoted the mechanized export agriculture sector and caused great economic problems for the pastoralists of Sudan.

In 1970, the Sudanese state introduced the Unregistered Land Act that declared that all land, occupied or unoccupied, belonged to the state and entitlement could no longer be acquired by long use. Only about one per cent of crop and grazing land is privately owned. The subsequent distribution of `state land' to absentee-landlords, encouraged the reorientation of agricultural production for export purposes. This move was sanctioned by the international "market forces" (foreign capitalists) who favor cash crops for export rather than food for the local people.

This development destroyed the social fabric and the natural environment and made it almost impossible for the Sudanese people to sustain a livelihood even though they lived in a resource-rich country. The ecological and social stress caused by large-scale mechanized agriculture created three types of conflict:

1. Conflicts between traditional farmers and owners of the big agribusiness farms. Small farmers were forced

> to sell their labor cheaply; pastoral nomads were driven out of the best areas of their traditional pasture; and agro-pastoralists were forced to abandon one of two activities and change over to agricultural labor for low wages and a lower standard of living.
>
> 2. Conflict among local people in the vicinity of the agribusiness farms because of scarcity of cultivable land, obstruction of animal herding routes or in the search for fresh grazing land.
>
> 3. Conflict between the state, as major backer of the agribusiness farms, and the small farmers and pastoralists. This is the most serious of all as the state has often opposed the spontaneous resettlement of such people when stricken by drought.

The internal conflicts which have exploded in the Sudan from the mid-1970s until the present reflect the growing resistance of millions of dispossessed people against the new economic regime based on export of resources.

Coercive acts against traditional cultivators and pastoralists were swift and brutal. When victims of famine and drought moved into the wetter zone in search of survival alternatives, they were often intercepted by the army. The only way left open for survival was to move to towns and relief centers to eke out a degraded existence dependent on begging, charity, and petty labor or theft and prostitution.

In the towns, these people were treated as third class citizens. The police were mobilized in arbitrary round-ups known as *kasha*, which sought to repatriate the migrants to their homelands, despite the fact that the land could no longer sustain them. These uprooted and homeless people are collectively known to the authorities as *Shamasa*, literally 'those who have no roof but the sun'.[184]

After decades of IMF and multinational corporate imposition of large-scale agribusiness on Sudan, so as to create a capitalist economy based on export, the environmental problems in the Sudan have uniformly worsened and now represent a major threat to Sudan's food security.[185]

According to the Food and Agriculture Organization of the United Nation's Plan of Action (2015 to 2019), the country faces the following threats:

- Rapid population growth;
- Continuous deterioration of the country's natural resource base;
- Escalation of political and tribal conflicts;
- Eighty-five percent of animal resources depend on pastoral and trans human systems of production;
- Increasing risk of transboundary plant and animal pests and diseases;
- Climate change and variability (i.e. dry spells, droughts, heavy rainstorms, floods, etc.);
- Institutions and procedures that control public funds not suitable for financing agriculture;
- Trade barriers of developed countries/potential markets limit access to Sudanese exporters; and
- Soaring prices of food and agricultural inputs.[186]

A report, published by the London School of Hygiene and Tropical Medicine and financed by the English State Department, revealed that about half of the dead were killed in fighting between ethnic rivals as it spread across the country, and the other half died from disease, hunger, and other causes exacerbated by the conflict.

During the 1960s and 1970s, Southern Sudan remained generally unscathed because it had largely remained isolated from colonial rule and the earlier civil wars. Since the 1970s, however, when the global capitalism system began going through structural changes, the rich countries of the northern hemisphere's efforts to adjust to these changes had a negative impact on the poorer countries in the southern hemisphere. This negative impact took the form of unfavorable terms of trade, including servicing and repayment of foreign debt, structural adjustment programs of the IMF, and a general worsening of the economic situation.

This pressure, in turn, was transferred by the domestic capitalist elite to the poorest people and the natural environment within their countries. Just as the poor nations began to export more and buy less, so the rural peasants and pastoralists were forced to produce more and buy less in the local market.

In Sudan, this meant a new drive to exploit hitherto less accessible resources, mainly in Southern Sudan. A number of schemes were started, based on the exploitation of oil, water, and land, and all in the name of "development." The profits, however, went mainly to the Northern Sudanese elites. Although the civil war had been halted in 1972 with the Addis Ababa Agreement, which recognized the autonomy of the South, Southern Sudanese were denied their share of the economic benefits of the peace.

As we have seen, civil war broke out in Sudan a mere eleven years after President Nimeiry abrogated the accord he himself had signed by redividing the South in order to include the Bentiu region where oil had been discovered into the North.

The first counter-attacks by the newly formed Sudan People's Liberation Movement (SPLM) under its leader Colonel John Garang, were directed against the installations of the Jonglei Canal and oil-exploration companies.

The Jonglei Canal was a canal project that was proposed to divert water through the vast Sudd wetlands of South Sudan to deliver more water to northern Sudan and Egypt for use in agriculture.

The Sudan Peoples' Liberation Movement *(SPLM)* grew to enjoy the support of a large section of the rural poor and dispossessed in the North, since it addressed the fears of marginalized peoples. This marked a fundamental transformation of the original north-south division of the country, because now economic urgency replaced the traditional ethnic divisions. For example, the ethnically mixed urban poor of Khartoum, the capital of North Sudan, were able to say, "Inshallah, John Garang will liberate us".

The war within the south between rival SPLM factions resulted from opposing political approaches to the reality of diminishing resources. In the search for a lasting peace it is necessary to understand this new dimension to the old conflict.

With the renewal of the civil war in 1983, a small but significant number of Southerners were drawn into the domestic capitalist class, while the majority of Southerners suffered the accelerated breakdown of social structures that had already been weakened through years of neglect.

Human and animal life depend on a delicate balance of the soil, water, and flora that support them, and disruption of any one of these vital elements creates havoc. Factors such as climatic change, land exhaustion through over-use and misuse, population growth or displacement, disadvantageous changes in land tenure, warfare, and export of resources cause lasting damage to the people, as well as the animals and the environment. In the last few decades, the following disasters have affected Sudan:

- Micro- and macro-climatic change (the practically continuous Sahel drought since 1967);
- Diminishing and erratic rainfall and accelerating desertification (the floods and torrential rains of 1988);
- More than tripling of the population in less than a half century (15.4 million people in 1970 to 25.4 million in 1990, to 52 million in 2018);
- Displacement, both internal and external, of some six million people;
- Doubling of livestock numbers within twenty years;
- Deforestation on a massive scale;
- Renewed civil war in the South, which is now encroaching on the west and east; and
- Aggressive expansion of legal and illegal rain-fed mechanized farming, from 1.38 million acres in 1967 to over 18.5 million acres in 1989.[187]

The Revolution Continues

What is tragic, however, is that the capitalist press and its institutions, as well as the institutions of the UAE and Saudi dictators are covering up any knowledge of the true cause of the chaos in Sudan and South Sudan. In doing so, they have also covered up any information about the true heroes of democracy, who, if they were acknowledged, would give pro-democracy forces throughout the world great inspiration.

Take the example of Sarah Mouawia who was arrested by al-Bathir's secret police and imprisoned in a covert "ghost house" known as the Refrigerator because of its permanently freezing temperatures. The police taunted her and beat her, but they could not stop her. For the next six months, up to the time of the Khartoum massacre, this twenty year old medical student kept marching for democracy. She managed to survive the massacre by avoiding the bullets of the RSF, who opened fire on the demonstrators. It was her bravery and those of hundreds of thousands like her who forced the military junta to the negotiation table.

No amount of repression or violence has been able to crush Sudan's democracy movement so far. After every attempt to silence them, the protesters have bounced back with bigger demonstrations than before.

In an interview with The *Globe and Mail*, a British newspaper, Sarah exclaimed "We can't be broken. The revolution will overcome, no matter how long it takes. I believe in this, more than ever. The revolution, she said, is not yet complete."[188]

Young people such as Ms. Mouawia have been the backbone of Sudan's extraordinary protest movement, which has gathered strength for much longer than anyone had predicted.

"Those who created the revolution from the beginning were the youth," says Amjad Farid, spokesman for the Sudanese Professionals Association, a key organizer of the protests. "These young people are the leaders and the decision-makers. The rest of us are their servants."

Even after the massacre in the capital city, the youth returned to the streets despite more gunfire and tear gas. Protest leaders told the marchers to "prepare their shroud"—a reference to the shroud that would cover them in death. "It is ready," the youth replied.

Violence will not destroy the protest movement, Sarah says. "We were born traumatized. And we're fearless now."[189]

It is not just the youth who have led the revolution, it has also been the women of all ages. Women have led the chants for freedom, justice, and peace. From the December 2018 protests to the great marches in the present, female demonstrators have led the occupation of the streets in the face of arrest, torture, sexual assault, teargas, and live bullets. Even after security forces killed people and raped dozens in the Khartoum massacre on June 3, women from a diversity of classes, generations, religions, educational levels, and ethnicities have continued to stand together to demand change even as the opposition talks with the Transitional Military Council have stalled. The women are clear about why they want a revolution. They demand justice for the sexual violence to which they have been submitted across the decades.

Many protesters, especially younger ones, also want the freedom to make both big and small choices in their lives.

Many protesters regard these laws and decrees as oppressive and contrary to the dignity of women. They object in particular to the use of religion to justify repressive state measures. "I think religion is about your relationship with God. My relationship with God might be better than theirs and it is not important to show that through the hijab," said one protester.

The specific issues of sexual violence and the imposed curtailment of women's freedoms under the onus of male-dominated religion come under the broader issue of gender equality.

A coalition of women known as MANSAM, which consists of eight political women's groups, eighteen civil society organizations, several youth groups, and various activists and academics, was one of the signatories of January 1 Declaration of Freedom and Change, which called for the end of al-Bashir's rule.

However, the coalition has not been represented in any of the negotiations with Sudan's Transitional Military Council. The continued exclusion of women in talks has prompted calls for equal gender representation in all official interim bodies. MANSAM has even put together a list of professional women who can participate in a transitional civilian government. Other groups such as the Strategic Initiative in the Horn of Africa (SIHA) and No to Women's Oppression, have made similar demands for female inclusion.

MANSAM has also developed a draft bill of rights that it wants to see included in a new Sudanese constitution. It supports equal gender rights and, if adopted, the draft bill would rectify gender discrimination in many areas, including family and public order law, and also codify rights to sexual orientation and gender identity, a radical position in Sudan.

In the words of one female protester: "This revolution is women's revolution."[190] According to another female student: "The Revolution is on the Curriculum."[191] It will be youth and women of the world like these fighters in Sudan who will lead the Nuclear Revolution to bring unity among the people of the world.

Lessons Learned from South Sudan

The main lesson from the Sudan is that, after the success of the revolution, the women and youth continued the fight for democracy while the men fought over control of political power and resources. This is why women and the young must play a primary role in nuclear revolution.

The revolution in South Sudan faced multiple challenges, including being colonized by the British and Egyptians, internal divisions between North and South Sudan, economic oppression by the IMF and international capitalists, religious oppression by the proponents of Islam and finally, tribal warfare. Fighting against these challenges took many years, destroyed much of the natural environment, and took millions of lives. In this situation, we see how the colonialists used a divide and conquer strategy to split the North and South and then through international disruption of the indigenous economy, destroyed the natural environment and turned tribe against tribe in order for them to survive. Add to this oppression, the pernicious impact of religious fundamentalism and faulty revoutionary leadership and you have a receipt for hell on earth.

President Nimeiry began as a socialist warrior who nationalized the banks and large industries and instituted land reform on behalf of the people, but then, in due course, his greed took over. At that point, he invited foreign corporations into the country to exploit oil and agricultural lands. The capitalists took the land and did nothing for the people. Nimeiry also got involved with the Muslim Brotherhood who promoted religious colonialism within Sudan. Things went downhill from there. Ten years later Nimeiry was taking money from the Reagan administration to allow the US to build military bases and a CIA spy station on Sudan soil.

Nimeiry was followed by a brutal dictator who set the stage for a civil war between the north and the south.

The civil war deprived the people of health care and education and displaced, starved, and deprived millions of people in the South. There were almost two million deaths. This is a heavy price to pay for freedom, but unfortunately the revolutionaries did not see things clearly. Immediately after gaining independence, they submerged the people in inter-tribal wars and allowed the people to be exploited by a local warlord, Joseph Kony, who turned their children into marauding child soldiers.

Thus, we see that the people of South Sudan faced many enemies that prevented a lasting revolution despite their fervor and love of freedom. While the people had to fight off the oppression of the colonizers, dictators, and religionists, their revolutionary leaders then failed them by not maintaining a service mentality, but instead by reverting to self-interest and group psychology. This is why it is so important that the revolutionary leadership be grounded in a universal ideology and a service mentality otherwise hard-fought gains come to naught and the people's suffering cannot be overcome.

History has taught us that all revolutions bring warriors to the fore. If these revolutionaries are not spiritual warriors, the great sacrifices made by the people come to naught. In this instance, it was the youth and the women who assumed the position of being spiritual warriors.

In the Middle East: Greater Kurdistan

The revolution in Kurdistan also has much to teach revolutionaries about the struggle for democracy. The Kurdish revolution has continued for almost a century and is spread over four different countries. Although there has never been a nation-state called Kurdistan, the Kurds are a homogeneous people with their own language, culture, and history. Thus, for purposes of this analysis, we will speak of Greater Kurdistan as the homeland of the Kurds, whose land has now been divided into four separate nation-states. Greater Kurdistan covers around 73,360 square miles in Turkey; 48,263 square miles in Iran; 25,097 square miles in Iraq; and 4,633 square miles in Syria, with a total area of approximately 151,352 square miles.

How did the Kurdish homeland get divided into four separate countries?

Before WWI, Greater Kurdistan had been occupied by the Ottoman Empire. Afterwards the Allied Powers, who were the victors of the world war, struggled among themselves to see who would claim its territories. The British and French, in a backroom deal called the Sykes Picot Agreement, created the countries of Iraq and Syria as their own colonies and incorporated much of Greater Kurdistan into these new countries.[192] Today the Kurdish territory in Iraq is called Southern Kurdistan, while the Kurdish territory in Syria is called Western Kurdistan. At the same time that the Allied Powers were carving up Greater Kurdistan, the Turks, who had been the heart of the Ottoman Empire, were also trying to expand their territory. In a compromise with the Allies, they were granted control over another area of Greater Kurdistan, which is now called Northern Kurdistan in Turkey.[193]

Eastern Kurdistan, or Kurdistan in Iran, developed by a slightly different trajectory. A portion of it had long been occupied by the Persian Empire, which today has become the country of Iran. In 1639, the Persian Empire was defeated by the Ottomans, who then formalized the partition of Kurdistan. For the next five hundred years, the Kurds in Iran have struggled for independence.

Therefore, in all these countries, the Kurds find themselves a persecuted minority group that is seeking justice or independence from the countries in which they reside. As such, this divided people are conducting revolutions against four separate countries at the same time. Sometimes the fight is for autonomy within the host country, at others, it is for complete separation.

Eastern Kurdistan (Kurds in Iran)

When Reza Shah came to power in Iran in 1925, he created a centralized state in which minority communities were expected to assimilate into the Persian culture. Revolting against oppression by the Iranian government, Kurdish leader Ismail Agha (Simko) tried to establish a Kurdish state in Iran. He was invited to a conference with Iranian military leaders and was assassinated. The assassination of Kurdish leaders became the method of choice to intimidate and weaken the Kurdish nationalist movements. Under the current Islamic government, Iranian Kurdish leaders in exile were sought out by Iranian agents and murdered.[194]

During the early twentieth century, there had been a growing sentiment of Kurdish nationalism and political activism in Iran. While the Kurds were unable to secure independence from Iran after World War I, a Kurdish state was created with support from the Soviet Union in the region of Mahabad, in northwestern Iran, after WWII. The Mahabad region had played a central role in Kurdish history for over one thousand years. Many of the Iranian Kurds who currently inhabit the area are descendents of the Murki tribe, who claim to speak the most ancient and unadulterated form of the Kurdish language.

Mahabad was home to the first Kurdish nationalist movement. The Republic was short-lived, however, because when the Soviets pulled out of Iran in 1946, the Republic of Mahabad was absorbed by the new Iranian government, and its leaders were executed.[195]

The continued rule of the Pahlavi dynasty in Iran was particularly brutal for the Kurds, and consequently Kurdish activists became active supporters of a regime change during the 1979 revolution. Although the Kurdish people and Kurdish organizations took part in the 1979 revolution on behalf of the Islamists, Ayatollah Khomeini declared Jihad ("Holy War") against the Kurdish people due to their demands for greater democracy in Iran and autonomy for Iranian Kurdistan. The Iranian Kurds had long fought for improved governmental representation and the protection of their basic human rights through the creation of their own federal state.[196]

Armed conflict between the new republic and the Kurds broke out as Khomeini tried to establish governmental control in the Kurdish regions. Since then, there has been a sustained military, economic, and psychological war waged against the Iranian Kurds, which has led, according to the Kurdistan Peace and Development Society, to a 'systematic genocidal campaign.' This phenomenon has claimed the lives of tens of thousands of innocent people.

In a speech in December 1979, Ayatollah Khomeini declared that the concept of ethnic minorities was contrary to Islamic doctrines. He pointed his finger at those "who do not wish Muslim countries to be united" by creating the issue of nationalism among minorities. His views were shared by many in the Shia Muslim clerical leadership. The exclusion of Kurds from participating in the new government of Iran, while simultaneously chastising them for wanting to form their own government, stoked the fires of Kurdish-Iranian conflict.

On December 2, 1996, the death of a prominent Sunni clergyman, Mulla Mohammed Rabiei, in Kermanshah, the largest Kurdish-speaking city in Iran, led to violent clashes between Sunni Kurds and the Iranian security forces who were Shia. Mulla Rabiei was the prayer leader in the Al-Shafe'i mosque in Kermanshah. The protests continued for three days and spread to neighboring towns in the region.

Iran is predominantly composed of Shia Muslims, while the great majority of the Kurds are Sunni Muslims. This constitutes another major fault line between the government of Iran and the Kurds.

Since the Iranian revolution that overthrew the capitalist-supported Pahlavi dynasty, the number of Sunni mosques in Kermanshah has more than quadrupled. At the time of revolution, there were around ninety-five Sunni mosques in this province. Today the number of mosques with Sunni Imams as prayer leaders is four hundred and ninety-one.[197]

In 1997, Sunni Kurds, like many other Iranians, took part in the presidential election. The new President Khatami, who received seventy percent of the votes, was little known until that point. Khatami had run on a platform of liberalization and reform and during his two terms as president, Khatami advocated freedom of expression, tolerance, and support for civil society. He advocated constructive diplomatic relations with other states including those in Asia and the European Union, and proposed an economic policy that supported a free market and foreign investment.

Khatami praised the glory of Kurdish culture and history. The Kurds wanted to use their own language, which had been outlawed in Iran.

They also wanted the appointment of top-level Kurdish officials. In his first term, Khatami appointed Abdollah Ramezanzadeh to be the first Kurdish governor of the Iranian province of Kurdistan. He also appointed several Sunni and Shia Kurds as his own or cabinet members' advisors. In his second term, Khatami had two Kurdish cabinet members; both of them were Shia. The increased presence of Kurdish representatives in the sixth parliament led to expectations that some of the voters' demands would be met. After the first round, in which eighteen Kurds were elected, one candidate said that he expected there would be more Kurdish instruction at the university in Sanandaj, and he called on the Khatami government to have more Kurdish officials. Subsequently, a forty-member parliamentary faction, representing the predominantly Kurdish provinces of Kurdistan and Kermanshah, was formed.

However, there were many other Kurdish activists who did not join the reform movement. Mohammad Sadiq Kaboudvand was among the latter. He started an independent human rights association to defend the rights of the Kurdish people. The territories comprising Iranian Kurdistan were among Iran's poorest and least developed.

In February 1999, Kurdish nationalists took to the streets in several cities such as Mahabad, Sanandaj, and Urmia and staged mass protests against the government and in support of Abdullah Ocalan, a Kurd living in Turkey, who was the founder and leader of the Party for a Free Life in Kurdistan (PJAK) and the Kurdistan Workers' Party (PKK).

The Iranian government viewed this movement as "trans-nationalization." In other words, the movement was identifying with the Kurds and not Iran. Consequently, these protests were violently suppressed by the government forces.

Despite the fact that more than one million Sunnis live in Tehran, many of them Kurds, not one Sunni mosque exists to serve their religious needs. In a rare public protest, eighteen Sunni parliamentarians wrote to the authorities in July 2003 to criticize the treatment of the Sunni Muslim community and the refusal to allow construction of a mosque in Tehran that would serve that community.

On July 9, 2005, a Kurdish opposition activist, Shivan Qaderi, and two other Kurdish men were shot by Iranian security forces in Mahabad.

For the next six weeks, riots and protests erupted in Kurdish towns and villages throughout Eastern Kurdistan such as Mahabad, Piranshahr, Sinne (Sanandaj), Sardasht, Oshnavieh (Şino), Baneh, Bokan, and Saqiz (which also inspired protests in southwestern Iran and in Baluchistan in eastern Iran) with scores killed and injured, and an untold number arrested without charge.

On March 13, 2006, Saleh Nikbakht, a well-known Iranian human rights lawyer, who was Shivan Qaderi's lawyer, announced that Qaderi's murderer was a member of the police, who shot the victim illegally. He also added that the murderer and the one who ordered the act were under investigation and the judiciary system had been cooperative up to that point.

In a report released in 2008, Amnesty International said that Kurds, who have suffered a long history of discrimination in Iran, have been a particular target of the Islamic Republic of Iran, which has repressed their social, political, and cultural rights and their economic aspirations. Anyone who questioned the government's violation of Kurdish human rights soon found that his or her human rights were also suspended.

In 2009, nation-wide street protests took place after the presidential election of incumbent Mahmoud Ahmadinejad. The people complained of electoral fraud and chanted "Down with the dictator," "Death to the dictator," and "Give us our votes back." They called for support of the opposition leader, Mir-Hossein Mousavi Khameneh, who was a reformist politician, artist, and architect.

Many Kurdish activists, writers, and teachers were arrested and sentenced to death during the government crackdown that followed the protests. Even before the elections, however, the Kurdish rebel group, Party for a Free Life in Kurdistan (PJAK), had taken up arms against the state.

In November 2009, Iran executed the Kurdish leader, Ehsan Fattahian, despite an international campaign calling for his release. Authorities accused Fattahian of carrying arms for an "illegal organization" and sentenced him to several years in prison. Fattahian attempted to appeal, and when he did so, he was sentenced to death for "enmity against God." His execution was condemned by human rights groups and activists internationally.

In January 2010, Iran executed a second Kurdish political prisoner, Fasih Yasamani, for "enmity against God." Like Fattahian, Yasamani was tortured and authorities tried to force him to confess, but he refused. He was also denied a fair trial.

Without notifying the families or lawyers of the political prisoners, Iranian authorities ordered the execution of five more Kurdish political prisoners—Ali Heydarian, Farhad Vakili, Mehdi Eslamian, Shirin Alam Hooli, and Farzad Kamangar. The political prisoners suffered severe torture at the hands of Iranian authorities and were also forced to confess their memberships in an illegal organization—namely PJAK. Amnesty International described the executions as "a blatant attempt to intimidate members of the Kurdish minority." All activists denied any links to PJAK, and the leader of PJAK also denied any links to the five activists.

Despite repeated international calls demanding the release or retrial of these political prisoners, all were executed without any prior notice or warning. Following the executions, Iranian authorities refused to return the bodies of those executed to their families.

As of May 2010, there were at least sixteen other Kurdish political prisoners on death row. Not one case has been reported as having received a fair trial.[198]

Due to their "otherness," Iranian Kurds are the frequent victims of human rights abuses. The Iranian government routinely accuses them of engaging in terrorist activities. Kurdish activists, including individuals with no ties to PJAK or other militants, are often detained and executed by the regime on terrorism and sedition charges. As a result, many Kurds in Iran see themselves as victims of a genocidal campaign directed by the ethnic Persian-dominated Shia Islamist state. Having exhausted all attempts to achieve self-rule for their community in Iran through peaceful means, Iranian Kurds, in the eyes of PJAK, have been left with little choice but to take up arms and engage in revolutionary struggle.

PJAK's stated goals, however, do not include calls for the unification of the region's disparate Kurdish populations into a single country or the secession of Iran's Kurdish regions from the Islamic Republic of Iran. Instead, PJAK claims to be fighting for the rights of Kurds in Iran as victims of an oppressive state. For PJAK, the establishment of an Iran that would guarantee Kurds substantial autonomy would satisfy the demands of Iranian Kurds. Nevertheless, many of its members would like to see a "Greater Kurdistan" down the road.

While tribal and clan-based kinship networks tend to politically divide the Kurds in Iran, their historical persecutions unify the Kurdish populations under a banner of resistance.

PJAK is modelled after the Kurdistan Workers Party (PKK), which is a political and militant organization in Turkey and Iraq. As such, PJAK has assumed the role of armed guardian of the Kurds in Iran. Relying on bases it shares with the PKK on the Iraqi side of the Qandil Mountain range, as well as on its own cells in Iran, PJAK is waging an asymmetric campaign that combines guerrilla-style insurgency operations with terrorist attacks targeting Iranian security forces, particularly the Iranian police and members of the elite Iranian Revolutionary Guard Corps (IRGC).

PJAK's tactics include small unit ambushes such as the August 12, 2010 attack against a convoy of IRGC officers traveling through the city of Urmia in West Azerbaijan Province that left three officers dead and one wounded.[199]

PJAK has also used improvised explosive devices (IEDs) and rocket-propelled grenade (RPG) attacks against Iranian security forces. The militia claimed responsibility for downing two Iranian military helicopters with RPGs in February and August 2007 that killed more than thirty Iranian soldiers.

In addition to targeting security forces, PJAK also attacks representatives of Iran's legal system, including judges and prosecutors, in retaliation for the large number of arrests and executions of Kurdish political activists.

PJAK will also target religious and political officials for assassination, including ethnic Kurds who are seen as collaborators with the regime. While PJAK is careful to avoid civilian casualties, legitimate targets include civilian collaborators, such as paid informants, or others assisting Tehran's efforts to root out the group are considered. Critical infrastructure, including energy pipelines, has also been targeted by PJAK in an attempt to disrupt the Iranian economy.

PJAK has broad support among the Kurds as indicated by the organized political network consisting of human rights and lobbying organizations,

Iranian Kurdish political parties operating in exile, and independent supporters who advocate on behalf of Iranian Kurds in the diaspora. In Europe, the group counts on the sizeable Kurdish population there for political, financial, and material support.

PJAK also operates a sophisticated network of websites to foster human rights and democracy discourse, to claim responsibility for its attacks, and to publicize its position on events and issues. It also uses the internet to memorialize fallen fighters and to demonstrate solidarity with the plight of Kurds everywhere.

Women assume prominent positions in all aspects of PJAK's work, including fighting on the front lines. Women are estimated to compose about half of PJAK's ranks.

Despite the fact that PJAK is made up of local people trying to defend themselves, primarily from government forces, the government of Iran seeks to portray the organization as the tool of foreign intelligence services, namely the CIA, but also the states of Israel, Saudi Arabia, and Great Britain. These, of course, are Iran's greatest foreign enemies.

In response to Iran's claim that the US was behind PJAK, the United States designated PJAK a terrorist organization on February 5, 2009 in order to appease Iran during its discussions with them concerning its nuclear program.

To this day, the relationship between the Kurds of western Iran and the Iranian government remains tense. The Party for Free Life in Kurdistan (PJAK) remains an active political and militant organization working for Kurdish autonomy particularly in the Iranian provinces of Kurdestan, Kermanshah, and West Azerbaijan.[200]

At present, The Kurds represent the third largest ethnic group in Iran, after the ethnic Persians and Iranian Azerbaijanis, comprising more than ten percent of the country's population.[201]

While the government of Iran continues to use missile strikes against PJAK, ultimately, in order to alleviate the long entrenched grievances and mistrust that Kurds feel toward the Iranian state, it will require a sustained commitment by the government to improve the lives of Iranian Kurds by implementing far-reaching social, political, and economic reforms in Iranian Kurdistan.[202]

Southern Kurdistan (Kurds in Iraq)

Southern Kurdistan or Iraqi Kurdistan is the Kurdish-populated region incorporated into Iraq. Most of the geographical region of Iraqi Kurdistan is governed by the Kurdistan Region (KRG), which is an autonomous region recognized by the Iraqi constitution. The capital city of the autonomous region is Erbil and the parliament has one hundred and eleven seats.

The region comprises 18,093 square miles and has a population of 5.9 million people. It is composed of four governorates, Duhok, Erbil, Silemani, and Halabja.

The establishment of the Kurdistan Region dates back to the Iraqi–Kurdish Autonomy Agreement of 1970 after the First Iraqi–Kurdish War that lasted from 1961 until 1970. The struggle was led by Mustafa Barzani on behalf of the Kurds' attempt to establish an autonomous Kurdish administration in northern Iraq. During the war, eighty percent of the Iraqi army was engaged in combat with the Kurds. The war, which took between seventy-five thousand and one hundred and five thousand lives, ended with a stalemate and led to the Iraqi–Kurdish Autonomy Agreement of 1970. The agreement failed to be implemented, however, and led to the Second Iraqi-Kurdish War.

In 1973, the US, under Richard Nixon, made a secret arrangement with the Shah of Iran to fund Kurdish rebels through the CIA, in collaboration with Israel's Mossad, against Saddam Hussein's Iraqi government.

In 1974, the Iraqi government retaliated with a new offensive against the Kurds and pushed them back to the border with Iran. On May 12, 1974, Hussein sentenced Leyla Qasim, a Kurdish student attending the University of Baghdad, to death for supporting the Kurdish cause. She was the first woman in modern Iran to face the death penalty. She arrived at the gallows wearing the traditional Kurdish dress. Today she is referred to as the Bride of Kurdistan and remains an inspiration to the Kurdish people.[203]

Hussein informed the Shah of Iran that he was willing to satisfy other Iranian demands in return for an end to its aid of the Kurds. Iran and Iraq reached a comprehensive settlement in March 1975 known as the *Algiers Pact*. The agreement left the Kurds helpless and Iran cut supplies to the Kurdish movement in Iraq. The Kurdish leaders left the country and many Kurds surrendered *en masse*, ending the revolution and the Second Iraqi-Kurdish War.

To begin the 1980–88 Iran–Iraq War, Saddam Hussein attacked Iran just after its revolution, fearing that the new Shiite-controlled government in Iran would cause Shia Muslims in his country to revolt against his Sunni-minority government. In this war, he also used the Iraqi Army to devastate the population and environment of Iraqi Kurdistan.

From 1987 until 1989, he carried out a genocidal campaign against the Kurds using chemical weapons, utterly destroying two thousand villages, and slaughtering around fifty thousand rural Kurds. The large Kurdish town of Qala Dizeh (population seventy thousand) was completely destroyed by the Iraqi army. Saddam Hussein's campaign also included a program to drive Kurds and other ethnic groups out of the oil-rich city of Kirkuk and replace them with Arab settlers from central and southern Iraq.[204]

On August 2, 1990, the Iraqi army invaded neighboring Kuwait to gain control of its oil reserves. This created international condemnation by the UN Security Council and played into the hands of President George H. W. Bush of the United States, who deployed US forces into Saudi Arabia, and urged other countries to send their own military forces to stop Hussein. Several nations joined his effort and the coalition formed the largest military alliance since World War II. The great majority of the coalition's military forces were from the US, followed by Saudi Arabia, the United Kingdom, and Egypt. The war against Iraq came to be called the Persian Gulf War. It was code-named Operation Desert Storm by the Americans.

At the same time, a majority of the Iraqi people did not support Hussein's wars. This led to domestic uprisings in 1991. The Kurds played a major role in the rebellion in the north, charging Hussein's government with systemic social repression and genocide. They agreed with other rebel groups that Hussein's government had become vulnerable to regime change. The Iran–Iraq War and the invasion of Kuwait had occurred within a single decade and had devastated the economy and population of Iraq.

Within the first two weeks of the uprisings, most of Iraq's cities and provinces fell to rebel forces. The rebel groups were a mix of ethnic, religious, and political affiliations and included military mutineers, Shia Arab Islamists, the Kurds, and far-left groups.

Following initial success, however, the revolution floundered due to internal divisions among the rebel leaders, as well as the absence of anticipated American support. Saddam's Sunni-dominated Ba'ath Party, which advocated Arab nationalism and anti-imperialism, also called for the unification of the Arab world into a single state. It managed to maintain control over the capital of Baghdad and soon largely suppressed the rebels in a brutal campaign conducted by loyalist forces led by the Iraqi Republican Guard.

During a brief, one-month period of unrest, tens of thousands of people died and nearly two million people were displaced. After the conflict, the Persian Gulf War Coalition established no-fly zones over northern and southern Iraq. The Iraqi no-fly zones protected the ethnic Kurdish minority in northern Iraq and Shiite Muslims in the south by denying Iraqi aircraft from flying inside these zones. The policy was enforced until 1998. During the time that the no-fly zones were in force, the Kurds were finally able to establish Iraqi Kurdistan as a Kurdish Autonomous Republic. The republic developed its own flag and national anthem.

The Iraqi Kurdistan's military forces, the Peshmerga, also succeeded in pushing out the main Iraqi forces from the north.

In 1992, the major political parties in the region, the Kurdistan Democratic Party (KDP) and the Patriotic Union of Kurdistan (PUK), established the Kurdistan Regional Government. Initial elections, held in the same year, however, produced an inconclusive outcome, with the assembly divided almost equally between the two parties and their allies.

During this period, the Kurds were subjected to a double embargo; one imposed by the United Nations on Iraq and one imposed by Saddam Hussein on their region. The severe economic hardships caused by the embargoes fueled greater tensions between the KDP and the PUK, over control of trade routes and resources. Relations between the PUK and the KDP became dangerously strained from September 1993 and this led to an intra-Kurdish civil war in Iraq between 1994 and 1996.

After 1996, thirteen percent of the oil sales of Iraq were allocated for Iraqi Kurdistan and this led to relative prosperity in the region. Saddam had established an oil smuggling route through territory controlled by the KDP, with the active involvement of senior Barzani family members. The taxation of this trade at the crossing point between Saddam's territory and Kurdish controlled territory and then into Turkey, along with associated service revenue, meant that whoever controlled the Dohuk governate and its main city Zakho had the potential to earn several million dollars a week. This was the major bone of contention between the KDP and the PUK.

Direct United States mediation in the Kurdish civil war led to a formal ceasefire between the KDP and the PUK that was codified in the *Washington Agreement* in September 1998.

In preparation for the US invasion of Iraq in 2003, the first Central Intelligence Agency (CIA) team entered Iraq in July 2002. Their efforts consisted of persuading the commanders of several Iraqi military divisions to surrender rather than oppose the invasion, and identifying all the initial leadership targets through reconnaissance missions.

They also met with the Kurdish Peshmerga and began to organize them to become the northern front of the invasion. The Peshmerga forces defeated al-Qaeda's Ansar al-Islam in Iraqi Kurdistan before the invasion and then defeated the Iraqi army in the north.

Ansar al-Islam was a Sunni Muslim militia group operating in Iraq and Syria that was established in Iraqi Kurdistan by former al-Qaeda members in 2001 as a Salafist Islamist movement that imposed a strict version of Sharia on villages it controlled in Iraqi Kurdistan near the Iranian border. Following the 2003 invasion of Iraq, the group fought against the Kurdish government, American led forces, and their Iraqi allies. The group continued to fight the US puppet government in Iraq following the withdrawal of US troops from Iraq, and sent members to Syria to fight Assad's government following the outbreak of the Syrian Civil War. Years later, the leaders of Ansar al-Islam officially announced that the group had merged with the Islamic State of Iraq and the Levant (ISIS).[205]

In the early hours of the morning on March 20, 2003, the US military made a surprise invasion of Iraq. There was no declaration of war. The invasion of Iraq was led by General Tommy Franks, under the US code-name "Operation Iraqi Freedom." Approximately forty other governments that President Bush termed the "Coalition of the Willing," participated by providing troops, equipment, services, security, and special forces, with two hundred and forty-eight thousand soldiers coming from the United States.

The invasion was premised on the Bush administration's false claims that Saddam possessed weapons of mass destruction and was tied to al-Qaeda whom they blamed for organizing the destruction of the World Trade Center in New York City.

In the US takeover of Baghdad, Saddam's Ba'ath party was disbanded and new political elections were held. Saddam went into hiding, but was captured on December 13, 2003 and tried for crimes against humanity. On November 6, 2006, he was found guilty and sentenced to death by hanging. He was executed on December 30, 2006.

The 2003 invasion of Iraq, led to subsequent political changes including the ratification of a new constitution in 2005. The new constitution defined Iraq as a federalist state consisting of Regions and Governorates. It recognized both the Kurdistan Region and all laws passed by the Kurdistan Regional Government since 1992. While there is a provision in the constitution for Governorates to create, join, or leave Regions, no new Regions have been formed, and the Kurdistan Regional Government remains the only regional government within Iraq.

In 2006, PUK leader Jalal Talabani was elected President of the Governing Council of Iraq, while KDP leader Masoud Barzani became President of the Kurdistan Regional Government.

Since the downfall of Saddam's regime, the relations between the Kurdistan Regional Government and neighboring Turkey have been in flux. Tensions reached a high stage in late February 2008 when Turkey unilaterally took military action against the PKK, which was a Kurdish militia charged with protecting the Kurdish people. We will discuss the relationship between the Kurds and Turkey in the section on Northern Kurdistan below.

Tensions between Iraqi Kurdistan and the central Iraqi government, however, mounted through 2011 and 2012 on the issues of power sharing, oil production, and territorial control. In April 2012, the president of the Kurdistan Regional Government demanded that Iraqi officials agree to their demands or face the prospect of Kurdish secession from Iraq by September 2012.

In September 2012, the Iraqi government ordered the Kurdistan Regional Government to transfer its authority over the Peshmerga militia to the Iraqi government. Relations became further strained by the formation of a new command center (Tigris Operation Command) for Iraqi forces to operate in a disputed area over which both Baghdad and the Kurdistan Regional Government claimed jurisdiction. On November 16, 2012, a military clash between the Iraqi forces and the Peshmerga occurred.

It was at this time (2011 to 2013) that the Iraqi insurgency, later referred to as the Iraq Crisis, escalated. There was wide-scale violence between local militia groups against the new Iraqi government as well as violence between Iraq's religious groups, particularly the Shi'a and the Sunni.

The insurgency was a direct continuation of events following the US-led invasion of Iraq in 2003. Sunni militant groups stepped up attacks against the country's majority Shia population to undermine confidence in the Shia-led government and its efforts to protect people as the US withdrew its resources. Among the most powerful of these Sunni groups was al-Qaida that was joined by many ex-military men from Saddam's army after the fall of his government. This group eventually evolved into ISIS.[206]

In 2014, the insurgency escalated dramatically following the conquest of Mosul and major areas in northern Iraq by the Islamic State in Iraq and Syria (ISIS). ISIS is a fundamentalist Islamic group that declared itself an international Islamic empire and while fighting in Iraq, it also entered Syria to fight the Shia government of Assad.

With the escalation of the Iraqi Crisis and fears of Iraq's collapse, Kurds have increasingly debated the issue of independence. During the 2014 Northern Iraq offensive, which began on June 4, 2014, ISIS began a major offensive in northern Iraq against the Iraqi government and against the Kurds.

ISIS and allied fighting forces captured several cities beginning with an attack on Samarra followed by the seizure of Mosul and Tikrit all in the month of June. As Iraqi government forces fled south in the face of this attack, the Kurdish forces stood against ISIS and took back control of the oil hub of Kirkuk from ISIS, which was also part of the territory in dispute with the Iraqi government.

Despite the security crisis, the Iraq's parliament did not allow President Maliki to declare a state of emergency. Many Sunni Arabs and ethnic Kurdish legislators boycotted the session because they opposed expanding the prime minister's powers.

On July 1, 2014, Masoud Barzani, President of the Kurdish Regional Government, announced that "Iraq's Kurds will hold an independence referendum within months." In response, the Iraqi government threatened the Kurds with "dire consequences" if they did not relinquish control of the Kirkuk oilfields to Iraq's control.

In September, Kurdish leaders postponed the referendum to focus on the fight against ISIS. In November, Ed Royce, Chairman of the Foreign Affairs Committee of the United States House of Representatives, introduced legislation to arm the Kurds directly, rather than continue working through the local governments.

In August 2014, the US began a campaign of airstrikes in Iraq, in part to protect Kurdish areas from ISIS.

In February 2016, Kurdish president Barzani stated once again that "Now the time is ripe for the people of Kurdistan to decide their future through a referendum," citing similar referenda in Scotland, Catalonia, and Quebec. On March 23, Barzani officially declared that Iraqi Kurdistan would hold the referendum some time "before October" of that year.

On April 2, 2017, the two governing parties of the Kurds released a joint statement announcing they would form a joint committee to prepare for a referendum to be held on September 25.

On the day following the referendum, September 26, 2017, President Barzani announced that the referendum had been a success in seeking independence, citing that ninety-two percent of Iraqi Kurds supported secession. He called on neighboring countries to support the new government and be open to future dialogue.

The Kurds in Iraq were not to achieve their freedom so easily, however. In November, 2017, the Iraqi Supreme Court ruled that the referendum on Kurdish independence was unconstitutional.

Immediately Iraq's government seized disputed territory from the Peshmerga, including the oil-rich province of Kirkuk. It also took control of Iraqi Kurdistan's only border crossings with Turkey and Syria.

The Kurdish referendum was also opposed by Iran and Turkey, both of which have sizeable Kurdish minorities. The US and the UN also opposed the referendum. These setbacks led President Massoud Barzani to step down from office.

Since October 2019, there have been violent protests in Iraq. The Iraqi people are demanding that the government disband itself because of corruption. The government will not disband itself and the Iraqi people will not accept anything less. In January of 2020 they expanded their list of grievances to include the expulsion of the US military from Iraq after a US drone strike killed Iranian general Qassem Soleimani near Baghdad.[207] While the Kurds, who are centered in the North can identify with the grievances of the people, they have not engaged in the protests because they cannot risk losing their autonomy if a new government comes to power that is unfriendly to them. They cannot jeopardize what little progress they have made toward autonomy.[208]

Northern Kurdistan (Kurds in Turkey)

Prior to World War I, the Kurds were under the Ottoman Empire and the economy of Turkey was still largely feudal and controlled by indigenous small capitalist land owners. The Kurds, at that time, were mostly sharecroppers, shepherds, or artisans, whose clan and tribal structures were beginning to break down.

Therefore, when WWI broke out in 1914, Ottoman Turkey still had a predominantly pre-capitalist economy. The defeat of the Ottoman Empire in the war led to the founding of the capitalist regime of Turkey under Kemal Ataturk in 1923.

As a consequence, the young Turkish Republic's economy was integrated into the capitalist world economy. This resulted not from internal dynamics of Turkish society, however, but from outside pressure. The world capitalists first used Turkey as a market, then as a place for investment. This led to the destruction of the Turkey's handicrafts market, because of "a flood of cheap imported Western goods." The comparative underdevelopment of Turkish Kurdistan was a result of this process.

While the Turkish economy remained subordinate to those of the capitalist industrialized states, it did learn industrial techniques. But the same was not true for Turkey's Kurdish region which remained at a subsistence level throughout the 1920s while the people continued to labor under the local landlords.

In what is known by the Kurds as Northern Kurdistan, located in east and south-east Turkey, agriculture is still today the principal economic activity. And while Kurdish peasants in the mountains now usually own the land they farm, this has not brought prosperity because the plots of land were ridiculously small and unable to support a family. Until as recently as the 1960s, Kurdish peasants were still paying the landowner a fixed proportion of the crop, which could vary from ten to eighty percent.

Many sharecropping agreements ended with the gradual introduction of mechanization to agricultural production that began in the 1950s. A surplus labor force began to appear, as landlords freed peasants they could no longer usefully employ. Former sharecroppers in Turkey's Kurdish region were now increasingly transformed into seasonal agricultural workers earning even less money, or were forced to migrate either to Kurdish cities or to western Turkey or out of the country altogether.[209]

The Kurds have always been a proud people and did not take this usurpation of their land and labor lying down. They had been conducting a long-running separatist revolution in Turkey beginning as are far back as the 1920s and 1930s. These insurrections were forcefully put down by the new Turkish government, who then declared Turkish Kurdistan to be a closed military area from which foreigners were banned. The ban lasted from 1925 to 1965. The use of Kurdish language, dress, and folklore were banned and the words *Kurds* and *Kurdistan* were erased from dictionaries and history books. The Kurds were only referred to as *Mountain Turks*.[210]

During the 1950s and 1960s, the Turkish government put a dominant emphasis on the development of private enterprise. This effort was funded by government and international capitalist investments with loans to both industry and agriculture. Politically, this signaled a clash between the old military-bureaucracy alliance that had founded Turkey and the domestic and foreign capitalists who were reacting to the statism of the old regime.

As soon as the capitalists gained power, however, they instituted their own version of complete state control. State agencies were privatized and parceled out.[211]

These new agencies failed to make a profit, however, and this only worsened the domestic capitalists' debt to the foreign capitalists and the World Bank. The government's initial response was deficit financing and more external borrowing.

As a result of these machinations, working people and poor farmers throughout Turkey suffered, but the Kurdish area was particularly affected, because the average Kurdish household only earned seventy-five percent of the average Turkish household. From here, things only got worse for the Kurds.

During the 1970s, in less than a decade, Turkey's foreign debt grew almost five-fold, going from $3.3 billion in 1973 to $15.3 billion by 1980. The result was a severe economic and political crisis which resulted in a military take-over and a World Bank structural adjustment program that demanded a shift to export-led growth similar to what we just witnessed in our look at the Sudan.

In 1978, a group of Kurdish students, under the leadership of Abdullah Ocalan, formed the Kurdistan Workers' Party (PKK). Since 1984, the PKK has been involved in an armed struggle with the Turkish state to achieve an independent Kurdistan. In the last few years, it has modified its goal to equal rights and Kurdish autonomy in Turkey.

The group's initial ideology was a fusion of revolutionary socialism and Kurdish nationalism. Its stated reasons for insurrection were the oppression of Kurds in Turkey and capitalist exploitation.

Following the military coup of 1980, the Turkish military officially prohibited the Kurdish language in public and private life. Many who spoke, published, or sang in Kurdish were arrested and imprisoned. The PKK began to arm themselves in an effort to establish the basic human and civil rights of the Kurdish people that had for so long been denied by the Turkish government and the international capitalists.

Abdullah Öcalan was captured and imprisonment in 1999. This event led him to move away from Marxist–Leninism and adopt a new political platform he calls "democratic confederalism." This new idea resulted from his reading of Western social theorists such as Murray

Bookchin, Immanuel Wallerstein, Fernand Braudel, and Friedrich Nietzsche.[212]

Democratic confederalism, much like the progressive socialism advocated in this book, is a "system of popularly elected administrative councils, allowing local communities to exercise autonomous control over their assets, while linking to other communities via a network of confederal councils."[213] Decisions are made by communes in each neighborhood, village, or city. Everyone is welcome to partake in the communal councils, but political participation is not mandated. There is no private property, but rather "ownership by use, which grants individuals usage rights to the buildings, land, and infrastructure, but not the right to sell and buy on the market or convert them into private enterprises." The economy is in the hands of the communal councils, and is thus (in the words of Bookchin) 'neither collectivised, nor privatised; it is common.'[214]

In 2005, Ocalan announced his "Declaration of Democratic Confederalism in Kurdistan," in which he advocated for a Kurdish implementation of Bookchin's *The Ecology of Freedom* via the creation of municipal assemblies in a democratic confederation of Kurdish communities across the state borders of Syria, Iran, Iraq, and Turkey. He called for agreement to the shared values of environmental defense, self-defense, gender equality, and tolerance for the coexistence of many religions, political parties, and forms of cultural expression.

In May 2007, former members of the PKK helped form the Kurdistan Communities Union (KCK), an umbrella organization of Kurds in Turkey, Iran, Iraq, and Syria.

In this same year, the President of Turkey, Tayyip Erdogan, was in the middle of his term as Prime Minister. Erdogan was the co-founder of the Islamic-rooted Justice and Development Party (AKP).

In June 2011, the AKP won by a wide margin in the parliamentary elections, securing a third term for Erdogan as Prime Minister.

In May 2013, anti-government demonstrations began to target Erdogan's policies, which included the Islamization of Turkey and the imposition of religious intolerance. A large demonstration was held in Gezi Park, the central park of the capital city of Ankara. This demonstration was triggered by Erdogan's plans to cut down the trees in the park and turn the Park into a mall. He also wanted to recreate Ottoman-era barracks. After this incident, protests spread to dozens of cities and the issues broadened to include Erdogan's autocratic style of government.

The protests remained peaceful, but thousands were injured when Erdogan's police descended on the demonstrators. In a scene similar to the demonstration held in Maidan Square in the Ukraine, the people, who came together to fight for their human and civil rights, were faced with violent government oppression. After five people had died in the police assault, Erdogan went public to describe the protesters as trouble-makers and said the government had reached the "end of its patience." Despite domestic resistance and international criticism of his "harsh measures" against peaceful protesters, Erdogan instructed his police that "we cannot allow lawbreakers to hang around freely in this square . . . We will clean the square."

Despite his warnings, new protesters of all ages continued to arrive in the park's camp, insisting they would not be frightened away. Among them was Kerim Ozken, sixty-three, a retired bank worker and writer. "I think the police might attack again tonight," he said, reflecting the fears of many in the camp, which was surrounded in places by makeshift barricades. "Tayyip [Erdogan] said it will be over tonight. He thinks it is a war. It is idiotic. Really idiotic. He thinks he can change people's minds by force."[215]

Over the next few years, Erdogan remained at odds with the Turkish people and with the PKK Kurdish militia who fought back against assaults by Turkish security forces.

In July 2015, the Turkish government conducted air strikes against the PKK's Mount Qandil headquarters inside Syria. This was at the same time that the Kurds were in a life and death struggle with ISIS in that country.

The air strike against the Kurds effectively ended a two-year truce between the Erdogan government and the PKK. After the air attack on their headquarters, the PKK issued a statement saying that the cease-fire is now off. "This truce has no meaning anymore," it said.

But instead of going on the offensive, the PKK organized a peace demonstration. In October, 2015 their peace rally, near the main train station in Ankara, was rocked by two explosions. The attack occurred three weeks before the national elections. More than ninety people were killed and two hundred and fifty wounded.[216]

Turkish government officials rushed to put the blame for the bombing on the rally organizers. Likewise, ISIS, with whom Turkey was accused of collaboration in the international media,[217] also pointed their fingers at the Kurds. While the Turkish government, ISIS, and the foreign capitalist establishment rushed to blame the Kurds for the attack, none of it made sense.

First, the peace rally was in support of the Kurds and the Kurdish organizations' desire for peace. HDP, a Kurdish political organization that had helped to organize the rally said that the bombs went off before the peace march had even begun. The Kurds had no reason to bomb their own peace rally.

On the other hand, the government had many reasons to do so and circumstantial evidence is compelling. The government, for example, immediately banned media from covering the attack, citing "security reasons." Social media such as Twitter was also closed. In this way, the government cut off any investigation by the media and the public to know the truth behind the bombings.

Witnesses at the scene said ambulances could not immediately reach the square and that the police obstructed the quick evacuation of the wounded from the square.

The HDP said in a statement that it had specifically been targeted by the bombs. Several HDP members and parliamentary candidates were among the victims of the attack. Their statement is given weight by the fact that in 2014, a rally for the HDP was also bombed ahead of the general election in that year.

Selahattin Demirtaş, co-chair of the HDP party, said: "This attack is not targeting our state and national unity, it is perpetrated by the state against the people. We are witnessing a massacre here. A cruel and barbarian attack was carried out. The death toll is high." Demirtas added that he did not expect those responsible for the bombings would be brought to justice.

Again, the Kurds showed restraint in the face of being attacked and killed. Immediately after the bomb attacks, the PKK declared a unilateral ceasefire and called on its fighters to halt all guerrilla attacks in Turkey.

At the same time, the US and Germany rushed to judgment in response to the Turkish government's proclamation that the bombs were a terrorist attack. President Barack Obama phoned Erdogan to say that the US would "stand with Turkey and its people in the fight against terrorism and other security challenges in the region." "So keen was the Obama administration to secure Turkish cooperation against the Islamic States that it would be unlikely to object to Turkey also taking on Kurdish militants," said Soner Cagaptay of the Washington Institute for Near East Policy.[218]

The German chancellor, Angela Merkel, expressed "sadness and dismay" over the attacks, while Russia's Vladimir Putin also offered condolences to Erdogan.[219]

Since then, Erdogan's administration has kept up a steady attack against the Kurds both within Turkey and in surrounding countries.

The former co-leaders of the HDP have both been jailed since 2016 on terrorism charges, with several other prominent party members accused of supporting terrorism over what the government says are links to the PKK.

Beginning in March 2019, Erdogan has systematically been replacing elected Kurdish mayors with loyalist trustees in more than half of the roughly sixty-five municipalities won by the HDP.

In June, 2020, the government detained four more mayors in Kurdish-majority areas as part of what it called terrorism related investigations.[220]

On May 28, 2019, The Turkish Army and Intelligent Secret Service conducted a joint operation, called Claw 1, on the Mount Qandil headquarters of the PKK in Syria, killing Diyar Garip Muhammad, a member of the group's presidential and executive council.[221]

Two months later, in July 2019, in the second phase of the attack against the Kurds, called Claw-2, Turkish fighter jets again struck Mount Qandil and killed at least three PKK fighters. In four months time, the Turks had killed a total of seventy-one PKK members since their assault against the Kurds in Syria began in May.[222]

After the second attack in July, the Kurds responded by killing a Turkish diplomat who was eating in a Turkish restaurant inside the city of Erbil, the capital city of Iraqi Kurdistan. Erdogan condemned the "hateful" attack.

Erdogan has been able to keep the US and European governments on his side by cutting deals with them in the international arena. For example, in December, 2019, he dropped his objection to a NATO plan to bolster the defense of Poland and Baltic states that are neighbors of Russia, but warned allies that he expected support in his fight against terrorism.[223]

Even so, Erdogan's authoritarian tactics have not gone without notice. Calling the Kurds terrorists is one thing, but calling journalists terrorists is another. The European Court of Human Rights has ruled against Turkey more than any other country in the world, even though its admonitions are routinely ignored. Surprisingly, the Turkish Supreme Court also ruled against Erdogan's two-year ban of Wikipedia in December, 2019, claiming that it was a violation of freedom of expression.[224] Over the last four years, Turkey has been experiencing a deepening human rights crisis with a dramatic erosion of its rule of law and democracy framework.[225]

Western Kurdistan (Kurds in Syria)

The Syrian Civil War, which began in March, 2011 as part of the wider wave of the Arab Spring protests, grew out of discontent with the Syrian Shia government of Bashar al-Assad. It quickly escalated to an armed conflict after protests calling for Assad's removal were violently suppressed by government troops.

The on-going war is being fought by the Syrian government's armed forces (SAF) and its international allies, including Turkey, Iran, and Russia against a loose alliance of mostly Sunni opposition rebel groups (including the Free Syrian Army, which consists of officers of the Syrian army who broke with Assad), Salafi jihadist groups like al-Qaeda, the Islamic State of Iraq and the Levant (ISIS), with a number of foreign nations, including the United States, supporting different factions of the rebels. The Kurds also joined the forces fighting against the Assad government as part of the mixed Kurdish-Arab Syrian Democratic Forces (SDF).

In some ways, the civil war in Syria has turned into a proxy war between the US and Russia. Russia has been conducting airstrikes and other military operations against rebel units since September 2015. The US-led international coalition, established in 2014 to counter ISIS, has conducted airstrikes primarily against ISIS, but also against government and pro-government targets. They have also deployed Special Forces and artillery units to engage ISIS on the ground.

The war in the north of Syria, even though it has put the Kurds under direct attack by ISIS, has benefitted the Kurds to some extent because it has provided the opportunity to establish their own autonomous region.

Since 2015, the US has supported the Syrian Kurd's region called Rojava materially, financially, and logistically.

Turkey has also been directly involved in operations against the Syrian government since August 2016, not only participating in airstrikes against ISIS, alongside the US-led coalition, but also actively occupying large swaths of northwestern Syria while engaging in ground combat against the Kurds, ISIS, and the Syrian government.

Additionally, even though officially remaining neutral, Israel has also conducted airstrikes against Hezbollah and Iranian forces, whose presence in southwestern Syria they view as a threat.

International organizations have accused virtually all sides involved, including the Syrian government, ISIS, opposition rebel groups, Russia, and the US-led coalition of severe human rights violations and massacres. The conflict has caused a major refugee crisis with millions of people fleeing the violence. Over the course of the war, a number of peace initiatives have been launched, but fighting continues. [226]

The war was greatly exacerbated by the most intense drought ever recorded in Syrian history. It lasted from 2006 to 2011. The drought, which is linked to man-made global warming, resulted in widespread crop failure, an increase in food prices, and a mass migration of farming families to the big cities. This migration strained the cities' infrastructures that were already burdened by the influx of some 1.5 million refugees from the Iraq War that occurred as a result of the US invasion of that country. An adequate water supply continues to be an issue in the ongoing civil war and the water supply is frequently the target of military assaults.[227]

In 2012, Assad withdrew government troops from northern Syria to fortify his southern defenses, leaving the Kurds to face the invasion of ISIS alone. In this fight, the Kurd PKK militia came from Turkey to support the Syrian Kurds, who were organized as the Democratic Union Party (PYD), which was an offshoot of the PKK in Syria.

In 2013, the PYD clashed with the various Islamist insurgent groups in Northern Syria and occupied a large swath of territory in the northern

Kurdish-majority provinces.[228] In the occupied territory, the Kurds created a semi-autonomous region in 2016 that they call Rojava.

In September 2014, PKK leader Abdullah Ocalan called for the Kurds to start an "all-out resistance" in the fight against the Islamic State, but later that month, the Kurdish-controlled town of Kobani was besieged and eventually captured by ISIS. This resulted in the exodus of a hundred and thirty thousand Syrian Kurds to Turkey. The ensuing battle for Kobani resulted in more than sixteen hundred deaths, but the Kurdish-led SDF forces eventually regained control of the city in January 2015. The SDF also liberated the strategic Syrian city of Manbij from the Islamic State in August 2016. In this battle, the Kurdish People's Protection Units (YPG) also clashed with Turkish-backed rebels attempting to gain control of the city.[229]

After the Kurds and their allies (SDF) consolidated control over the territory that they had captured from the Islamic State in northern Syria, the government of Turkey and Turkish-backed Syrian militias, including the Free Syrian Army (FSA), moved to recapture Kurdish-held cities and expel the Kurds. Turkish troops and the FSA launched an assault on the Kurdish city of Afrin in January 2018, eventually capturing the city in March 2018. Turkey continues to threaten assaults on other Kurdish-held areas inside Syria, including Manbij, and despite sharing a common enemy, many of Turkey's air strikes have targeted Kurdish fighters rather than Islamic State militants.[230]

At the same time that the Kurds are engaged in a life and death struggle with the combined forces of Turkey, ISIS, Turkish-backed Syrian militias, and Assad's government, they are arguably building the most advanced democratic state in the world. Rojava has become an example for the world's people of self-reliance and democratic process. It promotes a decentralized economy, gender equality, environmental sustainability, and pluralistic tolerance for religious, cultural, and political diversity. These values are reflected in Rojava's constitution, society, and politics. As such, Rojava serves as a model for any post-revolutionary democratic struggle.[231]

The people's freedom did not come easily. Under Syrian rule, the Kurds were subjected to enforced Arabization. They received little investment or development from the central government and there were laws that discriminated against Kurds owning property, driving cars, working in certain professions, and forming political parties. Education that used the Kurdish language was forbidden and hospitals lacked equipment for advanced treatment and instead patients had to be transferred outside the region.

In many instances, the Syrian government arbitrarily deprived ethnic Kurdish citizens of their citizenship. Over 120,000 ethnic Kurdish citizens saw their citizenship arbitrarily taken away and became stateless.

In 1973, the Syrian authorities confiscated two hundred and ninety square miles of fertile agricultural land that was owned and cultivated by tens of thousands of Kurdish farmers, and gave it to Arab families brought in from other provinces. In 2007, another two hundred and thirty square miles of farmland was granted to Arab families, while tens of thousands Kurdish inhabitants of the villages concerned were evicted. These and other expropriations were part of the so-called "Arab Belt initiative" which aimed to change the "demographic fabric" of the resource-rich region.[232]

In his report to the twelfth session of the UN Human Rights Council, entitled *Persecution and Discrimination against Kurdish Citizens in Syria*, the United Nations High Commissioner for Human Rights held that, "Successive Syrian governments continued to adopt a policy of ethnic discrimination and national persecution against Kurds, completely depriving them of their national, democratic, and human rights—an integral part of human existence. The government imposed ethnically-based programs, regulations, and exclusionary measures on various aspects of Kurds' lives—political, economic, social, and cultural."[233]

The de facto autonomous region of the Kurds in northern Syria is called the Autonomous Administration of North and East Syria (NES) or Rojava. It was officially declared in 2016. The legislature of Rojava is the Syrian Democratic Council of which the Movement for a Democratic Society is the ruling coalition. The leading party in the coalition is the Democratic Union Party (PYD) mentioned above.

Rojava itself is divided into several self-governing regions. Afrin, Jazira, and Kobanî cantons were first established in January 2014 and were envisioned as sub-divisions for a "future decentralized system of federal governance in Syria." Article 7 of the Constitution of the de facto autonomous region states:

> All cities, towns and villages in Syria which accede to this Charter may form Cantons falling within Autonomous Regions.[234]

The cantons of Afrin, Jazira and Kobani were later reorganized into three regions with subordinate cantons, areas, districts, and communes.

In September 2018, the regions of Raqqa, Tabqa, Manbij, and Deir ez-Zor were incorporated into the de facto autonomous region as well. Elections for the local councils of the Jazira Region, Kobani Region and Afrin Region were held in December 2017.

While the region is not officially recognized as autonomous by the government of Syria or any international state or organization, the supporters of the region argue that it is an officially secular political entity with direct democratic ambitions that, as stated above, promotes decentralization, gender equality, environmental sustainability and pluralistic tolerance for religious, cultural, and political diversity, and that these values are mirrored in its constitution, society, and politics. As such, it desires to be a model for a federalized Syria as a whole and does not seek outright independence.

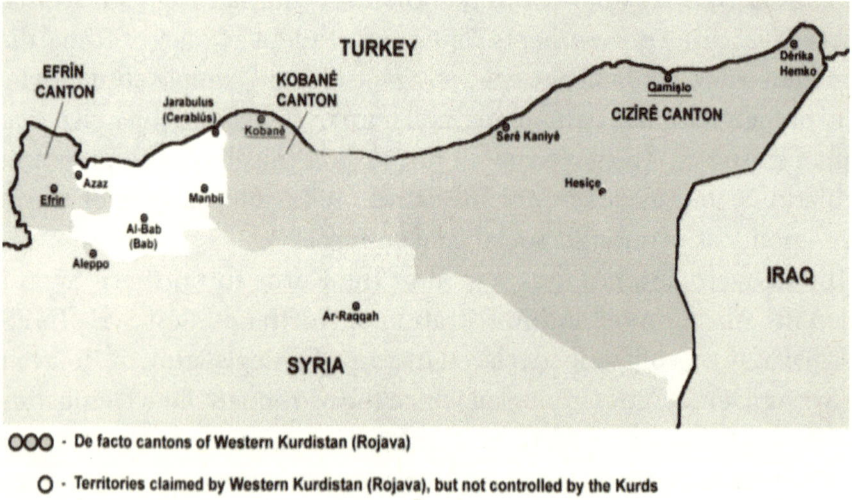

The Decentralized Economy of Rojava,

While the political and military considerations have always taken precedence in Rojava's revolution, the people cannot survive without their basic needs being met. As such, the creation of a local economy has provided the mechanism for uniting the Kurds in Syria and creating their new lifestyle and culture.

While the basic principle of the economic policies of the Assad regime in Rojava was to keep the people poor and deprived in order to maintain their dependence, the Jazira and Kobanî cantons served as the breadbasket of Syria. Before the revolution, forty percent of the wheat consumed in the country came from Rojava, and agriculture here is still people's primary source of income.

Sixty percent of the petrol used in Syria also came from the Jazira region. Nonetheless, Rojava is a region that was left impoverished despite its natural riches.

Today, the Economic Academy of Rojava is the central institution for economic life in the autonomous enclave. The board members of the academy describe the politics of the Assad regime this way: "The Syrian regime saw the resources of Rojava as its warehouse. Wheat was cultivated here and purchased by the state, which then had it processed in a different region and only then sold to the people. No factories or workshops that would enable the processing of the agricultural goods that grow in the Jazira Canton were allowed."

This, then, is the basis of poverty and deprivation on which the Rojava Revolution sought to establish an economy. Of course, it is also necessary to consider the war that has been on-going since the first day of the revolution. Seventy percent of the Kurd's budget has to be reserved for defense.

A third debilitating factor is the embargo imposed on the Kurds by the governments of Turkey, Iraq, and Syria. The border is closed between Rojava and Turkey.

Therefore, the economy of Rojava is geared towards the local market and for providing for the poorest members. Its basic principle is that everyone participate in production. In the words of a minister of economics: "If a single loaf of bread is manufactured in Rojava, everyone will have contributed to it."

This model is defined as a *social economy*.

The communes of Rojava are the building blocks of its political structure. Communes are basically neighborhood groups who organize among themselves to meet their own needs and contribute to the well-being of the entire region. They have regular weekly meetings to discuss the problems they face. Each commune also has a representative in the Legislature and in the neighborhood, village, or town where they are based.

Below is the definition of the commune from the Movement for a Democracy Society's (Tev-Dam) manifesto:

> Communes are the smallest cells and the most activist in society. They are formed practically in society and there is freedom of women and ecology and the adoption of direct democracy.
>
> The Communes form on the principle of direct participation of people in the villages, on the street and the neighborhoods and the towns. These are the places that people willingly organize themselves with their opinion, create their free will and initiate their activities in whole residential areas and open the door for discussion about all the issues and their solution.
>
> Communes work on developing and promoting the committees. They talk and search for solutions of social matters, political, education, securities and self-defending & self-protection from its own power, not from the state. Communes create their own power through building organization in the form of agricultural communes in the villages and also communes, cooperatives and associations in the neighborhoods.
>
> Forming the Communes on the street, villages and towns with participation of all the residents. Communes have a meeting every week. In the meeting, Communes make all its decisions openly by people who are in the Commune and are older than 16 years-old.[235]

In other words, the communes are basically neighborhood groups that are building a lifestyle in which everybody participates, not only in the roles of the economy, but also in the roles of governance.

The communes, or community governing groups, are responsible for setting up workers cooperatives that serve to meet their basic needs.

The question of how and where the cooperatives are founded depends on the needs of the families that live in the communes.

The local Economic Committee, which is attached to the Movement for a Democratic Society, prepares a plan. Then the concept is presented to the communes near the location of the proposed cooperative, or to a council of communes. The communes or the councils have the task of finding poor families and needful families to take part in the cooperatives. The commune determines the employees of the cooperative. In this way, the cooperatives are connected to the commune. They are locally created, autonomous, and democratically-run enterprises that aim to meet the people's basic needs while providing the community with affordable alternatives.

After a cooperative is established by the local people, the Economic Academy of Rojava gives half of the needed money and the seeds to the cooperative as a loan. Each cooperative is thus founded with the means of the revolution. All these steps toward the foundation of a cooperative are done by the Economic Academy and the Autonomous Administrations in collaboration. The cooperatives keep seventy to eighty percent of the income when they sell their products. Only twenty to thirty percent is given to the Economic Academy of Rojava.

In the Jazira Canton, in 2018, there were eighty-seven cooperatives with thirty thousand members. The cooperatives provide work opportunities to all the participants, some of whom work in turns by taking three-month shifts. The cooperatives have been playing a big role in providing supplies for the local markets, stabilizing prices, and overcoming monopoly. They are the reason that local people meet their basic needs of food, shelter, clothing, health care, and education.

As such, the cooperatives are diverse and are involved in livestock, agriculture, industry, and commerce.

Agricultural cooperatives include those that are specialized in planting vegetables and grains, greenhouses, and animal feed. They have 10,593 participants, and the price of shares cost between 10 and 70 thousand Syrian Pounds ($20 to $136). The agricultural cooperatives have been successful in Jazira and have caused the price of potatoes, onions, tomatoes, and cucumbers to drop. In 2018, the price for a kilo of potatoes was 225 Syrian Pounds (43 cents), then it dropped to 45 SYP (9 cents). The price of onions was 70 Syrian Pounds (14 cents); they dropped to 29 SYP (5 1/2 cents).

Livestock cooperatives specialize in breeding cattle and they produce dairy products. There are 6,796 participants in total, and the price for a share varies between 50 to 100 thousand SYP ($97 to $194).

There are many women working in dairy factories, and they make yogurt, milk, and cheese. A dairy factory produces a ton of yogurt, in 470 containers, and distributes it in local markets. Fifty-three gallons of milk are sold every day from the Shilêr women's cooperative, and the market price for them is 225 SYP (44 cents) a container, instead of 250 SYP (48 cents).

Commercial cooperatives provide consumer goods, buildings, building materials, and food products. They have 11,574 participants, and the price for a share ranges between 10 to 20 thousand SYP ($20 to $40).

Industrial cooperatives include factories that manufacture tahini and halva, water bottling, sewing workshops, ovens, electricity generators, fruit drying factories, and handmade food. There are more than a thousand participants, and the price for a share ranges between 25 to 300 thousand SYP (5 cents to $582).

The special thing about electricity generators is that after paying the entire price for the electricity generator, the participants can own it. Seventy-five percent of the profit from sewing workshops are distributed to the workers, and the remainder goes to the cooperative. Cooperative workers receive monthly wages that vary from 50,000 to 60,000 SYP ($100 to $116).

The local cooperatives have a very positive impact on stabilizing market prices, especially the food product cooperatives. While they are based on profit, they aim to prevent monopoly and the raising of prices to the local people.[236]

The cooperatives are built on land that was previously nationalized by the Syrian regime, but which has been reclaimed since the start of the revolution. This land makes up eighty percent of the autonomous territory.

The first step in creating the decentralized, self-reliant economy was to establish the Economic Academy of Rojava. Its outposts were created in every town with the participation of engineers and economists with relevant expertise, along with other volunteers. Seed companies were set up to improve the agrarian economy and enable the villagers to cultivate the soil.

The administrators at the Academy describe the model they want to create: "We reject the capitalist economy, but we are not adopting the economic model of real socialism either. Ours is a communal economy based on cooperatives. We do not block private initiatives but, we are following a policy that prevents the formation of monopolies."

According to the Rojava Social Contract:[237] "Everyone has the right to the use and enjoyment of his private property. No one shall be deprived of his property except upon payment of just compensation, for reasons of public utility or social interest, and in the cases and according to the forms established by law." Article 42 states:

> The economic system in the provinces shall be directed at providing general welfare and in particular granting funding to science and technology. It shall be aimed at guaranteeing the daily needs of people and to ensure a dignified life. Monopoly is prohibited by law. Labor rights and sustainable development are guaranteed.[238]

Basically, the means of production in Rojava, its factories, land, forests, water, and underground and above ground resources, are the property of the Democratic Autonomous Administration.

While the Rojava Revolution is not opposed to private property, the communes and cooperatives, which are the political, social, and economic organization of the people, and the main power of the people's revolution, are given precedence in the use of the resource base of Rojava.

The more the people develop their own decentralized economy, the more they are able to oppose the monopolization of resources in the name of private property and one of the first measures against monopolization have been the cooperatives.

While the Kurds struggle to develop a local economy dedicated to meeting their basic needs that is in harmony with the natural environment, the Turkish government continues to threaten their lives. Aside from its air raids, the Turkish government has also been expanding dam projects along rivers running south into Rojava, turning rivers into dust and making it harder for people to grow food and be self-sustainable.

Turkey has also built massive walls along its border, carving miles of scars through the region's ancient forests.[239]

On October 22, 2019, Recep Tayyip Erdogan and Vladimir Putin signed a 10-point agreement ratifying Turkey's presence in northeast Syria that forced the Kurds self-defense units to withdraw from Turkish-held territory. Since then, Turkey has been accused of driving out the Kurdish population and replacing them with two million Sunni Arabs who had fled from elsewhere in Syria to Turkey. Before Turkey invaded in 2018, eighty-five percentof Afrin's population was Kurdish. Now it is twenty percent.[240]

While Turkey invaded from the north, Assad loyalists and the Russians rushed in from the south, promising protection. But they have demanded a high price: They want the Kurds and their allies to pledge allegiance to Assad's army. We do not yet know whether the Syrian Kurds as a whole will accept that.

Regardless of the macro-politics playing out all around them, the project for Kurdish autonomy in Syria has left an impression that will not soon go away. Because the Kurds had devolved power to the local level and did not try to build a nation-state, the revolution has planted deep roots into both Kurdish and non-Kurdish communities in Northern Syria.

Under the prevailing circumstances of war and uncertainty the democratic autonomy project of Rojava has offered a successful strategy "for coping with security issues, economic chaos, and internal disunity and dispersal." The Kurds' system of decentralized self-rule has survived so far because it works.[241] Particularly, the Kurds have built a system of local governance that is based on the principles of environmental sustainability, gender equality, and pluralistic tolerance. As such Rojava stands as a beacon of light for all of humanity.

Environmental Sustainability

The local economy, insofar as it is based on food and farm products, is dedicated to environmental sustainability. The local communes and cooperatives have made protection of the ecology part and parcel of their work. Their campaign addresses issues related to food cultivation, reforestation, renewable forms of electricity, limiting fossil fuel usage, preserving the water supply, and developing recycling and other waste management solutions.

Over the last few decades, many wildlife and plant species in the region have been wiped out by deforestation and monoculture practices. Many wolves, foxes, wild pigs, and birds found refuge in the Hayaka forest, one of the last forests near Derik in Jazira. In 2014, the democratic self-administration of Rojava declared the Hayaka forest a natural reserve in order to preserve and restore the biodiversity of the region.

Developing urban agriculture is one tier in the ecological campaign to expand food security and autonomy in Rojava. Monocultures, the act of growing a single crop, had long been the policy supported by the Assad government, which has controlled Syria since 1971. Decades of monoculture has not only resulted in the loss of ecological diversity across the region, but the use of artificially supplied water and chemical fertilizers have degraded the soil and made cultivation of food impossible in some areas.

In Afrin, for example, the Syrian regime cut down ancient forests to grow a monoculture of olive trees that degraded the topsoil.

An internationalist commune, made up of volunteers from around the world, helps raise money for projects to realize the vision of the ecological campaign, including, for example, a tree nursery that grows fruit trees for distribution and also offers training to farmers on the different aspects of fruit tree production. The Internationalist Commune also plans to create rooftop gardens and to plant vegetables on undeveloped land within cities, which will further decentralize agriculture across Rojava. The return to traditional organic fertilizer-based agriculture is also a solution being put into practice by the Internationalist Commune.[242]

Gender Equality

The Revolution in Rojava has been attempting to create gender equality as a fundamental tenet of its work. This goal expresses itself in the equal opportunities of men and women in the political process where women play equal roles in the Democratic Self-Administration of Rojava (DSA), Rojava's legislature, and the Democratic Union Party (PYD).

Women are heavily involved in every aspect of the government committees, groups, and communes. In comparative terms, women comprise only nineteen percent of the US government and twenty-two percent of the United Kingdom government, but they comprise thirty percent of the Kurdistan Regional Government. Both the PYD in Syria and the PKK in Turkey require co-chairs of one women and one man. This is part of an intricate quota system to encourage greater gender equality in Kurdish politics.[243]

In the military, the head of any military section must include women. In addition to this, the women have their own armed forces. Kurdish women lead the world in female fighters. Whereas the Chinese military is made up of 7.5% women, the Russian military, 10%, the American military, 14% and the Israeli military 33%, the Kurdish military is comprised of 40% women.

In the economy, aside from the general cooperatives, women's cooperatives are also affiliated with women's communes. This ensures that women are integrated into every aspect of social production.

This is not to say that Kurdish society is not patriarchal, and while Kurdish women are increasingly being educated and represented in the workplace and office, they still have a long way to go to achieve equality.

Kurdish women are still victim to female genital mutilation and honor killings, and female suicide rates often go uninvestigated, yet the psychology of Kurdish society regarding women is already radically changing as women play a larger role in society.

Currently, recruits to Rojava's six thousand-strong police force receive their weapons only after two weeks of feminist instruction.

In the war against ISIS, Kurdish women soldiers fueled a glamour about the Kurdish independence movement and its gender equality. Often times pictures and videos of female fighters were used in the Kurdish militia groups' internet activism, and arguably helped recruit foreigners to join the fight against ISIS. The women present themselves as tough and scrappy and willing to do anything to protect the people. If women today stopped working or withdrew from the above groups, Kurdish society would well collapse.[244]

Pluralistic Tolerance

The Kurds have a well-deserved reputation for being the most tolerant people in the Middle East. While Islam remains a "background presence in the law and in the conservative culture of wider Kurdish society," the Kurds respect the separation of religion and state. They view the intrusion of literalist and ultra-conservative versions of Islam as particularly toxic.

The Kurds are also tolerant of other ethnic groups. In a recent Christmas address, the President of Iraqi Kurdistan strongly stated: "What differentiates [us] from most of the countries around us is religious and ethnic tolerance. Accepting and defending each other's rights strengthens the principle of humanity in this country, particularly in difficult times."

In Kurdistan, the main political parties are secular, with seats set aside for Turkmen, Assyrians, and Armenians. There is also a minimum quota of thirty percent of the seats for women. A recent report by the Foreign Affairs Committee of the UK Parliament described Kurdistan as 'a beacon of tolerance and moderation in a wider region where extremism and instability are on the rise.' It examined the Kurdistan Regional Government's (KRG) self-image as 'a haven of tolerance and moderation in the wider Middle East' and found it largely confirmed in the evidence received.

Examples of religious tolerance can be seen in their treatment of the Yezidis and Arab Christians. The Yezidis, who were condemned and subjected to a genocidal campaign by ISIS and whose women were used as sex slaves, are the oldest faith group in Kurdistan. In an effort to strengthen the resolve of the Yezidis, a group of Kurdish men and women converted to the Yezidi faith. This move was supported by the Kurds in general. Neither the Kurdish government nor the people condemned them, nor found them guilty of apostasy (heresy).

Arab Christians fleeing from the emergence of ISIS also sought sanctuary in Kurdistan and were warmly welcomed. A Kurdish Christian Deputy Prime Minister received a papal knighthood for his efforts.

Kurdistan once hosted many Jews, who formed seventeen percent of the population in Slemani.

When ISIS terrorists killed one hundred and twenty-eight French citizens in Paris in November 2015 as the city was enjoying a soccer match between France and Germany, the Kurdish leadership rallied in solidarity with the French people. The Foreign Minister of the Kurdistan Regional Government participated in the solidarity demonstration in Paris and condemned the "Obscurantist extremism" of ISIS.

The Kurds also express great tolerance toward foreign visitors. Kurdistan had already accepted a quarter of a million Syrian refugees, who have now been joined by well over a million internally displaced people living in makeshift camps, schools, and half-finished construction sites as well as mosques and churches.

Caring for these people is a priority for the Kurds who understand the plight of those who flee oppression.

They are indeed the harbingers of a nuclear revolution.

As we have seen in Rojava, the provision of full rights for Kurds as equal citizens and the establishment of a form of autonomous government need not be a threat to Syrian Arabs or to Syria's unity. Kurds themselves are insistent that democracy and liberalization must be created for all citizens in Syria and the world, not just for their own community.[245]

Lessons from Greater Kurdistan

Of all the revolutions we have reviewed, the Kurds provide the greatest example of fighting against multiple enemies at once. These include the governments of Iran, Iraq, Turkey, and Syria, as well as Russia and ISIS. They have also been betrayed by the United States, as well as by internal forces.

As such, the Kurds have had to stage revolutionary movements in multiple countries just to establish their basic civic and human rights as a people. Their efforts constitute the most advanced revolutions on the planet in relation to the goals of a nuclear revolution. As such, revolutionaries around the world would benefit by taking the Kurds struggle for democratic freedoms as an example of the work that needs to be done to create a nuclear revolution.

The most advanced stage of the Kurdistan revolution are the efforts to establish semi-autonomous Kurdish areas in Bakur Turkey and Rojava in Syria.

It is here that the Kurds have established the greatest concrete example of how to create a nuclear revolution against all odds. The basic building blocks of the Kurds participatory democracy are: (1) a universal world view; (2) an army for self-defense; (3) an autonomous basic needs economy based on workers cooperatives; (4) protection of the natural environment, (5) gender equality, (6) pluralistic tolerance, and (7) a dedication to serve the poorest first.

The first lesson they learned is that when a more powerful government is dedicated to the repression and oppression of your people, you must create an armed insurrection. This can be seen in the creation of the Kurdish Peshmerga in Iraq, the PJAK in Iran, the PKK in Turkey, and the YPG in Syria.

Armed insurrection by the Kurds in these countries proved critical for the continued survival of the Kurds as a people. This is a lesson that must be learned and factored in to any revolutionary struggle.

When Turkey was founded as a country, its constitution denied the existence of distinct cultural minorities. As a result, any expression by the Kurds of their ethnic identity was harshly repressed. For example, the use of the Kurdish language was made illegal. To this day, any talk of Kurdish nationalism is grounds for imprisonment.

The Turkish government has consistently thwarted attempts by the Kurds to organize politically. Kurdish political parties are banned and party members are harassed and imprisoned for "crimes of opinion."

Aside from being culturally and politically suppressed, the Kurds in Turkey are also economically oppressed. The Turkish government systematically withholds resources from the Kurdish region, turning it into a third world economy.

The repression and oppression on the part of the Turkish government led to the Kurds forming an armed separatist movement, the PKK, in 1984. It was the only force fighting for broader Kurdish cultural, political, and economic rights. The situation of the Kurds in Turkey is a clear example of the need for armed struggle when those in power will stop at nothing to deny a people human and civil rights.

Of course, the Turkish government responded with even greater repression and oppression of the Kurdish people. It deployed three hundred thousand troops to the Kurdish area and instituted a system of "village guards," by which it hoped to split the Kurds by paying and arming Kurds to keep the PKK revolutionaries out of their villages.

Kurdish villages that refused to participate in the guard system faced being torn down by the Turkish military, while those that capitulated faced reprisals by the PKK for being traitors. In this way, the Turkish government succeeded in splitting the Kurdish people.

So, the lesson to be learned here is that while armed separatism is required to face unremitting repression and oppression, the revolutionaries must expect a brutal backlash in the short term and do everything in their power to stick together. To date, the Turkey government has destroyed over three thousand Kurdish villages, creating more than two million refugees.[246]

Abdullah Öcalan, the founder of the PKK, was more than a warrior; he was also an intellectual visionary. While initially a Marxist-Leninist, Ocalan became familiar with the writings of Murray Bookchin while in the Turkish prison system wherein he had been incarcerated since 1999.

Bookchin was a New York socialist who argued that capitalism's fatal flaw is not the exploitation of the working class, but rather its conflict with the natural environment, which, if left unopposed, would lead to the dehumanization of people and the destruction of nature.

In his book, *The Next Revolution*, Bookchin argued that the most fundamental message of his vision, which he called *social ecology*, is that the very *idea* of dominating nature stems from the domination of human by human. For an ecological society to develop, first the inter-human domination must be eradicated.

According to Bookchin, "capitalism and its alter-ego, 'state socialism,' have brought all the historic problems of domination to a head," and the market economy, if it is not stopped, will succeed in destroying our natural environment as a result of its "grow or die" ideology.

Inspired by these ideas, Ocalan began to see the creation of an independent Kurdish state as just another continuation of the status quo. He argued that "the system of nation-states has become a serious barrier to the development of society and democracy and freedom since the end of the twentieth century." In his view, the only way out of the crises facing humanity was to establish a democratic confederal system "that will derive its strength directly from the people, and not from globalization based on nation-states." He argued that neither the capitalist system nor being pressured by imperialist forces will lead to democracy. It will only serve the interests of the ruling elite. "The task is to assist in developing a grassroots-based democracy . . . which takes into consideration the religious, ethnic, and class differences in society."[247]

The many letters and documents Ocalan sent from prison regarding these matters set the course for the socio-economy of the Kurds in Bakur, Turkey and Rojava, Syria. In these letters, Ocalan specifically argued for three strong points promoted by Bookchin. These are: (1) *Dual Power* that espouses the development of alternative institutions to the state in the form of popular assembles and neighborhood committees, which would in time "hollow out" state power from below and eventually make it superfluous; (2) *Confederalism*, which is a network of administrative councils whose members are elected from popular face-to-face democratic assemblies in the villages, towns, and neighborhoods of large cities; and (3) *Social Ecology*, which is the belief that human relations must be reordered in a post-capitalist world so that humanity can live in a protective balance with the natural world. In this regard, the emancipation of women is one of the most important aspects of social ecology because as long as men look to dominate women, they will treat the natural environment in the same way. They are just commodities to be exploited.

Thus, we see that along with the need for armed struggle, Ocalan also realized that the Kurdish people in Turkey had to create their own economy and protect the natural environment.

These same ideas were transferred to the Kurds in Syria. While it was the armed PKK that protected the Syrian Kurds from ISIS and from attacks by the Syrian and Turkish governments, it was the Mesopotamian Ecology Movement founded by the Turkish Kurds that transferred Ocalan's political and economic ideas to the Kurds in Rojava, Syria. It was here that the political and economic revolution among the Kurds took roots and advanced the farthest.

The Mesopotamian Ecological Movement at its inception in 2016 was composed of local political and environmental groups in Turkey who set themselves the following resolutions:

- To establish a strategic intellectual, organizational, and operational coordination with national and international ecology movements in order to enhance common discussions and actions against ecological destruction and exploitation;
- To struggle against the mental, physical, and ideological destruction of energy, water, forests, soil, cities, agriculture seeds, and technology; and based on the approved policies of the Mesopotamian Ecology Movement, to mobilize a struggle for the construction of a new life;
- To fight the system that demolishes urban settlements and burns forests in Kurdistan; to publicize the ecological devastation experienced in Kurdistan and to map the devastations occurring within the war;
- To plan actions, in coordination with other ecology movements, against the destruction of cities in Kurdistan; to ensure active participation in solidarity platforms that have been established in these cities;
- To continue struggles to preserve cultural and natural sites in Kurdistan that face extinction—such as Hasankeyf, Diyarbakır-Sur, the Munzur Valley, and "Gele Goderne"—due to the energy and security policies;
- To develop an ecological model suitable for Kurdistan;
- To build a greater and more regular presence in print and digital media and to establish ecology academies;
- To carry out legal struggles parallel to ongoing actions and campaigns; and
- To expand our own organizational structures throughout Kurdistan and the Middle East.[248]

They viewed their movement as a struggle for freedom that takes a stand against the system that jeopardizes nature, society, and the individual in the interests of capitalist profit and state hegemony.

In this way, the members of the Mesopotamian Ecological Movement, situated in the heartland of the ancient Garden of Eden that birthed Western Civilization, came to offer a second chance for human survival, and in so doing, helped build the socio-economy of Rojava in Syria.

Today, Article 9 of the Rojava Social Contract reads as follows: "Democratic, environmental, and societal life are the basis for building an ecological democratic society in order not to harm, abuse, and destroy nature."

In this Article the people of Rojava address head-on the two major challenges that address humanity as a whole. These are: (1) an unjust economic system that results in widespread poverty and war, and (2) a rapid state of decline in our planetary environment upon which all of us depend.

Capitalism can only be overcome through a participatory economic and ecological revolution that is global in scope. While this process will take time, Rojava has calculated this into its Social Contract. It allows for investment in private projects if they "take into account ecological balance" (Article 42). Likewise, the right to private ownership is guaranteed "unless it contradicts the common interest" (Article 43).

At this first stage, ecology is not seen in opposition to capitalism, but rather as a limit to capitalism destroying nature and human health.[249]

The Kurds in Bakur, Turkey and Rojava, Syria did not wait to replicate the model until a Greater Kurdistan could be created. Rather, they came to identify themselves with the greater humanity, while maintaining a feeling of pride among themselves for the universality of their vision.

They knew that their model needed to expand beyond the Kurds and they encouraged the exchange of ideas and activities between themselves and the people of other areas of Turkey and Syria as well as revolutionaries around the world.

These are vital lessons for all those who want to participate in a nuclear revolution. As such, it behooves the international revolutionary forces to support the current struggle in Rojava where the US Trump administration stabbed the Kurdish people in the back by withdrawing its support from them after they fought to contain ISIS for the US.

Even worse, the abandonment of the Kurds opened the door for Turkey's ethnic cleansing campaign against the Kurds and it has now marched its army into Rojava to kill the people and destroy their remarkable work.

As revolutionaries, we should support shutting down arms sales to the Turkish state, including those from Germany, England, and the United States. Secondly, we should staunchly oppose the economic blockade Turkey has imposed on Rojava. Thirdly, we should publish articles about the contribution of Rojava to the world's people's democratic struggle and demand that the mainstream press do the same. Fourthly, we should fight to get the US and Russia out of Syria and pressure the Assad regime to transition the country to Kurdish style confederalism. Finally, we should pressure the United Nations to oversee this transition as a means to bring lasting peace to the Syrian people and the people of the Middle East.[250]

The urgency here is that the foremost human struggle for local self-reliance, democracy, and environmental sustainability is on the brink of recolonization and extermination by oppressive governments that surround them.

Lessons Learned about Capitalism

The contemporary revolutions that we have analyzed have been piecemeal efforts based upon the geo-sentiment of nationalism or the socio-sentiments of religion, race, or culture.[251] While these revolutions have been defined by different events, different struggles, and different results, they have all been against the same enemy (the self-serving psychology of capitalism). This self-serving psychology dominates national leaders whether in the western capitalist countries, the Middle East capitalist countries, the African countries, or the Eastern communist countries of Russia and China. Therefore, the freedom-fighters in the struggles against these capitalists, anywhere on the planet, regardless of differences in political systems or cultures, are all heroes of humanity. They are our brothers and sisters in the fight for freedom and democracy. They continue to give their lives for the greater good.

By saying that the self-serving psychology of capitalism is the common enemy of humanity, we mean that the capitalists' single goal is to maximize personal or group gain at the expense of others. This is a sanitized way of saying that their goal is to expropriate the natural resources of other people and exploit their labor. Capitalism is not about political democracy when it does not serve the immediate economic interests of the capitalists. And it is never about economic democracy.

Currently the major contradiction within the international capitalist class is between the western capitalists and the communist capitalists of Russia and China. The people of the world are caught between these two forces in an attempt to gain their human and civil rights. When people in the West, as exemplified by our study of Cuba, Chile, and Nicaragua, want to realize their rights and freedoms, they must ultimately free themselves from the domination of the western capitalist class and their paid political and military underlings.

In their struggle against the western capitalists, particularly the US capitalists, the revolutionaries come to believe that people in the communist countries have a better life and they try to enlist the enemy of their enemy in their revolutionary struggle. This may have a short-term tactical advantage, but in the long run it is fruitless.

Just as the people under the thralldom of the western capitalist class look to the eastern communist-capitalists for support, the people in thralldom to the eastern capitalist Russia and China look to the western capitalists for support. The people under the oppression of either empire believe that the grass is greener in the other's system. In fact, they are the same exploitive system.

There is, however, a slight difference between the western and the eastern capitalists. The western capitalists have long been in power and it is their global economic system that the eastern capitalists are trying to break into. Russia and China were, in their revolutionary days, opposed to capitalism. They formed so-called communist governments that purported to function on behalf of the working people. Because they could not overtly steal the resources and labor of other countries as readily as the western capitalists do, they were forced to employ a warrior psychology to protect themselves and secure their self-reliance. As such, they attempted to gain wealth and power through military dominance. This is in keeping with the law of the social cycle.

In time, however, not having developed their merchant psychology adequately to run their national economies competitively with the West, they wound up allowing domestic capitalists to control the people's wealth and labor, not for the welfare of the people as a whole, but for the personal profit of the capitalists and self-serving party operatives.

In this way, they came to mirror the capitalist system of the west. In Russia, President Putin made a backroom deal with the domestic oligarchs to allow them to exploit Russia's resources and people in exchange for the oligarchs' support of his government. In this way, Russia entered a long stage of counter-revolution. So also, in China, the Communist Party reverted to capitalism under the Deng administration. While remaining members of the Communist Party, the leaders reduced the ability of the state to make economic decisions, but took this power of the state onto themselves in order to line their own pockets and those of other people imbued with the capitalist psychology in China.

While the newly adopted capitalist-inspired policies brought short term benefits to the people in Russia and China, these did not last long because the leaders of these countries were not inspired by the virtue of selfless service, but rather selfish intent at the expense of others. Here is another example of the warrior and capitalist elites collaborating to exploit the people under the name of Communism.

Capitalists everywhere seek to maximize profits. They accomplish this by expropriating the natural resources of other people, as well as by exploiting their labor. This sums up capitalism at its core. Capitalism is not about democracy. We have witnessed this in Cuba, Chile, Nicaragua, and in South Sudan. We have also witnessed this in Kurdistan, in the Soviet Bloc, and in Hong Kong. Capitalists go to absurd lengths to lie about capitalism's fundamental nature. This is true of capitalists around the world, whether in capitalist, socialist, or communist countries. This is why it is fruitless for revolutionaries in capitalist countries to depend upon communist countries for long-term support and why it is fruitless for revolutionaries in communist countries to depend upon capitalist countries for long-term support. The only solution is to fight against capitalist-controlled governments and corporations everywhere in the world, while simultaneously building local, cooperative, environmentally-sustainable, self-reliant economies that begin by meeting the local people's need for food, shelter, clothing, health care, education, and physical security.

In the US, capitalism has meant turning the oceans into dead zones; filling the atmosphere with carbon emissions and other toxins that will render the climate unfit for humans; pumping chemicals and waste into the soil, water, air, and food supply so that cancer becomes an epidemic; buying off elected officials and judges to serve the exclusive interests of capitalists; and privatizing social services, including health care, transportation, education, and public utilities.

If, for the capitalists, reducing the cost of labor means forcing workers to remain unorganized while abolishing work, health, and safety regulations, if it means moving industry overseas where foreign workers toil like nineteenth-century serfs, if it means suppressing wages at home to force an impoverished population into debt peonage, what does it matter to the capitalists; this is the way profits are made.[252]

It is not accidental that the United States now has the worst income inequality since the 1920s. The Russian and Chinese capitalists are not far behind, but they are playing catch up.

It was the economist Milton Friedman who legitimized, in public perception, the most selfish impulses of capitalism by convincing the world that the sole responsibility of the corporation was to generate profit for its owners.

The lopsided emphasis on maximizing profits for the few has been responsible for the majority of social, environmental, and political problems in contemporary society, including extreme economic inequality, distortion of human needs, environmental destruction and climate crises, corporate tax evasion, and, above all, the commandeering of political power by the corporate elite. Interestingly, it was Milton Friedman who said that Hong Kong was the world's most free market economy, worthy of global emulation. "If you want to see capitalism in action," he said, "go to Hong Kong."[253]

In their book, *Global Reach*, Richard Barnet and Ronald Muller paint a vivid picture of what life is like in a typical third world country after decades of American imperialism. All these countries share the same symptoms of underdevelopment to a lesser or greater degree.

The countries are a contradiction of rags and riches. One out of every ten thousand persons lives in a palace with high walls and gardens and a Cadillac in the driveway. A few blocks away hundreds are sleeping in the streets, which they share with beggars, chewing-gum hawkers, prostitutes, and shoeshine boys. Around the corner, tens of thousands are jammed in huts without electricity or plumbing. Outside the cities, most of the population scratches out a bare subsistence on small plots, many owned by the few who lived behind the high walls. Even where the soil is rich and the climate agreeable most people go to sleep hungry. The stock market is booming, but babies die and children with distended bellies and spindly legs are everywhere. There are luxurious restaurants and stinking open sewers. The capital boasts late-model computers and receives jumbo jets every day, but more than half of the people cannot read . . . .

Nationalist slogans are prominent, but the basic industries are in the hands of foreigners. The houses behind the walls are filled with imported cameras, TVs, tape recorders, and fine furniture from the United States or Europe, but the major family investment is likely to be a Swiss bank account. There appear to be three groups in the country distinguishable by what they consume. A tiny group live on a scale that would make a Rockefeller squirm. A second group, still relatively small in number, lives much like the affluent middle class in the United States—the same cars, the same Scotch, the same household appliances. The vast majority eat picturesque native foods like black beans, rice, and lentil soup—in small quantities. The first two groups are strong believers in individual development for themselves and their family, but they see no solution for the growing plight of the third group. So they fear them, and their walls grow higher. For the third group disease, filth, and sudden death are constant companions,

> but there is an air of resignation about them. Life has always been full of pain and uncertainty and it always will be. The only development they see is the same journey from cradle to an early grave that their fathers and their grandfathers took." [254]

It is argued by capitalist economists that third world countries lack the infrastructure to create more wealth—roads, communications systems, schools, technology, etc. In other words, they lack the "capital stock" to develop their economies. The traditional analysis is that the countries never produced the wealth necessary to generate the savings required to invest in their own economic growth. As we have already seen, however, this economic analysis is disingenuous because finance capital, under modern capitalism, has nothing to do with savings. It was accumulated by enslaving people and by stealing their lands and resources. Today the capitalists use this "capital stock" to control the world's people. The global banking system has the ability to create paper out of thin air—paper that is used to control people everywhere because it poses as a legitimate and legally binding media of exchange and store of value. Of course, this worthless paper is backed by the CIA, Special Forces, and, if necessary, the US army. As far as Africa, South America, and Central America are concerned, we must also remember that the Europeans and then Americans decimated the labor force of these continents by enslaving their populations.

If the majority of the countries in the world are poor, even though they are rich in natural resources, it is because the foreign corporations, who control the wealth in those countries, take immense profits from them instead of investing in the local people. The finance capital generated by the natural wealth of these "underdeveloped" countries is not used to develop local businesses, schools, and other structures required to meet the local people's basic needs. Rather it is siphoned off to the "developed" world, first as plunder, then in the more respectable form of dividends, interest on loans, stocks, technical fees, etc. In other words, the wealth is used to finance the luxurious amenities of the capitalists and to expand continued opportunities for profits. Today these profits are used for the most part to gamble on the stock market, invest in derivatives, wager on currency wars, and store wealth in unmarked bank accounts in foreign tax havens like Hong Kong. The capital that is left in the countries goes to the small class of locals who cooperate with the foreigners in the expropriation of the local people's wealth. Thus, over the years, the wealth of the "undeveloped" countries has been systematically depleted and the people impoverished. This process has gone on regardless of whether a Democrat or Republican administration was in power in the US government.

Revolutions occur when the people seek to replace the existing power structure. Counter-revolutions occur with the existing power structure acts to suppress the people's desire for change. Revolutions usually occur when a majority of the people require greater support to meet their basic needs and want greater freedoms in their lives. However, as we have seen in our study of how societies move, a successful revolution usually results in a new ruling class coming to power, which then, if not guided by a universal ideology and a service mentality, seeks to subject the people to their new rules and authority built on self-interest and limited group psychology. In time, this new class begins to suppress the very people whom it had fought to liberate. We have seen this clearly in the cases of Nicaragua and South Sudan. The Ortega revolutionaries were continually attacked by counter-revolutionary capitalist forces even after they had overthrown the old capitalist government. As such, they became paranoid of their own people and began to suppress even the slightest criticism of the Ortega administration's rule. In South Sudan, the revolutionary leadership split into tribal rivalries based on self-interest and group psychology and once again plunged the people into war and chaos. Thus, revolutionaries can turn into counter-revolutionaries in time.

All of the revolutions that we looked at suggest that a simple political-economic revolution is not enough to bring people real happiness, nor are such revolutions a guarantee that the leaders will continue to stand with the people after they have assumed power.

In order to address the people's real happiness and the continued benevolence of the leadership, we need a nuclear revolution. Only a nuclear revolution is able to transcend national boundaries, religious exclusivism, and the ideologies that create boundaries between races, genders, and classes. A nuclear revolution also addresses personal and interpersonal dynamics. In order to produce such a comprehensive and unifying revolution, the revolutionaries must be inspired by a universal ideology and be guided by a spiritual force that will allow them to overcome the divisive tendencies of their own egos as well as the divisive forces in society that are driven by their own self-interests.

A nuclear revolution is based upon principles that go beyond liberal and conservative ideologies, or capitalist or communism ideologies. As we have seen in our look at contemporary revolutions, conservativism and liberalism are not to be confused with capitalism or communism. Nor are they to be conflated with specific political parties. Support for either can be revolutionary or counter-revolutionary. In the revolutions fought by the people in Eastern European countries, that we have examined, capitalist ideas were revolutionary because they fostered a democratic movement against the conservative, autocratic rule of communist leaders. The people overwhelmingly wanted to join the European Union. In contrast, the people's revolutions in Latin America believed communist ideas to be liberating as they struggled to overcome capitalist exploitation and their autocratic puppets.

Human civilization has now reached a critical stage in its trajectory. In the decline of the capitalist/communist system, exploitation of one human being by another has assumed alarming proportions. At critical junctures in the past, when exploitation had reached the zenith point, history witnessed the emergence of mighty personalities who were able to overcome the problems in society. Today also, the guidance of mighty personalities with a comprehensive ideology is required to lead humanity away from the pit of disaster towards a glorious future. The emergence of such personalities is an indispensable necessity of history.[255]

Nuclear revolution can only be brought about by spiritual warriors (sadvipras) who reside in the nucleus of the social cycle. Spiritual warriors are those who, by their moral and spiritual integrity and all-round endeavor, will mobilize and guide the people on a revolutionary path. This struggle and mass upheaval will liberate society from exploitation and usher in a new era of peace and prosperity.

About one hundred years ago, in 1917, amidst the disaster of World War I in Europe, Rudolf Steiner, an Austrian philosopher, scientist, and spiritual thinker, formulated a unique approach to addressing the cultural, political, and economic needs of society in an effort to prevent another destructive world war. Observing that the ideals of the French Revolution were *liberté* (freedom), *egalité* (equality), and *fraternité* (brotherhood), he defined these human values in a completely unique manner. He said that *freedom* is achieved by the cultural and spiritual behavior of the people. *Social justice* and *equality* are achieved in the political realm insofar as the government deals with questions of rights, laws, and agreements. Finally, the goal of *brotherhood*, he said, can only be achieved in the economic realm when all people have their basic needs met and are no longer fearful for their lives. Taken together, this trinity of practices constitute the foundation for organizing and empowering a self-governing citizenry that will find its authority, responsibility, and accountability through active civic engagement.[256] It is the intent of this book to build on this realization by Rudolph Steiner and on the Universal Ideology of P. R. Sarkar to educate potential revolutionaries about the qualities of a united human society and in the knowledge and competencies to bring such a society into concrete reality.

In this struggle for democracy, the well-wishers of human unity should support any group fighting for their civil and human rights, whether or not the overlords are US, Russian, or Chinese capitalists. They all oppress local people in order to confiscate their labor and resources.

The activities of protesters are not lost on young people around the world. For example, as tens of thousands of anti-government protesters formed human chains across Hong Kong on Friday, August 23, more than one hundred supporters joined hands in Lithuania to show solidarity. Hong Kong's demonstration was held on the anniversary of the "Baltic Way" in 1989, when two million people in Lithuania, Estonia, and Latvia held hands to form a one thousand and ninety-four-mile human chain, spanning the three nations to push for independence from the former Soviet Union.

The movement inspired Hong Kong activists to launch their own version, calling it the "Hong Kong Way."

Lithuanian lawmaker Mantas Adomenas said he was impressed with Hong Kong's struggle for liberty. "They are at the frontline of fighting for freedom and democracy."[257]

Vytenis Fuks, who attended the event said, "I believe it is a must to defend democracy and human rights. For Lithuanians, it is written in blood. We support other nations who now have to protect themselves just like we had to do it during Soviet occupation."[258]

The problem we all face, whether in capitalist or communist countries is the psychology of self-interest that seeks wealth and power at the expense of social equality. This selfish psychology dominates our collective consciousness instead of a psychology of selfless service that would bring us together as a human society and heal the planet.

The solution to our collective problem is to end the capitalist age, which is based upon exploitation and greed. Both the capitalist and communist empires suffer from the same psychology because they are founded on materialism instead of a universal spirituality that promotes human unity. As such, the leaders of these countries have no higher pursuit than to enrich themselves and lord it over others.

All the great spiritual masters of history, regardless of time or place, or personal style possessed the psychology of selfless service. They taught us that if you serve others, the loving, divine Being will serve you and you need not worry about your own mundane problems. More than ever, we need a leadership among all people that embodies the psychology of selfless service. We need the young of the world to adapt this psychology and fight collectively for the welfare of humanity and the planet as a whole. The future of all nations and all peoples, regardless of the divisive religious and political ideologies that control our societies, depends upon the development of a mentality of social equality instead of personal pleasure, if we are to survive as a species in a world that will permit us to live.

We need a vision of ourselves as one human society, living in one world under the protection of an absolute loving consciousness. The revolutions that we have analyzed reach beyond capitalist or communist ideologies, as well as beyond the media-hyped divisions between the Left and the Right. Everywhere, the struggle of the people is a struggle for democracy and the right to make their own decisions regardless of whether the overlords are communists or capitalists. We need an ideology that is bigger than either of these ideologies and we need a method that allows us to respect and help each other in the face of oppressive governments and political-economic systems the world over.

The authors apologize for the long and complex analysis of these revolutions around the world, but it is our hope that the reader will get a strong sense of what organizing and living through a revolution actually entails, as well as what is required to lead a revolution and what push-back to expect. Hopefully, too, the reader will appreciate the need to be mentally centered and have a spiritual core if we are to succeed in bringing about the positive changes we envision.

Chapter Seven: The Socialists' Dilemma

The Left and Right in America

Before we look more deeply into the various components of a nuclear revolution, it would benefit us to look at the nagging question of whether we can solve our common problems through reform measures, or is revolution the only way out. It is an important question and our revolution can only be understood in the context between the two great progressive camps of reformers and revolutionaries. These camps still exist today. Whether or not we can depend upon these parties and organizations in our nuclear revolution will be determined by whether they are guided by reformers of the capitalist system or by revolutionaries.

In the United States today, both the conservative Republicans and the liberal Democrats promote capitalism. Thus, there is no question about revolution in their minds. The Republicans support corporate welfare policies, free of government intervention or social service programs. The Democrats want to see a mixed economy which supports capitalism and the capitalist class, but also wants to mitigate the capitalists' extreme exploitation of the people by the means of government regulation and domestic social service programs. The underlying reality here is that both Republicans and Democrats support capitalism and thus the right of capitalists to exploit American people, the world's people and the environment at will.

This reality is demonstrated by the capitalists' willingness to fund politicians of both parties.[317] In donating to the political campaigns of both sides, and by conducting lucrative backroom deals, the lobbyists of the capitalists are able to convince both the Democrats and Republicans to support their initiatives instead of those of the people. This is essentially why wages remain weak, women and racial minorities are kept in check, college students are in debt, people do not have adequate health care, many of our best jobs have been transferred overseas where labor is cheaper, and the minimum wage is not enough for people to live on.

At the same time, the capitalists' mainstream news media continues to play on the apparent difference between the Republicans and Democrats. While they support both parties, the capitalists want the people to be divided in their allegiance to one or the other. Let us take a closer look at this dynamic. In the motion of any society, the contradiction between conservatives and liberals is a natural phenomenon. The conservatives look to the past for solutions to the present, while liberals look to the future. The conservatives want to go back or maintain the status quo. The liberals want to change the status quo and move forward into greater social justice and equality, but only insofar as their geo-sentiment or socio-sentiment extends. These contradictory positions are inherent in the movement of society as it chugs along in a pulsative movement, alternately contracting and expanding. During the periods of contraction, the conservatives are in power, during the times of expansion, the liberals hold sway. The Capitalists require both the Left and the Right to remain in power.

This natural dialectic, however, becomes aggravated when people face extreme economic insecurity. As they look for a reprieve from their suffering, they alternately turn to the Republicans and the Democrats, only to find them engaged in mutual blame and finger pointing, which only further polarizes the country. As economic pressure increases, the Right moves toward fascism and totalitarianism, and the Left toward socialism and democracy. If Republican totalitarianism wins out, so much the better for the capitalists. If Democratic social democrats win out and it costs the ruling class a few more quarters, they can also live with that.

If we review the actors on the Right, we find a range of organizations that espouse various political, economic, or cultural issues. The *Libertarians*, for example, promote cultural issues, particularly individual liberty while the *Tea Party* also promotes conservative Christian values based on control of a woman's body and suppression of LGBTQ rights. The *Neoconservatives* want to control the government to advance economic colonialism and fight insurgencies that stand in the way of capitalist exploitation. The *Wall Street Republicans* want to free Wall Street capitalists from any economic restraints or oversight. At the extreme right are the fascists, neo-Nazis, and white nationalist organizations like the KKK, American Nazi Party, the Fascist League of North America, *Qanon, Proud Boy, Oathkeepers*, etc. whose ideas and methods are adopted by the moderates when they feel unusually threatened by the changes proposed by the Left.

These extreme right-wing organizations are essentially white nationalists who eschew any gains made by women, or racial and religious minorities. They promote white male supremacy as a solution to present day problems. Because right wing parties and organizations are essentially counter-evolutionary and counter-revolutionary, rather than evolutionary or revolutionary, they are not a subject of discussion in this book. This being said, revolutionaries may still find individual allies among those conservatives whose primary attachment to the past is the US Constitution, and a strong sense of community and family.

Moving to the Left beyond liberalism, we encounter a variety of social democrats, democratic socialists, communists, and anarchists. Most of these groups today believe that it is within their power to reform capitalism and in so doing create a democratic society that is run by the people as opposed to the capitalist oligarchy. Few have thought how to achieve this, however. Some advocate the replacement of capitalism. Although there has been some strategic thinking about how to achieve a democratic society, there has been little activity beyond organizing political and cultural campaigns. Where we do find concurrence among some leftist factions, is in the mutual building of local economies to create greater self-reliance among local people in meeting their basic needs.

Among progressive forces in the United States, we find campaigns to fight climate change, secure universal health care, oppose government suppression of immigrants and racial minorities, etc. All good causes in their own right, but woefully inadequate as isolated activities to address the problems facing humanity as a whole.

It is important that people, who are involved in the struggle for greater democracy, understand that the division between the Right and the Left is actually supported and promoted by the capitalist class because the conflict between the two sides keeps the masses of middle class and lower class people preoccupied with the "faults" of each other, rather than the reality of capitalist exploitation. Because this "feud" has also been a major preoccupation of the capitalist-controlled mainstream media, it has become a central preoccupation of the US national consciousness in our attempts as Americans to envision a better tomorrow. No president in recent history has exploited this feud as dramatically as Donald Trump.

Egberto Willies, writing an article for *Daily Kos* entitled "How do we avert the inevitable Plutocrat driven civil war?" explains this divisive ploy with great insight.

> The difficulty is keeping the different poles that comprise the masses balanced. They must be preoccupied with one another and not with that which is inflicting pain on them, that which is bleeding them.
>
> The plutocracy depends on the polarization, and so far they have been successful. Trump is not the genius of the plutocracy. He just magnified the hate at the appropriate time. He is but the puppet of the system which chooses its protagonists not only in American but throughout the world.[318]

We are falsely led to believe that if we can reform the government, then we can reform capitalism. This false premise can be seen in the behavior of the capitalist class and by the historic failure of the left to curb their rapacity.[319] But because people do not understand their history, they do not understand what they are up against.

As long as the capitalists are able to foment political polarization between the Left and the Right; that is, between the forces of expansion and contraction, they are able to mask the true division in American society between the oligarchs and the American people as a whole, specifically between the 1% and the 99% of American society resulting in much of the 99% being kept in a state of subservience and, in many cases, poverty. By poverty we mean the inability of people to consistently meet their basic necessities of food, shelter, clothing, health care, and education.

In conclusion, therefore, we cannot depend upon the Republicans or the Democrats to solve the crises endemic to capitalism. The only way to benefit humanity and protect the natural world is to deny the capitalists their political and economic power and put it in the hands of *moral leaders who seek the well-being of all.* This is the only way to help people improve their lives as well as the planet as a whole. This shift in power relations cannot be achieved by accommodating the capitalists.

In regard to the principles of Universalism, there can be no compromise with exploitive predators. Only a shift in power relations can solve the problem of exploitation and this can only occur through struggle, not through negotiations and compromise.

As such, the reform of capitalism is not an option in solving the problems that now face humanity and the planet. Given the quick deterioration of the natural world, neither is evolutionary change an option. Only revolution will do and the revolution must be a nuclear revolution.

The Left and Right in the World

Although there are differences between the Left and the Right on the international level, specifically between the so-called Capitalist and Communist countries, this too has become little more than a media fabrication. Today, the major "communist" countries have become little more than Capitalist countries. This is especially true of Russia and China. Much of the Russian economy today is controlled by capitalist oligarchs that gained prominence with the collapse of the USSR. Similarly, in China, the Capitalists have taken over the Communist Party and have assumed greater influence of over China's enconomy than the government. Neither the people living in capitalist countries nor those living in in these former communist countries should look upon each other as enemies. We are all being exploited by the international capitalist elites. And while there may be competition among them for a bigger piece of the pie, their values are self-serving capitalist values. Although some capitalists may support the environment and spend their resources on social programs, and should be applauded for doing so, these actions will not provide sufficent relieve from the exploitation that the capitals system continues to impose on the world's population.

The conclusion that we reach from this understanding is that political engagement, on the part of the people, is not enough to overcome the capitalist-initiated crises. We also require an economic revolution that takes power away from the multinational capitalists wherever they have set up their base, and return it to ourselves, the local people. One way to do this is to build an integrated network of local economies, under our local control, and using our own local resources to meet our own basic needs. This is a way to free ourselves from capitalist control.

Let us never forget that as the people on the Left and the Right, in their ignorance, raise their fists to each other, the real criminals remain above reproach. The people fail to understand that under a capitalist system, there are two main contradictions: (1) the conflict between the capitalist class and the rest of humanity and (2) the conflict between the system's continuous consumation of resources and the earth's ability to continually provide those resources. The former conflict is also referred to as class conflict by the socialists and communists, or as a conflict between the oligarchy and democracy by the social scientists and philosophers. By down-playing the contradiction between the capitalist class and the people, the American politicians allow the capitalists to continue their exploitation of diminishing resources, while the people fight among themselves. Despite their opposition to Western capitalism, Russia and China also allow the capitalists to exploit their own people. This is a fundamental dilemma that must be faced by honest socialists around the world. The truth is that there are very few examples of true socialism in this world and fewer true socialist political-economies worth copying.

The Need for Viable Socialist Economic Models

In general terms, a capitalism economy, by whatever name people choose to call it, moves in a pulsative movement like everything else. In its contractive phase, it throws millions of workers out of jobs and destroys the environment. Historically, socialist revolutions, including those in Russia and China, despite their original intent, have not been sucessful and eventually resulted in leaving control of the means of production in the hands of the corporate capitalists. While other socialist revolutions, like Cuba and Western Kurdistan, although having some success, continue to struggle against the capitalist system.

Consequently, while the world's socialist organizations reject totalitarian systems, neither the "command economies" of the communist countries, nor the "mixed economies" of the capitalist countries are able to provide a viable economic model to address the abuses and devastation created by global capitalism.

The critical problem for the socialists remains how to socialize the means of production and thereby overthrow the dominance of the capitalist class for good. Fortunately, there has been a growing interest among progressive socialists in ideas like solidarity economy where the essential needs of the community are controlled and met at the local level. Besides the Rojava model in Syrian Kurdistan, other alternative models exist like the Emilia-Romagna coop federation in Italy, the Timbaktu Collective in India, and the Mondragon coop federation in Spain. As socialists, we should be supporting the people who are building these models and learn from them what lessons we should apply within our own communities.

Traditional Socialists Do Not Distinguish Between Religion and Spirituality

Another problem for traditional socialists is that their materialist worldview prohibits them from understanding the difference between spirituality and religion. This situation confines them to a relative world, limited by the duality of mind. Not being able to distinguish between the mental distortions of religions and pure Absolute Consciousness also robs them of the shelter of a greater consciousness and only serves to bolster their egos as the only means to succeed. As such, they will not be able to fulfill their thirst for limitlessness, nor, as we have seen in earlier examples, will they have the strength to keep from exploiting the people once they have seized power through revolution.

When the object is subtle, the mind becomes subtle, and when the psychic structure becomes subtle, the thinking power is developed and higher ideas are created. If one meditates on materialistic ideas, the mind becomes crude and materialistic.

A socialist spiritual warrior must make his or her mind more subtle and, in doing so, more expansive and loving.

Socialists require a new economic model and a spiritual path. These are the contributions that progressive socialism (PROUT) makes to the history of socialism. This will bring socialism into greater alignment with the majority of the world's people, all of whom we wish to serve.

In this material world, only a unified socialist movement, acting according to the principles of progressive socialism, has the potential to create a nuclear revolution. If the revolutionary forces are to succeed, we must work in unity with progressive democrats, progressive individuals, democratic socialists, the greens, sustainable development activists, issue-centered activists, Gene Sharp revolutionaries, local economy activists, "value-oriented intellectuals," leftist students, and anarchists, but only insofar as they stand as true revolutionaries and are not merely proponents of Capitalist reform.

We need to begin our nuclear revolution by creating mutual decision-making bodies among revolutionary socialists at the national and regional levels, as well as on local level. For a more detailed description of the function of these bodies, see chapter five of *Universal Ideology: The Thought of P. R. Sarkar*

For the nuclear revolution to be successful, however, we will expect that the leadership positions will be filled by those who are competent, have a high degree of morality, are versed in the tenets of a Universal Ideology, and have set the attainment of Supreme Consciousness as their personal goal. May we succeed in bringing peace and harmony to humanity and this wonderful planet Earth.

Chapter Eight: Unpacking a Nuclear Revolution

In this chapter, we look more in-depth at what distinguishes a nuclear revolution from the political, economic, and cultural revolutions we have examined. Our new revolution is called "nuclear" for three reasons. First, a nuclear revolution is based upon universal spiritual values (not to be confused with religion). Because the Divine Consciousness exists in the nucleus of the universe, a revolution which has this Divine Consciousness as a goal, can also be called a nuclear revolution. Secondly, because a nuclear revolution is not imposed top-down, but is generated in the heart of every participating individual, the revolution is called a nuclear revolution because it begins in the nucleus of every participating individual. Thirdly, because a nuclear revolution touches every aspect of individual and collective life and reformulates it, we can call it a nuclear revolution because life itself becomes changed at a nuclear level. For all these reasons a nuclear revolution differs from the political, economic, and cultural revolutions of the past.

Universal Spirituality

The philosophies that human beings have followed can be divided into four categories: (1) the dogma-centered philosophies espoused by religions, (2) the matter-centered philosophies espoused by communists,

(3) the self-centered philosophy espoused by capitalists, and (4) the God-centered philosophy espoused by spiritual masters and mystics.[320]

People who follow dogma-centered philosophy exploit others in the name of providence for their own self-interest. For example, the proponents of dogma often claim that they have been blessed with divine revelation. They say that they had a dream in which God appeared before them and commanded them to do particular work, and on this pretext, they exploit others to the full.

In matter-centered philosophy, collective enjoyment of material objects is the main motive, as in contemporary Marxism.

In self-centered philosophies, individual selfishness is the motive, as in capitalism.

On the other hand, those who follow God-centered philosophy serve humanity and the entire living world with equal love and affection. While there may be differences between individual expressions, the center is divine. In God-centered philosophy, God is the nucleus, and the radius of one's selfless love and devotion is determined by the distance between one's self and the divine nucleus.

The work done by God-centered people elevates the standard of human beings. In the past, they elevated humanity; in the present they are elevating humanity; and in the future they will continue to elevate humanity. But until now, the world has not experienced any God-centered philosophy that includes a specific socio-economic model for collective fulfillment.

Assuming this perspective, let us begin by saying that a nuclear revolution is grounded in a God-centered or universal ideology. Spirituality is the core of our being. It is the source of our eternal quest for limitlessness. Without understanding our spiritual nature, we will not make progress as human beings in this world. We will merely bob up and down on the waves of life without moving closer to our source of unity and our ultimate goal.

Unless we are able to grasp the full potential of our individual and collective humanity and imbibe the message of love given to us by the world's spiritual masters, we will remain victims of our largely programmed, groupist sentiments, and the social programming imposed upon us by our history and current institutions. Under these conditions, we remain unable to envision self-realization as a common goal of humanity. Therefore, a nuclear revolution is based upon spiritual values rather than material values.

"Spirituality" does not mean "religion." Although religions may reflect some aspects of universal spirituality, for the most part, many religions are fundamentally dogmatic and as such, divisive. This is true of Christian fundamentalists, Jewish fundamentalists, Hindu fundamentalists, Muslim fundamentalists, etc. Spirituality, on the other hand, means universal love. It speaks to our highest human potential and provides us with a rationale for human unity and respect for all of life. It links each of us with the Divine, with nature, and ourselves. While it is often the goal of religious leaders to divide people, it is the goal of spiritualists to unite us. This is important to know, because the leaders of a nuclear revolution will all be spiritualists. In fact, they will be spiritual warriors.

Universal Morality

In the sphere of society building, there are two distinct psychologies. One is service psychology, which inspires people to promote collective interest. The second is group psychology, which only tries to promote the limited interest of a small group. Those who are guided by service psychology do not like to separate politics from morality. Their thoughts and ideas remain far above narrow group interests. On the other hand, those who are guided by group psychology want to establish the authority of their group and impose their interests on others. This leads to interpersonal and inter-group conflict. Only the synthetic approach leads to unity and cohesion amongst numerous individuals and groups. The followers of the analytical path often become vocal revolutionaries, and become extremely active to establish their raj (kingdom) entirely without niiti (morality). Thus, out of these two distinct types of psychology two social outlooks arise. P. R. Sarkar.[321]

According to P. R. Sarkar, those who are guided by service psychology have a synthetic outlook, and those who are guided by narrow group interest or self-interest adopt an analytic outlook.

The second component of a nuclear revolution is therefore a universal morality. By "universal," we mean that it extends beyond any particular group. All people, regardless of gender, race, age, class, religion, or nation are due the same respect as human beings. Further, all plants, animals, insects, etc., have the same existential status and require the same respect for their existence. In the post-Capitalist world, our universal morality will provide the basis for a world constitution. New laws and regulations will replace the hodge-podge of sins and crimes of different groups, which are based upon local sentiments or religious dogmas. In addition to universal moral cardinal virtues Sarkar also provides ten universal principles. These are: (1) Seeing everything as forms of Divine Consciousness, (2) using discriminating truthfulness, (3) practicing non-harmfulness, (4) non-stealing, (5) living simply, (6) practicing personal cleanliness, (7) maintaining peace of mind, (8) self-sacrifice, (9) seeking the shelter of Divine Consciousness, and (10) understanding spiritual issues. Further discussion of these principles can be found in *Universal Ideology: The Thought of P. R. Sarkar.*

If the idea of a "universal morality" sounds too grandiose or abstract, perhaps a simple story will bring its meaning down to earth. I was in a discussion with a local farmer about our town mayor who wanted to sell the town's water department to a private company and pocket a "finder's fee" from that company. I was particularly angry about the mayor's latest attempt to take advantage of the local people for personal gain. I told the farmer that the mayor was "a typical capitalist." In response, the farmer said to me, "Any system that operates without a conscience is bound to fail." I thought this observation summed up the situation nicely, and I told the man so.

Morality is not the goal of spirituality, but it is the foundation of spirituality, and a nuclear revolution cannot succeed without it. Spirituality exists in the hub of the revolution where it influences every thought, word, and deed of the revolutionaries. The existence of a Universal Ideology, based upon the highest principle of morality, may further polarize society between those who are committed to the welfare of all and those who are committed to personal and group aggrandizement. It will create a battle between the forces of selflessness and the forces of selfishness. In this battle, the service-minded people will support the revolution and will remain with us. The self-centered people will oppose us and will continue their efforts to block our progress in the future. We are sure to be victorious since we are not alone; the Divine is with us. This is because it is the intention of the Divine to see all people move along the path of unity. The path of unity, according to all the great spiritual masters who have ever lived, is realized by love for the Divine and love for neighbor. Jesus Christ said that this is the foundation of his teaching. A Universal Ideology founded on universal spirituality and universal morality will keep the revolutionaries forever inspired and energized.

As important as it is to build the nuclear revolution on a strong core, there is much more to a nuclear revolution than a universal spiritual sentiment and moral rectitude. A wheel requires more than a hub to move. It also requires spokes and a rim. In our analogy, the spokes are the different aspects of human society. These aspects are: (1) personal transformation, (2) interpersonal relations, (3) culture, (4) history, (5) sociology, (6) politics, and (7) the economics of society. The rim of the wheel is the social cycle wherein one age rotates into the next. All of these social dynamics will change in the course of a nuclear revolution. The changes will occur naturally as the revolutionaries create a worldwide revolutionary force to protect the natural world and build decentralized economies to secure people's basic needs.

Personal Transformation

A nuclear revolution will require a new kind of leader. These leaders will not come about through training in any existing organization, be it military, corporate, or bureaucratic. Nor will they come about through the exhortation of politicians or political activists. Rather leadership by spiritual revolutionaries will only come about by the spiritual practice, personal discipline, and spiritual training of many individuals worldwide. As stated above, one of the reasons that we call our revolution a nuclear revolution is because it begins in the core of every individual. The aspiration to become a spiritual revolutionary by many individuals around the planet will create a new understanding of personal potential and place the individual in an exceptional status within human social formations.

P. R. Sarkar, who first articulated the idea of nuclear revolution called the leaders of this kind of revolution "sadvipras" or spiritual revolutionaries. He said that these new leaders are those whose every effort is directed towards the attainment of divine bliss. Divine bliss is achieved when the love for the Divine is so strong that one actually becomes one with the Divine.

P. R. Sarkar defined spiritual revolutionaries as "Those who strictly adhere to the principles of morality, are ensconced in sacrificing service, and are ready to wage war against immoralists."[322] Therefore, it becomes the prime duty of everyone to support these spiritual revolutionaries. See *Universal Ideology: The Thought of P. R. Sarkar* for guidance in this endeavor.

Interpersonal Relations

Spiritual practice, especially meditation, will be necessary to not only improve one's indivdual self, but also to improve one's relationship with others. As people begin to look at the capacity of individuals in a new light, this will also change interpersonal relationships. Aashna Jain, in an article entitled "Can meditation heal your relationship? 5 powerful tips that will solve your problems," talks about how the spiritual practice of meditation will improve relationship problems by creating patience, awareness, happiness, forgiveness, and gratefulness.[323]

When we talk about the spiritual practice of meditation improving relationships, we are talking about making a commitment to our neighbors to support each other's well-being. This commitment is achieved by local people creating local socio-economies based upon the production of goods and services that meet everyone's basic necessities of food, shelter, clothing, health care, and education. These are called *people's economies*. Whereas capitalism is based upon the isolation of individuals and a competition to gain the most wealth and material possessions, progressive socialism is based upon the cooperation of individuals for mutual benefit. When this spiritual psychology becomes dominant within communities, the nature of interpersonal relations will change significantly.

Cultural Transformation

According to progressive socialism (Progressive Utilization Theory), all of human society shares the same culture. While there are local variations in the mode or state of cultural expression, culture, nonetheless, is universal. While everybody eats food, some eat by hand, others by chopstick, and still others by spoon. Some prefer wheat, while others prefer rice. In another example, American dance may be based on rhythm, while Indian dance is based on dance postures, but they do not represent two different cultures.

Local modes of cultural expression are called customs. These differences in culture tend to diminish if there is a close blending between different groups of people. In Nepal, there is a blending of Hindu and Buddhist customs. In Bengal, there is a blending of Aryan and Dravidian customs. In the United States, there is a blending of black and white customs. Thus, local modes of expression that bear local or group specialties are customs, but human expression itself is culture.[324] Human society is one and our culture is one. Human society is like a bouquet of flowers. There is one bouquet although the flowers that compose it may be many and varied.

When a particular group, which is motivated by its group-sentiments to exploit others, tries to destroy the customs of other groups, it forcibly imposes its own language, dress, and ideas on these groups, and thus paves the way for exploitation by paralyzing those people psychologically. This is one of the ways how people guided by divisive socio-sentiments perpetuate exploitation in cultural life.

Exploitation in the cultural sphere is accomplished by the propagation of psycho-economic exploitation. For example, one way method to exploit people is the propagation of a consumer culture by the capitalist ruling class in their campaign to break the local people. Psycho-economic exploitation first weakens and paralyses people psychologically in various ways, and then exploits them economically. Some of the methods of psycho-economic exploitation are: (1) pseudo-culture, (2) the suppression of the indigenous language and cultural norms of the local people; (3) the imposition of numerous restrictions on women; e.g., forcing them to be economically dependent on men; (4) an unpsychological educational system with frequent political interference by vested interests; (5) the suppression of various forms of spiritual practice in the name of secularism; (6) creating divisions among the people into castes and groups; (7) damaging women by the use of unnatural and harmful methods of birth control; and (7) placing the control over different mass media such as newspapers, radio, and television, in the hands of the capitalists. This exploitation is a great danger to our entire humanity today.[325]

To counteract this situation, a revolutionary organization will need to create a powerful popular sentiment for the liberation of intellect. In order for this to succeed, the first requisite factor is that the revolutionary leadership must keep their own intellects pure and unblemished. They must identify with the issues of the people, assist in their development, and extend their support to all anti-exploitation movements. This approach will help ensure that exploitation is rooted out, that the social structure is stabilized, and that the intellectual horizon of the human mind is expanded. If human society follows such a path, we will move forward to a brilliant future with rapid steps. The spiritual revolutionaries should use the arts, literature, education, trade, commerce, industry, agriculture, morality, and social relations to get their message across to the people. This will free the masses of people from their inferiority complex. Creation of a powerful popular sentiment is a vital component of nuclear revolution.

Politics and Decision-Making

Regarding the subject of Politics, a nuclear revolution is not based on divisive party politics. Party politics only serves the capitalists by keeping the people divided while the capitalists continue to impoverish us all. Rather, a nuclear revolution is based upon informal and formal political systems. The informal political system is made up of those interactions and activities of all the local people as they build trust among each other and strengthen their socio-economy in informal ways. The formal political systems will be established by the revolutionary leadership and be based upon local social boards, socio-economic units (samajs), federated states, and ultimately a world government (see chapters 4 & 5 of *Universal Ideology: the Thought of P. R. Sarkar*).

A successful political system will also depend on the leadership of spiritual warriors. This universe belongs to all. Every dust particle of the universe is the common inheritance of each and every one of us, so it is not at all proper to allow a particular social class to perpetuate its rule at the expense of everyone else. As such, as the cyclical evolution of the social cycle continues. Spiritual warriors will have to fight against the degeneration of each social class as it gains dominance.

Nobody can stop the rotation of the social cycle, not even the sadvipras. They stay in the nucleus of the social cycle, vigilantly observing the process of rotation. One after another, one Age follows the next: Warrior after Capitalist, Intellectual after Warrior, Capitalist after Intellectual. The sadvipras cannot stop this process, but they can accelerate the natural movement of society when an Age enters its degenerative phase.

After establishing the post-revolutionary Warrior Age, the sadvipras will have to keep a vigilant eye on the warriors so that they only rule and do not descend into exploiting the other classes. When they show signs of beginning to exploit, the sadvipras will immediately create an antithesis to end the Warrior Age. They will do the same thing in the Intellectual and Capitalist Ages. In other words, they will not allow society to follow the process of natural degeneration. They will shorten the period of degeneration and bring about social revolution whenever necessary.[326] In so doing, they will act to minimize the amount of pain and suffering people have to endure when the society enters the period of decline.

Society belongs to all, but it will be under the guidance of the sadvipras. Although sadvipras will require the assistance of warriors, intellectuals, and capitalists of strong moral character in the leadership of society, the responsibility for leading society cannot be left solely in the hands of the Warriors class, or Intellectual class, or Capitalist class, because all three groups, as their age begins to decline, will begin exploiting those of the other groups. The toiling masses (Shúdras) do not have the capacity to undertake the leadership of society.

Therefore, the responsibility for leading society can only be entrusted to the sadvipras because they are well-established in universal morality and imbued with spiritual ideation. They are also trained to incorporate all of social psychologies within themselves so as to be able to do the greatest good regardless of circumstance. The social cycle will surely rotate, and as a rule, the dominance of the warrior, intellectual, and capitalist will take place in succession. But if sadvipras control the nucleus of society, these social classes will attain some degree of prominence in social life, but they will never be able to become the absolute exploiters that we have seen in history.[327]

How will spiritual warriors come to lead society? There will be many factors at play. First, they will be driven by their mission to serve humanity and the natural world as part of their compact with the Divine. Secondly, they will have the skill set to lead society and to serve society by whatever means are required, physical force, mental force, financial force, and the willingness to get their hands dirty and stand with the lowest of the low within society. Thirdly, the Sadvipras will arise from the local people everywhere in the world and the local people themselves will demand their leadership. Sadvipras will never have any rest. They will continue to fight tirelessly. This fight is the natural course of living beings. In the absence of this fight, human progress will cease to exist. Because the spiritual warriors will utilize the skills of the intellectuals, warriors, merchants, and workers, their leadership will mean the victory of every social class It will mean the victory of human society as a whole.

Economic Democracy and Decentralization

The economic base of a nuclear revolution will consist of decentralized, democratic economies built by local people to meet their own basic needs which consist of food, shelter, clothing, health care, and education. These local socio-economies, called People's Economies, will permit local people to withstand the forces of economic globalization that exploit local labor and local resources and pit us against each other. Members of these local communities will gradually increase their strength by developing a network of communities. In this way, a network of people's economies will be built and integrated in a movement toward a global network, which ultimately will influence the drafting of a global constitution and the laws and regulations based upon this constitution. See *Universal Ideology: The Thought of P. R Sarkar* for the attributes of a global constitution and the means by which it will come into existence. While politics will tend toward centralization, the local economy, which is dedicated toward meeting basic needs, will remain decentralized to ensure that local people will always have the means of survival under their own control. However, besides meeting the basic needs of the local community, Prout communities will also ensure all aspects of human growth are attended to including scientific study, arts and literature, music, entertainment, cultural events, and time for spiritual practices and activities.

Chapter Nine: What Is Required to Create a Nuclear Revolution?

Wide-Scale Exploitation

CERTAINLY, THE FIRST CONDITION for a nuclear revolution to occur is the presence of wide-scale exploitation. This condition presently exists. It is characterized by the prevalence of extreme poverty; social insecurity; injustices against the common people; lack of purchasing capacity to acquire the minimum requirements of life; huge economic and social disparities between social classes; the irrational distribution of wealth; the destruction of the natural world due to over-exploitation, etc.

When a people are long oppressed, they become weak and static and are unable to resist their aggressors. Under these conditions there may be occasional protests or demonstrations, but these actions are insufficient to overcome the people's habit of subservient obedience to the regime. Under such conditions, those with a fighting spirit will want to resort to violence, thinking that this is the only way to bring about revolutionary change. The problem here is that violence is the main weapon of the ruling elite and that despite the bravery of the people, they are seldom a match for the military might of the dictatorship. Although there are exceptions to this as we saw above in Cuba an Nicaragua, in most cases, the use of violence only leads to more oppression, and the warriors find themselves alienated from the people for whom they are fighting. This is not to say that violence is unnecessary. It is only to say that revolutionary leaders must be willing to explore and to use every means necessary to bring about revolution.

The fact that widespread exploitation has not already triggered a nuclear revolution indicates that additional conditions are still required. The lesson of history teaches us that the common people will not automatically rise up against oppression. They require leaders who can provide a vision and inspire them to take action.

Universal Ideology

The second condition, therefore, is the wide acceptance of a Universal Ideology that can provide a common vision of human capacity and human unity, while also charting the course toward self-reliance and political resistance against corporate and governmental exploitation. This Ideology must, at minimum:

1. Address the inability of mainstream ideologies to solve our collective problems;
2. Provide a comprehensive view of history and the characteristics of class psychologies;
3. Provide a clear analysis of human nature and the consequences of acting from the instinctive, sentimental, rational, and intuitive perspective; and
4. Point out the nature of psychic exploitation and categorize those group sentiments that are used to exploit the people.

As discussed in our book *Universal Ideology: The Thought of P. R. Sarkar,* the authors also believe that a revolutionary ideology should also contain the following:

1. Describe the workings of devotional sentiment;
2. Describe the process by which the absolute consciousness is changed into forms;
3. Explain the science of god-realization;

4. Examine the process of evolution and the means to psychic liberation;
5. Explain how the law of karma works and the reason for death and rebirth;
6. Explain the connection of the brain and the mind;
7. Present an argument for the goal of humanity;
8. Define spiritual practices;
9. Present the tenets of a universal morality and cardinal human values;
10. Define the politics of progressive socialism;
11. Present the factors required to ensure the success of society;
12. Provide guidelines for government;
13. Define a government structure that can ensure optimum freedom for the individual;
14. Present an understanding of criminal minds;
15. Describe the economics of progressive socialism;
16. Define a local socio-economy;
17. Define the principles of a local socio-economy;
18. Provide planning guidelines for creating a local socio-economy; and
19. Provide a methodology for analyzing social change.

The Universal Ideology must be based on practice, not on theory. Having said this, if there are any defects in the practical implementation of the philosophy, then it can be rectified, but if there are fundamental defects in the philosophy itself, the philosophy can never be implemented, nor can it be rectified.

If there are any defects in the philosophy or if it is not comprehensive, there is the danger that the revolutionaries will lose control of the leadership of society. This will be extremely harmful for the progress of society.

The theories of Marx and Gandhi are examples of defective philosophies. Marxists say that revolution is the only solution to capitalist exploitation. This is unquestionably a positive idea. But the concepts of dialectical materialism, the materialist conception of history, the withering away of the state, proletariat dictatorship, classless society, etc., are defective ideas which can never be implemented. That is why the post-revolutionary stage in every communist country has suffered from turmoil and oppression. There is not a single country in the world which is established according to these Marxist principles.

Gandhism is also defective. Instead of guaranteeing liberation from exploitation, it favors the interests of the exploiters. Hence, while certain elements of Gandhism, like nonviolent resistance, may be used tactically, we cannot base a revolutionary strategy upon it. Liberation from exploitation is impossible when the exploiters themselves find shelter in the philosophy. The co-existence of the exploiters and the exploited can never lead to a society free from exploitation No revolutionary organization can accept Gandhism as an ideal philosophy. If any organization does, then it will no longer be a revolutionary organization, and it will break apart within a short time. This is an historical inevitability. Thus, the revolutionary organization must adopt a positive philosophy, which is without defects.[328]

Qualified Leadership

The third required condition for a nuclear revolution is the creation of a qualified leadership. The ideal leadership will be that class of spiritual warriors discussed above. Generally speaking, these leaders will be intellectuals and warriors drawn from the middle and lower classes who are inspired by a universal ideology and who will lead the masses of people to take action. Those leaders that currently exist must come forward and unite to begin the revolution. Although anyone can become a spiritual warrior, this will not occur overnight. It will take time because it will require people to educate themselves concerning a universal ideology and to begin the process of self-discipline to become spiritual warriors. There are also many other books written about the Progressive Utilization Theory (PROUT) that will help with this venture.

The main task of revolution is to overcome the barriers of inertia in the collective psychology of the Capitalist Age and establish the next age in the order of the social cycle. Society moves according to the inexorable laws of the social cycle, yet those who are concerned for the collective welfare will always endeavor to accelerate the speed of the social movement by fighting relentlessly against exploitation so that everyone can move forward in unison as quickly as possible.

When the psychic balance in social life undergoes a revolutionary change, it means a collective upsurge against exploitation has been brought about by a revolt against the status quo. It means there has been an effective struggle of the benevolent forces against the malevolent forces, which has created a new collective psychology among the people.[329]

To ensure the all-round welfare of the society, the benevolent leadership of spiritual warriors (Sadvipras) is required. Sadvipras alone can represent the genuine aspirations of all the people.

The revolutionary leadership must do more than end the existing system. It must also be able to put a new socio-economic system in place after the revolution. It will not do to destroy the old system and leave the people without a new system to meet their needs. This new system must begin long before the revolutionary leadership sounds the call for physical revolution. It will begin exactly when the local revolutionaries arise and begin the process of creating model People's Economies where they live.

This "hands-on" experience will build our character and teach us how to love. The guidelines for creating a local People's Economy are included in the book *Universal Ideology: The Thought of P. R. Sarkar*.

Its success will require the systematic and rational application of progressive socialism (PROUT) by the collective effort of many highly intelligent and self-disciplined people imbued with a service mentality.

Although Democracy may be the best system currently existing, it cannot solve human problems because, in a democracy, one particular class usually winds up dominating society while the other classes are subjugated. Neither are religious autocracy nor communist party rule satisfactory. Therefore, PROUT supports the centralization of political power and the decentralization of economic power. The centralization of political power does not mean centralization in one person, or party, or institution. Rather, the spiritual warriors who lead the nuclear revolution shall work with other leaders to elect and form multiple boards consisting of persons who are conversant and experienced in the various branches of administration. By the time that the Sadvipras sound the clarion call for revolution, they will be organized on the state and international levels to administer the transition from a capitalist political economy, with its various multinational corporations, government bureaucracies, and international organizations, to an entirely new era. This transition will require the following highest administrative boards:

1. The Supreme Board of Sadvipras: The Supreme Board of Sadvipras shall be an elected body of Sadvipras. It shall be the supreme body for framing policy and supervising the functions of the other different boards in society.
2. The Legislative Board of Sadvipras: The Legislative Board of Sadvipras shall consist of those Sadvipras who are experienced in the work of legislation. It shall frame the constitution and the policies and laws according to the principles of Universal Ideology and the policies laid down by the Supreme Board of Sadvipras.
3. The Executive Board of Sadvipras: The Executive Board of Sadvipras shall consist of those Sadvipras who are experienced in the work of administration. It shall be responsible for materializing the policies and laws that are framed by the Legislative Board of Sadvipras. It shall supervise the work of selecting and nominating the personnel who shall be employed in the bureaucracy. It shall also supervise the work of the various subboards established for different branches of the administration.
4. The Judicial Board of Sadvipras: The Judicial Board of Sadvipras shall be elected by Sadvipras and consist of those Sadvipras who are experienced in the work of the judiciary. It shall frame the rules and procedures for maintaining the constitution. It will also oversee the appointment of judges and different personnel of the judiciary.

5. Sub-boards of Sadvipras for various branches of the administration: Sub-boards of Sadvipras shall be appointed by Sadvipras and shall consist of those Sadvipras who are experts in the respective branches of the government. The lists of names for the representatives of the sub-boards shall be prepared by the Executive Board of Sadvipras and forwarded to the Legislative Board of Sadvipras. The Legislative Board of Sadvipras shall recommend the list, after amendments are made if necessary, to the Supreme Board of Sadvipras. The Supreme Board of Sadvipras shall have final approval of this list.

However, if the Supreme Board of Sadvipras does not approve the list, it will send the list back to the Legislative Board of Sadvipras with amendments if necessary. In this case, the Legislative Board of Sadvipras shall either send the list back to the Supreme Board after accepting the amendments, or send their own opinion again to the Supreme Board.[330]

These higher administrative boards, existing at the global level, will have their roots at the local level from which the Sadvipras at the highest level of government will be drawn.

Sadvipras at the local level will start by helping local people plan their socio-economies and serving on boards that oversee the process of food production, clothing production, and housing, as well as education and health care. Sadvipras will also serve in various key industries, such as energy and transportation, which provide infrastructure services on a "no-profit no-loss" basis. They can also serve as administrators, teachers, healthcare workers, etc. When federated states evolve from the merger of local socio-economies (samajs), Sadvipras will also serve on boards at this level and, in this way, gradually expand their familiarity with different levels of government. Sadvipras constitute the ideal form of benevolent leadership.

Trained Cadre

The fourth condition for a nuclear revolution to occur is the creation of trained cadre.

Before the revolutionary leaders give the clarion call for revolution, there must be intensive preparation. In spite of the presence of all the conditions for revolution, it may not take place unless the exploited masses are psychologically prepared for revolution. If the people do not support revolution, the clarion call for revolution will not be heeded. Ideologically-educated revolutionary workers will have to channel the psychology of the world's people towards revolution and inspire them to undertake revolutionary struggle! Such workers must adopt a rational approach inspired by a positive philosophy, have a well-developed socio-economic-political consciousness, and be dedicated to uplifting the standard of the common people. The duty of these workers is to direct the people's frustration along the path of revolution. Through their dedication and dynamism, the cadre will bring the collective psychology in their favor. The first and foremost duty of the revolutionary leaders therefore is to create dedicated workers.[331] Just as it is not enough to create military officers without an army, so it is not enough to create a revolutionary leadership without an army. The spiritual warriors must continually pass on their experience to educate a cadre concerning the tasks at hand.

The trained cadre will come from the local level and like their trainers gain their stripes in the effort to serve the local people. The cadre will be educated with the intention of making them spiritual warriors as well. In our revolutionary army, spiritual warriors will be found at every level of organization. "Spiritual warrior" is not a rank; it is a state of mind that includes the willingness to do whatever is necessary to serve the people. As such, a spiritual warrior will strive to employ the skills of every social class in order to assist or lead the people regardless of circumstances.

Revolutionary Organization

The fifth required condition for a nuclear revolution is the creation of a revolutionary organization. The revolutionary organization is that network of sadvipras and their co-workers who are working to realize a post capitalist society. As the revolution progresses, the organization of the leaders will expand and become strengthened. Its vision and strategy will become more integrated and the speed of change will increase.

Revolution and war are almost the same. Revolution is a type of war. The difference between the two is that, in war, force is applied on behalf of a person or state, but in revolution, force is applied by a group of people to establish a society free from exploitation. A revolutionary organization is essential for revolutionary war.

During the psychic preparation for revolution, the disgruntled sections of society, that is the spiritual warriors and the cadre, will come to build a revolutionary organization which will create a congenial environment for revolution.

A diverse, multifarious organization is required to conduct a revolution. The responsibility of the organization is like that of a government. The revolutionary organization must be operational from the local level to the highest level possible. Local activists or coordinators will be connected to every level of the structure. As the revolutionary organization develops beyond the grassroots level, more activities of the revolution will come to be directed by the higher bodies.

If a revolution is started without building a proper structure, or if loopholes are left in the formation of the structure, the results may be disastrous.

The revolutionary organization must follow a Universal Ideology. A progressive, comprehensive ideology will be the invincible weapon of the revolutionary organization. It will counteract negative ideas in society and generate a powerful, positive psychic flow in the collective psychology. People will become revolutionary-minded. Even so, vested interests will determinedly try to resist positive change. Therefore, polarization will occur in the collective psychology. The duty of the revolutionary leaders is to create polarization through the propagation of a Universal Ideology.

A Role for Students, Youth, Women, and People of Color in Nuclear Revolution

Students and Youth

The younger generation usually has greater knowledge than the older generation of how circumstances will change in the future because it is their nature to look ahead and, consequently, they focus more attention on the future than their elders. It is also the younger generation that has a more universal vision when it comes to protecting the natural world and overcoming the divisive tendencies in human society. They understand more deeply how the present momentum affects the future.

While nobody can deny the value of practical experience, we should remember that the past does not always repeat itself. In this case, past experience is not always of much value in the present.

Experience helps us to determine the relationship between cause and effect, but in the absence of far-sightedness, experience cannot be effectively utilized. With a change of time and place, people must be able to maintain a consistency between past experiences and probable future events when they determine public policies.

When a chief goal of a local community is to formulate policies for social progress, experience cannot be the sole prerequisite for this work. Rather a combination of the past experiences of the older generation and the creative zeal of the young should determine the speed of social progress. We cannot afford to neglect either group. The human race must attain glory by giving due recognition and justice to all.[332]

While studying is the main occupation of students, being youthful and energetic, they can be engaged in various forms of social work and particularly in the moments of crisis. They can also be engaged in activitist campaigns like fighting against illiteracy, eradicating poverty, advocating social justice, fighting against corruption, etc. They can also come to the aid of the distressed during emergencies.

The students and youth of today lay the foundation of the future society. As such, society should give them the love and encouragement due to them, and see that their studies and work are not unduly hampered.

Women

The exploitation of women is not something that began with the capitalist age. Rather, it began far back in history at the dawn of patriarchy during the warrior age. It started with the creation of marriage as an economic transaction between the husband and the women's brother or uncle.

The enslavement of women by marriage continued throughout most of the evolution of Western Civilization and remnants of this long heritage still exists today in the form of the bride price, bride service, forced marriage, child marriage, exchange of females, gift exchange, dowry, or indirect dowry.[333]

The giving of a payment by a man to the women's family in order to marry her, essentially meant that the husband had bought the woman and her children and thus they became his property to do with as he wished. Following the enslavement of women through marriage, white Aryan tribes then began to enslave people of other races in the course of their empire building. At first, they enslaved just the women to be used as domestic workers or concubines, but they soon learned how to enslave men as well. See *The Untold Story of Western Civilization*, Volume 1 for an in-depth presentation of this development.

Although the enslavement of women and people of color has been discontinued in Western Civilization, this mentality remains today in the attitude of superiority that men have toward women and whites have toward people of other races.

Insofar as women are concerned, it is not only white men who oppress them. Most races of men oppress their women. Men's behavior toward women historically is a blight on our humanity and a fundamental reason that we have not progressed spiritually as a human race.

As such, we can see that the task of creating a balance in the relationship between men and women requires more than a revolution against capitalism and the international capitalist elite. It requires more than an uprising against any particular nation-state or organized religion. Rather it requires a revolution against male chauvinism in general and against those organizations and their policies, dogmas, ceremonies, and laws that serve to keep women in a subordinate and persecuted position in human society.

A nuclear revolution must free women from millennia of persecution if it is to be called a nuclear revolution. Insofar as freedom is seldom granted by others, but is earned by one's own actions, women must take leadership positions within any revolutionary movement. We say this not so much to convince women of their rights and responsibilities in this regard, but to assure that women will be taking charge, and it is up to all revolutionaries to encourage and support such a development.

Equal opportunity between men and women in every respect must be realized in order for a nuclear revolution to be successful.

The process of achieving equal opportunity for women requires eliminating all those means of subjugating women within every occupation, field of study, physical endeavor, religious pronouncement, and state and federal law.

How can men begin to support women in this struggle? Let us begin by looking at the family structure where the subjugation of women began. Domestic violence to this day remains a big threat to women's physical, mental, and spiritual well-being.

It is unfortunate to say that many men marry women, not out of love, but as a means to subjugate them by using physical violence. Such men need a woman to keep house, prepare their meals, and be a sex slave, and if she does not obey, he brutalizes her.

Such domestic violence must be terminated in a revolutionary society and it falls to the men to help eradicate it. Not only must we change our own behavior toward women, but we must help protect women from husbands who will not change. There is nothing sacred about a marriage in which one partner uses the bond of marriage to exploit and brutalize his mate.

Unfortunately, under present conditions, the hands of the police are tied from achieving this goal. Either they themselves blame the woman in a situation of domestic violence or legal means prevent them from terminating the domestic violence, placing the woman in ever greater danger.

Thus, it becomes the responsibility of revolutionary leaders to support its female community members and to act on behalf of battered women. Our actions should entail any means necessary to protect women from being brutalized or from any retaliation from her husband if she reports him, leaves him, etc. In this case, force must be met by force if there is no other alternative. If a man poses a threat to a woman by reason of verbal threat or past behavior, he should be removed from the woman's presence until it is determined by the woman and by the revolutionary leadership that he is no longer a threat. In this situation, whether in jail or mental health facility, the man should receive counseling and his behavior should be rectified before he is released.

If a man remains incarcerated over a period of time so that the economic welfare of the woman and her children is threatened, the revolutionary leadership must be responsible for addressing this problem by providing paid work and/or child care for the mother.

In many neighborhoods across the US, citizens participate in "Neighborhood Watch" programs to prevent theft from occurring in the neighborhood. In like manner, Neighborhood Watch programs can be set up to protect battered women in the neighborhood. Training for such volunteers should be offered by the revolutionary leadership. Wife beaters should be shamed by society at large and their behavior condemned by local community.

If some men think this may be going too far, consider the treatment of women even in this so-called enlightened American society. Women can be sexually assaulted and raped anywhere at any time across the land. They are sexually harassed on the job, in business, in the military, by the police, and in the universities. At the same time, child rapists who dress as religious leaders go free from punishment. These same religion leaders condemn women from their pulpits as inferior. They use religious scriptures to validate their male chauvinism, and use state laws to control women's bodies against their wills. Priests also attack nuns and Protestant ministers attack their female parishioners. In many states, child marriages and forced marriages are still legitimate. Seeking justice, women are persecuted for being raped and for wanting to abort their rapist's offspring. Women have to fear attack just walking down the street. A female jogger must suffer cat calls and face the prospect of rape and murder.

In less enlightened societies, women are forced to cover their entire bodies in black clothing. They cannot leave their homes without being accompanied by a male. They are not allowed to go to school or drive a car. Men use smart phone apps to track their every move. They are subject to female genital mutilation which is even being practiced in the United States.

Because of this long and constant history of oppression, a nuclear revolution cannot occur without the full and equal participation of women in the revolutionary leadership. Women must be represented in equal measure within all economic, cultural, and political organizations and on all local social boards that are responsible for creating the progressive socialist revolution at the grassroots level. It may take time to build a level of trust between people in the new leadership, and to balance various interests, but it is imperative that we do so.

People of Color

Regarding people of non-white races, a similar situation has also occurred. While all societites and races have been guilty of enslaving others, many of the white societies throughout history, have been built upon the slavery of others. Most European countries have been built through colonization and slave labor, and nothing has been easier for them than color-coding slaves for easy identification. No honest person can deny that racism was one of the major factors in the building of the United States. Although stealing of land was not an unusual occurrence among both Native Americans and Latinos, whites not only stole the land from Native Americans, including Latinos, but they also subjected them to slavery, brutality, rape, and death. Once they had the land, the whites brought black people from Africa, under the most horrendous circumstances, to work it for them. Though it is impossible to give accurate figures, some historians have estimated that six to seven million slaves were imported to the Americas during the eighteenth century alone, depriving the African continent of its most valuable resource, its healthiest and ablest men, women, and children.

While Africa suffered this loss, the wealth of the Southern states was procured through slave labor. But it also built the big banks, corporations, and universities of the North. The white dominated system kept blacks from voting until 1965, almost a full century after the Civil War, which ostensibly freed them from slavery. Yet even today, twenty years into the twentieth first century, members of the Republican Party have contrived every possible means to keep people of color from voting, included mass incarcerations.[334]

For this reason, many Native Americans and blacks in the United States will also join the revolutionary leadership to set the course of a new world. In a similar way, colonized people across the world will come forward to join the revolutionary leadership and overthrow the capitalist masters and thereby finally enjoy their freedom. While the people of non-white races will set their own course of freedom, it will be the responsibility of white revolutionaries to support them in their endeavor.

One way that whites in the US can support people who are not white is to oppose racial segregation. New public policies can play a central role in delegitimizing patterns of racial segregation in our neighborhoods, as well as fighting discriminatory practices in the workplace and local governments. Racial discrimination, information gaps, stereotypes and fears, and disparities in income and purchasing power all work together to perpetuate segregation, Because there are spiritual and mental benefits from living in a culturally diverse neighborhood, the public policies of the revolution must intervene to break the cycle of residential and economic segregation. To accomplish this, local revolutionary leaders will need multi-dimensional strategies that tackle the multiple causes of segregation simultaneously.

Neighborhoods matter to the well-being of children and families and we want all of our neighborhoods to be integrated in order to provide basic needs and essential services equally for all people. Quality food, housing, health care, and education, along with reliable child care, safe after-school activities, and healthy recreational facilities should be available to all. Our neighborhoods must transmit the norms and values that we espouse, so that our children know what we expect of them as they mature.[335] To create such a neighborhood is each of our responsibility, regardless of gender or race.

Revolutionary Strategy

The forces opposed to revolution possess immense military power. In spite of it, the revolutionaries will achieve victory. This victory will be due not only to a well-knit organization, progressive ideology, and exemplary leadership, but also to revolutionary strategy.

Without a revolutionary strategy, crucial decisions will be left to chance. This is unacceptable. Furthermore, without a strategy, the leaders can only react to the initiatives of the capitalists, putting themselves always on the defensive. Under such conditions, the best that can be expected are a few reforms. This is essentially the situation in which the progressive forces find themselves today. Because reformers, and not revolutionaries, are currently leading the resistance against capitalism, the people have no long-term revolutionary plans to accomplish their liberation. Today's strategies are focused primarily on reforms and not revolution.

A revolutionary strategy serves several purposes. It provides the most effective ways to bring down the exploitive capitalism. It allows us to determine when the political situation and popular mood is ripe for revolution, and it allows the leaders to choose how and when to make the call for revolution.

To create a revolutionary strategy, we need a realistic assessment of the capability of the people to achieve freedom. Strategic planning increases the likelihood that all available resources will be mobilized and employed most effectively to accomplish this end. Further, a grand strategy must take us beyond destroying the capitalist system because without a plan to do so, we risk just creating another oppressive social system.

A revolutionary strategy is required to transfer political and economic power from the capitalist class to the people and then to protect the interests of the people after the fall of the capitalists from power. To do this, we must understand wherein political power is derived. To answer this question, we look to Gene Sharp who wrote *From Dictatorship to Democracy: A Conceptual Framework for Liberation*, which has inspired revolutionary movements around the world since it was first published in 2002. According to Mr. Sharp, the political power of rulers is based upon several different factors. These are: (1) Authority, (2) Human Resources, (3) Skills and Knowledge, (4) Intangible Factors, (5) Material Resources, and (6) Sanctions. Although it may seem counter-intuitive, each of these factors by which the capitalists maintain their dominance requires the acquiescence of the world's people to remain operable.

For example, the authority of the capitalists is based on the people's belief that the exploitive capitalist system is legitimate and that the people have a moral duty to obey it.

The capitalists' human resources consist of the number and importance of the persons and groups which obey, cooperate with, and provide assistance to the rulers. When the people withdraw from participation in these groups, the power of the capitalists is weakened and ultimately destroyed.

The capitalists also depend upon specific skills and knowledge to perform specific actions. These skills and knowledge are supplied by cooperating persons and groups. For example, in Volume 5 of *The Untold Story of Western Civilization*, we saw how the capitalist class in the United States depends upon "policy-oriented intellectuals" to spread the capitalist ideology in universities and throughout society by using the mainstream mass media and political parties. We also saw how they depend upon warriors to carry out their imperialist wars and how the bankers, weapons manufacturers, and the fossil fuel industry make fortunes off their deaths.The capitalists also possess psychological and ideological factors that induce people to obey and assist the rulers. These factors will be eliminated by the Universal Ideology propounded by the revolutionary forces.

Material resources are another factor that the capitalists depend upon to maintain their political power over the people. Their power extends to whatever degree they have access to, or control, property, natural resources, financial resources, the economic system, the means of communication, transportation, etc. One means of breaking this power of monopoly is for people to build their own basic needs economies at the local level.

Sanctions are the bottom line by which the capitalists maintain power over the people. Sanctions are the threatened or applied punishments that the regime uses against the people for disobedience or non-cooperation, so as to ensure their submission. These sanctions are the reason that the people require a trained fighting force to combat the capitalists' use of physical force and intimidation against the people. The American Revolutionary War against the British was not only an intellectual revolution, it was also a physical revolution. The American revolutionaries trained militias in every town and village and it was these militias that eventually destroyed the British control over the United States.

So now we have seen how the capitalists' domination depends upon the submission of the people and how the revolutionaries must create their own factors of political power to strengthen the people and weaken the capitalists. Withdrawal of cooperation weakens the regime commensurate to the degree that the people become self-reliant.

As Gene Sharp tells us, "The degree of liberty or tyranny in any government is in large part a reflection of the relative determination of the subjects to be free and their willingness and ability to resist efforts to enslave them." [336]

It is the strategy of the revolutionary leadership therefore to diminish the power of the capitalists by noncooperation and by political defiance, while simultaneously working with the people to achieve self-reliance through building local economies under their own control.

In a nuclear revolution, leadership is not imposed from above, rather it is established through sincerity, dedication, ideological zeal, a fighting spirit, and all-round capacity of the local people. The revolutionary leaders will gradually acquire such capabilities step-by-step in their work with their neighbors at the local level, and, in so doing, become the heroes that the people will gladly accompany into battle.

A common base of integrity and unity amongst the exploited members of society will have to be established. This is essential because at the time of revolution there will be an all-out fight against anti-social activities and corruption. In this struggle, the revolutionaries will have to fight against three forces—outside exploiters, internal exploiters, and inner selfish tendencies. All these three forces are powerful. Yet the moral, psychic, and spiritual strength of the revolutionaries will be the cause of their victory, because physical weapons are not the only source of power. Moral and spiritual power will ultimately overcome physical force.

Although the primary duty of the revolutionaries is to drive out the exploiters, they also must ensure that these exploiters do not get new opportunities to capture power or undermine society in the future. The greatest achievement of a revolution is to eradicate exploitation and bring about a progressive change in the collective psychology through the minimum loss of life and property.

The Power of Sentiments

The revolutionary leadership will create a revolutionary strategy, keeping in mind that people are fundamentally sentimental. Therefore, a nuclear revolution will take place around strong common sentiments. If there is no strong common sentiment, a revolution cannot take place. Sentiment has always been stronger than logic.

Communism, for example, propagated sentiments like "workers of the world unite." Initially, people were attracted to such sentiments, but after some time they discovered that they were hollow, consequently intellectuals became dissatisfied with them. Communism is now unable to fight against the local sentiments that are coming up in different parts of the world because these sentiments are stronger than communist sentiments.

PROUT philosophy is based on a *universal sentiment* and the progressive socialist movement is systematically moving towards the implementation of this sentiment. Only progressive socialists can address all local sentiments, while at the same time leading everyone in the world to universalism by gradual stages.

Revolutionaries must know how to arouse the sentiments of the people and channelize their sentimental legacy towards universalism; that is, that all people and the entire planet need to be treated as manifestation of the divine with a pervasive sense of love and compassion. During the preparation for revolution, unstinting efforts must be made to arouse this sentimental legacy in the people, because sentiments inspire popular support for the cause of revolution, and infuse the revolutionary workers with tremendous power and conviction.

According to PROUT, there are two types of sentiments—positive sentiments and negative sentiments. Positive sentiments are synthetic in nature. They unite society, elevate humanity, enhance collective interests, and encourage progressive development. Negative sentiments are narrow in scope and divide society.

Some important positive sentiments include anti-exploitation sentiments, revolutionary sentiments, moral sentiments, universal sentiments, and spiritual sentiments. Some negative sentiments include provincialism, nationalism, lingualism, religious exclusivism, misogyny, and racism.

Sentiments should never be used to create artificial fissiparous tendencies in society. Rather, sentiments should always be used to bring unity amongst people. Hitler used racism in an effort to unite the German people and he succeeded in the short-term, but because he used negative sentiments only and had no positive sentiments, his approach resulted in devastation to the Jewish people, a world war, and the near destruction of Germany itself. The path of negativity is extremely dangerous and harmful for society. Positive sentiments are the real tools by which to build society. This must never be forgotten under any circumstances.[337]

Chapter Ten: What is the Grand Strategy of Nuclear Revolution?

The Grand Strategy to realize a nuclear revolution is composed of several campaigns. These campaigns address the multiple aspects of a nuclear revolution. They are the personal campaign, interpersonal campaign, socio-economic campaign, cultural campaign, environmental campaign, political campaign, and military campaign. The campaigns of the grand strategy are meant to challenge and destroy the capitalists' control over the social, political, economic, and cultural expressions of society, and lead to the creation of a better system for meeting the basic needs of all people. Each of these initiatives must be planned out and put in a timeline within a constantly changing field of struggles characterized by continuous interplays of moves and countermoves. Within each campaign strategy, specific tactics, and methods are necessary to realize the strategic goals.

The creation of a new era will entail making the people self-reliant while confronting their being exploited by the capitalists. Thus, the *revolutionary strategy* must be a polarization strategy, the purpose of which is to create the maximum distance between the exploiters and the exploited and to isolate the exploiters in the minds of the people. Through this polarization approach, the exploited people will learn who the exploiters are and how to defend themselves against exploitive practices.

Personal Campaign

In the personal campaign, individuals will seek to become spiritual warriors. They will familiarize themselves with the Universal Ideology and begin spiritual practices. Spiritual practices are those practices that link the human mind with the Divine Consciousness. In this process one's small ego will gradually expand to merge into and identify with Absolute Consciousness. This is called the process of enlightenment. Enlightenment is the state that Jesus spoke of when he said "I and my father are one." The process of enlightenment is not dependent upon religion. The goal of enlightenment is to experience the total oneness of consciousness, and not engage in divisiveness.

By observing progress in the fields of science, technology, etc. it appears that humans are making progress in the physical world. But is this true human progress? If human beings themselves remain stuck at the same place where they initially started their journey, what progress has there been? By designing a road, one does not advance along the path of progress, rather one will have to move along the road. As such, humans must make progress in the mental realm. And in the process of this movement, they will reach a stage when the mind ends in the spiritual realm. Unless and until the mind reaches that final supreme point, human beings will remain imperfect and unfulfilled in life.

Human beings are moving out of animality and into Divinity. This trajectory can be witnessed in our thirst for limitlessness. We want the Absolute, but we mistakenly try to achieve it by possessing finite objects. We always want more because finite objects never completely satisfy us. In tantric spiritual science, the spiritual path is called Bhagavad Dharma, the Path to the Divine.

The path of spiritual progress can be divided into four stages.[338]

In the first stage, the spiritual practitioner makes a sincere effort to attain the Divine. While one continues making efforts in this stage an unlimited time may pass away. A person may say, "I am trying," but if this trying lasts for three, four, five, twenty or twenty-five lifetimes it may lead to retardation. So, to say, "I am trying," is not enough. Nevertheless, the benefit of the first stage is that at least human beings make constructive efforts.

To progress from the first to the second stage requires firm determination. The only thought now is "I must succeed." This sort of firm determination is the precondition for success. The story is told that Lord Buddha performed spiritual practices for a long time, but still did not attain the final goal. So, when he sat down the second time, he made a firm determination:

> Unless I attain the supreme height of enlightenment, I won't move an inch from this posture, even if my body withers, and my bones, flesh and skin are destroyed.[339]

The second stage starts from the point when one sits down with this firm determination to attain the final success. What happens in this stage? To begin, one gains a certain degree of control over some psychic propensities for a while. Later, at some other time, one may gain a little control over some of the other propensities. This is what a devotee normally experiences in the second stage. It is often noticed that many saints and monks who give up hearth and home have a weakness for food. They are pleased with anyone who offers them delicious food and drink. Other saints and monks, however, who have overcome their weakness for food, get very angry if their visitors do not pay obeisance to them.

These are all psychic diseases or weaknesses of the mind. As spiritual warriors, we must start fighting these obstacles immediately. We must take a vow of "do or die" because this psychic imbalance should not continue for long.

In the third stage one gains full control over a particular psychic propensity. Once controlled, it will never return, and will never cause further degeneration. For example, if a person becomes agitated, causing his respiration to speed up and his fists to close tightly, the propensity of anger gets easily expressed. If this situation continues for long, what will happen? Among the propensities of lust, anger, greed, vanity, attachment, and jealousy, anger will easily dominate the other propensities.

There are some people whose only weakness is greed. They have an uncontrollable desire for money, or food, or sex. They do not even care what other people think about them because they have totally lost control over the instinct of greed.

To have control or lose control over a particular propensity may be temporary or permanent. Some people are totally free from greed. They have neither greed for money, nor greed for food, but they cannot control their other propensities.

In the process of psychic development, spiritual warriors reach a stage where a particular propensity either hinders their spiritual progress, or leads them towards extreme subtlety. In yoga philosophy, this stage is called *ekendriya*.

Next comes the fourth and final stage of spiritual progress. In this stage, all the psychic propensities are brought well within the control of the spiritual practitioner. As everything comes within his or her perfect control, he or she attains the perspective of divinity.

According to Intuitional Science, the attainment of the fourth stage of spiritual progress results in the attainment of eight occult powers. These powers become available through the practices of the spiritual warrior. These powers are listed and explained in Appendix A.

The spiritual path is the proper path for human beings. This is how human beings have progressed step by step in the past, are progressing in the present, and will progress in the future as well.

Now that we have analyzed the stages of spiritual evolution let us look at how this can impact the thinking and behavior of a spiritual warrior.

Our minds have to act within the periphery of time, place, and person. Their creation, maintenance, and destruction are dependent on these three factors. This is why opinions carry their highest value in a particular time, place, or person, and lose their value, or become altogether non-existent, when any of these three relative factors change. Opinions have value in both individual and collective life. But when one is dealing with the Absolute Entity, who transcends the boundaries of time, place, and person, does one's opinion have any value? No, in this case an individual opinion has no value whatsoever.

An opinion is a psychic object. The opinion of a particular person is, in effect, an object of the object of the Supreme Consciousness. The objects of the Supreme Conscious are minds and the objects of minds are opinions. So how can an opinion become the knower of the Supreme Subject? How can an opinion know the Knower of the known? One's opinion in this regard has no value. For anyone to have an independent opinion about Supreme Consciousness is meaningless.

Regarding the Supreme Consciousness, the individual mind can do only one thing: become one with Him/Her as the Supreme Subject. The unit mind must accept the Supreme Consciousness as its subject, and run towards Him/Her.

An opinion held by a particular individual or group in regard to the social sphere, the economic sphere, or the political sphere does have value, however. Thus, an individual mind may hold many opinions about the physical and mental worlds, but no opinion of value regarding the spiritual world. The path which leads to supreme bliss is a singular one; there is no second path. The path is one, and only one. It is the mystical path referred to by all spiritual masters regardless of religion or absence of religion. On this path one learns to love the One and love the many. And the first is equal to the second. The Supreme Consciousness exists within and without.

If this universe is envisioned as a circle, its nucleus is the bliss of perfect love to which every point on the circumference is connected. If any individual located on any point of the circumference wants to reach the Nucleus, he or she will have to move with cosmic ideation, floating on the divine waves of bliss. No matter what point on the circumference one measures the distance to the Nucleus, that radius will always be the same. This radius is termed the path of bliss. So, the path of bliss is the only path. The Divine Father/Mother cannot be attained through rational thought. Rather, it can only be attained through intuition, the highest human mind set. Therefore, a spiritual warrior must develop his or her intuition. He or she must learn to love everything as an expression of the Divine. We must worship the Divine within all forms. To do this a spiritual warrior must awaken his or her mind and fight against the negative influence of the static principle. We will have to struggle ceaselessly against psychic impurities and physical ailments in our individual life, and against social evils and economic disparity in collective life. A spiritual warrior is not afraid of the crushing load of unhappiness in life, and never surrenders to pessimism, but fights against the miseries of life with revolutionary zeal. This is the criterion of a spiritual warrior. Such dauntless fighting spirit will win the laurels of victory one day, a victory which is never possible simply by coaxing and cajoling others.

The individual mind progresses by the process of clash and cohesion. The mind finds natural problems that it must solve, such as procuring food, finding accommodation, and rearing children—not to mention simply staying alive. The more difficult these obstacles are, the more scope the mind gets to unfold. The struggle to overcome obstacles is the primary factor in the development of the mind.

It should be the aspiration of every human being, after having made considerable psychic progress due to natural clashes or due to contact with more-developed minds, to reach that stage of expanded consciousness, where one is able to control the psychic propensities instead of being subservient to them. A human being is a "mind-preponderant being."

As the mind becomes more developed, the physical structure becomes more complex. It grows in complexity in order to serve as a proper vehicle for the expression of a developed mind. When the psychic activities are channeled in different directions or engaged in combatting various obstacles, the brain, as the medium of the mind, is also bound to become more complex.[340]

Similarly, undeveloped human beings can make rapid psychic progress due to natural clashes or due to contact with great personalities. The greater the application of psychic energy to a human being, the more his or her physical stuff is transformed into mind-stuff. This helps one attain greater psychic expansion, and consequently become a greater reflection of Cosmic Consciousness.

All beings are made of the same consciousness. Although we may find ourselves at different stages of psychic elevation, we are all linked by the bonds of fraternity. If we remember this fact, if we work for the well-being of all, then undeveloped human beings will be able to develop their spiritual potentiality when they come in contact with true spiritual warriors. Likewise, every being, including plants and animals who come in close contact with spiritual warriors will be influenced by their spiritual vibrations.

Spiritual practices and education will bring about mental development in those humans who have degenerated to the level of animality, but it will be very difficult for them to develop spiritually. We must pay more attention to those people who are undeveloped, but who have not entirely lost their human sense. If we fail to elevate them, our social system will be futile—all our education, culture, and civilization will be futile.

In the most developed minds, our infinite thirst for limitlessness attains the supreme rank of Divine Consciousness, which is bliss personified. In this way, our infinite thirst for limitlessness is finally quenched. The sustained effort to attain this blissful state is called spiritual practice.

If human beings do not find the proper way to satisfy their thirst for limitlessness, they go astray. Even those who follow the spiritual path may perform harmful actions at any time. Even those whom society respects as intelligent or learned may, in reality, be no better than what society calls "monsters." Our modern society is full of such people. They may talk about human welfare like parrots, but they have no Universal Ideology in their individual and collective lives, and merely dedicate themselves to the deception of the human race as they strive to serve their petty self-interests by any means they can. We cannot expect any beneficial contribution from them.

The existence of individuals is bound up in action. We will have to act and to move. Life is a dynamic process from beginning to end. No one has come to this world to remain static; staticity is contrary to existence. Even the physical body changes every moment.

Human beings perform two types of action: original actions and reactive actions—actions prompted or goaded by karma. Original actions are performed under one's own initiative, and thus one is fully responsible for them. Every original action is a new action. It may represent an extension of the experience of the past, but it is not a reaction. And the actions which human beings are compelled to perform as reactions to their previous actions are called reactive actions. In other words, original actions constitute efforts, and reactive actions constitute the resultants of the original actions. This is a vital concept to understand.

When someone has the intent to harm others by robbing them, cheating them, or slandering them, they are committing original actions. When a dishonest corporate or government employee takes advantage of people, it is an original action, and when he or she becomes ill and has to be rushed to the doctor, it is a reaction to an original action. It does not serve at that point to pray, "I have not knowingly done anything wrong. Oh, Lord, why have you given me such severe punishment?" But God did not give this punishment—the illness was the result of past original actions.

The moment a spiritual warrior starts spiritual practice, he or she must surrender all original actions to the Supreme Consciousness so that they do not have to endure the reactions. This surrender is the most important aspect of spiritual practice. When you surrender the fruits of your actions to the Supreme, you are not performing actions, God is acting through you.

The Sufis tell a wonderful story about Ali, the chief disciple and general of Prophet Mohammed's troops. At the Battle of the Trench, the Muslims were fighting against the Quraysh tribe who were the oppressors that drove them out of Mecca. Ali proved to be an invincible warrior by killing Amr ibn wod al ameree who was one of the most feared warriors at the time. After Ali dropped Amr ibn wod al ameree from his horse, Amr spat at Ali. Ali got angry, and instead of striking Amr, he walked away for a moment to calm himself. Upon returning to Amr he told him, """If I had killed you then I would have satisfied myself and not God's will" and then after surrendered his action to the Divine, he killed Amr. The Prophet said "Ali's strike on the day of the trench, is worth the combined worship of all of mankind and Jinns." Ali's fame grew with every battle in which he fought due to his courage, valor, and God-consciousness. The Prophet Muhammad acknowledged him as the greatest warrior of all time.

The yogis put it another way: The action of offering is Brahma, the ghee offered into the sacrificial fire is Brahma, the fire is Brahma, and the person who offers is Brahma. Those who will maintain this spirit in every action will finally merge in Brahma.

Wherever there is struggle, action is bound to become predominant; likewise, one will have to acquire knowledge—the knowledge of how to struggle. Through knowledge, a spiritual warrior becomes fully aware that all his or her sorrows and afflictions are the results of past original actions. In order to be relieved of their affliction, they do not cry pitifully to God, but, displaying the spirit of valor, they say, "O Lord, give me strength to continue the struggle. I do not want to escape from affliction and suffering, I want to attain You in a joyful struggle against the affliction and suffering."

But how does one gain such knowledge? The mystics tell us that we gain such knowledge by performing spiritual practices. Here are five such complementary means to gain spiritual knowledge. These are: (1) self-care (proper diet, exercise, rest, etc.), (2) adopting a Universal Ideology, (3) selfless service, (4) meditation, and (5) acts of devotion for the Supreme.

In the final stage of spiritual practice, the individual mind becomes one with the Cosmic Mind. The moment before the final merger, the spiritual warrior realizes that the Entity who has come in the form of happiness is his dearest Lord, and the Entity who has come in the form of sorrow is also his dearest Lord. He feels the divine joy of the cosmic play. A spiritual warrior never retreats, for having passed through the initial stages of spiritual practice, he or she has acquired immense courage and valor. In the final stage of spiritual practice, the spiritual warriors offer their greatest treasure—their minds—to the Divine and expect nothing in return. With the absence of mind, they cannot enjoy the sweetness of the divine play any longer. At that supreme stage of surrender the spiritual warriors have no minds of their own, for they have already surrendered their minds to their dear Lord. This is the supreme surrender; this is the supreme attainment.

> Your abode is brimming over with gems and jewels. The goddess of fortune is Your housekeeper. What can I offer to You, O Lord? Oh yes, there is one thing You lack, as it has been stolen by Your devotees; it is Your mind. I therefore offer my mind to You. Please accept it.[341]

In this process the spiritual warrior surrenders his or her individual ego desires in order to better benefit of whole community. Therefore, it is important to create as many spiritual warriors as possible. Each new spiritual warrior will add to the glory of his or her local community and play a key role in guiding the people to a better life in all respects.

As mentioned above, the spiritual warriors also work to gain the skills of the four social classes—Worker, Warrior, Intellectual and Merchant. With a universal spiritual core and the skill sets of all the social classes, the spiritual warrior becomes the ideal leader for which humanity has always longed.

Interpersonal Campaign

The Interpersonal Campaign has three components. The first is for the spiritual warriors to improve their relations with their family and community by engaging in selfless service. By doing their spiritual practice, the spiritual warriors will become a model for their family and thereby create the ideal conditions for the family's well-being. So also, by becoming involved in the community's efforts to become self-reliant, they will be able to model the attitude of selfless service and to help bring others closer to fulfillment and happiness.

The second component of the Interpersonal Campaign is for the spiritual warrior to build relationships with other spiritual warriors. In order for spiritual warriors to lead a nuclear revolution, they must become aware of each other and organize with each other to create the required social transformation. In other words, spiritual warriors must develop a class-consciousness in order to take the reins of power. This class-consciousness will mostly emerge at the local level and fan out through networking with spiritual warriors in other communities. However, spiritual warriors may also begin to band together at the global level to help bring about a world government.

The third component of the Interpersonal Campaign is to create balance within society. Balance means equilibrium. It is referred to in PROUT as prama.[342]

Prama is required in our physical, mental, and spiritual realms, both individually and collectively. Just as pramá is indispensable in individual life, it is indispensable in collective life. The excellence of the social structure, culture, or civilization of a community of people is derived from the degree of pramá which that community attains in its individual and collective life.

Although human beings came onto the earth about a million years ago, and although human civilization started about fifteen thousand years ago, human beings could not bring about pramá in the three spheres of individual and collective life. We did not even feel the necessity of establishing this pramá in individual and collective life.

For example, though the western world has made some material progress by bringing about a certain degree of pramá in the physical sphere and while it has made some effort to bring about pramá through intellectual development, no sincere effort was made to establish pramá in the spiritual sphere, although some is effort being made to do so today.

If we analyze the history of different communities of people in the world, we notice that despite their tremendous physical, psychic, and spiritual potentialities, they did not utilize the opportunity they had to establish pramá in individual and collective life. This was due to their less developed ideas and practices and their imperfect social and economic systems. In the absence of pramá, they could not make their fullest possible contribution to human society by developing different branches of human knowledge, uplifting the standard of culture and civilization, and raising the level of intellect and intuition.

Nature has been bountiful to each and every region of the world, and has endowed us with enormous wealth, both on and under the surface of the earth. In fact, different areas of the world are full of agricultural, mineral, aquatic, medicinal, and forest resources, yet in many economic regions of the world there is utmost poverty, a low standard of living, and cultural and industrial backwardness. As a result, faced with the curse of an acute scarcity of food, clothes, and shelter, as well as lack of educational facilities, even today in the twenty-first century when material science claims to have made rapid progress, millions of people are fighting for their physical survival. Due to the blessings of nature, there is no shortage of physical resources in any economic region. But due to the lack of benevolent propensities within human beings, particularly the ruling class, those materials have not been utilized for social and economic development. Consequently, the people's basic physical necessities (food, clothes, shelter, medical treatment, and education) in many parts of the world have not been met. Obviously, there is a gross lack of pramá in the physical sphere.

Had the economic planning of the world been based on the principle of prama instead of capitalist exploitation and selfishness, the social and economic picture of the world would be completely different from what it is now.

Human society today has reached a stage of degeneration and, as a result, is lost in a wilderness of economic bankruptcy, social unrest, cultural degeneration, religious superstition, and environmental breakdown.

Once the society reaches the state of degeneration, prama cannot be immediately restored. It will take much effort to create balance between the three realms and between the individual and society. Although all three spheres of life—physical, psychic and spiritual—carry equal significance, the physical stage should be given greater importance in the initial stage. When pramá is lost in the physical sphere, the antisocial elements gain the upper hand in society and pollute the entire social environment. Consequently, pramá in the mental and spiritual spheres will also be lost, and the mental condition will further degenerate. So, the establishment of prama in the physical sphere is the foremost necessity.

Prama in the Physical Sphere

What should be done to restore balance in the physical world? First, we must divide the physical sphere into the following economic substrata: agriculture, industry, trade, and commerce. Within each of these sectors, we must determine what can be done to optimize our ability to meet people's basic needs at the local level in a manner that respects the natural world and is conducted in a manner that sustains it.

For example, in agriculture, this will entail rebuilding the topsoil, reducing pollution in the water and air, protecting eco-systems and wildlife, etc. It also means developing planting and animal husbandry strategies that will create the necessary balance between proteins, carbohydrates, fats, vitamins, and minerals for the local people. To do this, a proper irrigation system that does not use underground aquifers, and the use of heirloom seeds instead of GMOs will have to be introduced. Agricultural cooperatives and producer cooperatives will have to be started, and farmers' brigades may have to be formed. Agriculture should be conducted on the basis of the principle of consumption, and not the principle of profit. There should be a proper preservation and distribution of agricultural products. A proper balance in the field of agriculture will help establish balance in the physical sphere. In fact, an agricultural revolution is the basis of a progressive socialist economy.

Creating prama in housing will entail using renewable energy, reducing the quantity of energy used, building with green products like hemp, optimizing the use of passive solar, reducing waste, recycling water, using trees and shrubs for cooling instead of air conditioning, utilizing space efficiently, etc.

We must use the same methods of analysis to eliminate toxins, reduce energy consumption, and use natural materials in the production of clothing.

In health care, prama can be created by reducing the dependency of people on multinational pharmaceutical companies and insurance companies. This will entail research into the healing properties of food and herbs that can be grown locally. This research must include developing information on dosages, diagnosis of illnesses, contraindications, and delivery mechanisms. Other types of medical systems like acupuncture, acupressure, chiropractics, Rolfing, reflexology, etc., should also be explored.

In developing systems to restore balance in the physical stratum as a whole, the following four factors should be considered:

1. The physical demand at present and the physical demand in the foreseeable future;
2. The physical supply at present and the physical supply in the foreseeable future;
3. The maximum utilization of land; and
4. The Five Fundamental Principles of PROUT as they apply to the physical stratum:[343]

The Five Fundamental Principles of PROUT are:

1. No individual should be allowed to accumulate any physical wealth without the clear permission or approval of the collective body.

2. There should be maximum utilization and rational distribution of all mundane, supramundane, and spiritual potentialities of the universe.
3. There should be maximum utilization of the physical, metaphysical, and spiritual potentialities of the individual and collective bodies of the human society.
4. There should be a proper adjustment amongst these physical, metaphysical, mundane, supramundane, and spiritual utilizations.
5. The method of utilization should vary in accordance with the changes in time, place, and person and the utilization should be of a progressive nature.

To address these factors in a comprehensive manner, the local revolutionary leaders will need to develop a Master Plan for their area. See Appendix B for the elements to be addressed in a Master Plan.

Prama in the Psychic Sphere

Just as balance is required in the physical stratum, balance in the psychic sphere must also be created. As far as education is concerned, the science subjects come within the scope of the physical stratum as they are directly concerned with the material world. On the other hand, the arts and humanities (language, literature, history, philosophy, etc.) come within the scope of the psychic stratum.

The human being is predominantly a mental being. Our minds have two main functions: to think and recollect. The excellence of human beings lies in our thinking capacity, intellectual subtlety, and wisdom. In the process of expressing our creative faculties, we externalize colorful and varied ideas and feelings of our psychic world in a variety of ways: on canvas with colors and brushes, in poetry and literature with the stroke of a pens, and in sculpture with the subtle use of hammers and chisels. Our philosophical ideas, our scientific observations and experiments, and our study and analysis of various branches of knowledge are exclusively within the psychic realm of the human mind. But if there is a lack of pramá in the psychic sphere, many mistakes and defects are bound to enter into our art, architecture, literature, philosophy, science, and other branches of human knowledge. Dance may lose its rhythm, painting may lack proportion, music may lose the harmony of its melody and rhythm, and in the various branches of literature, the expressions of thought may remain immature and useless for humanity's progress.

Philosophy began in the distant past in an effort to understand the mysteries of creation. Various scholars in different branches of human knowledge established various schools of philosophy in different ages. Some of these philosophies were idealistic, some were materialistic.

The true purpose of philosophy should be to discover the link between the Creator, human beings, and the natural world. But philosophers, in spite of their sincere efforts, have not yet been able to build a bridge between the relative world and the absolute world. It seems philosophy has lost its way in a labyrinth of metaphysics. The defective conclusions of philosophers have made respective schools of philosophy merely a dogmatic intellectual exercise. For these reasons, it is essential to have a Universal Ideology to create balance in this world. See our book *Universal Ideology: The Thought of P. R. Sarkar* to understand the dimensions of a balanced philosophy that seeks the welfare of humanity and the natural world as one.

The local education system will also need to be improved by having the students imbibe a Universal Ideology that seeks the benefit of all people and all of life. By viewing their class subjects within this context, a sense of the whole will be created for students and they will more easily be able to think for themselves to solve problems.

Thinking analytically has its benefits, but it cannot be the sole manner of thinking. We must also learn how to think in terms of synthesis, to understand that the whole is greater than the sum of its parts and to think from this perspective.

Prama in the Spiritual Sphere

In the spiritual realm, the main purpose of spirituality is to discover the Supreme Consciousness who is lying quiescent in every human being, and to establish oneness between the Supreme Consciousness and individual consciousness.

Very often, while ignorant of real spirituality and goaded by religious dogma, people undertake long and hazardous journeys to places of pilgrimage, sometimes even selling their earthly possessions such as houses and cultivable land to make the trip possible. They hope to attain virtue by taking a holy dip in a sacred river. Needless to say, this not only causes a loss of energy, time, and money, but also causes much trouble and my bring little or no spiritual gain. Religious myths and dogma are a glaring examples of lack of pramá in the spiritual sphere.

Balance in the spiritual sphere can be accomplished when the leadership of the nuclear revolution engages in spiritual practices to achieve union with the Absolute Consciousness. By providing this example, the people will also take up this challenge. The current myriad of contradictory and belligerent religions only serve to maintain imbalance in the spiritual sphere. If these religions are to maintain any relevance in the future of human society, they will have to acknowledge that their current view of Absolute Consciousness is too small if it does not embrace the welfare of all of life in its purview.

When all the balances within the physical, psychic, and spiritual strata collectively form the final prama, its central point will coincide in a perfect balance with the central point of the Supreme Consciousness. A perfect state of balance will be established between individuals and the collectivity, and between the collectivity and the Cosmos. This supreme balance in all spheres will bring a perfect state of pramá everywhere. It will create the stage of all-round welfare, progress, and perfection for all humanity and for the entire living world.

Socio-Economic Campaign

In solving any problem, great or small, the solution must involve a genuine love for humanity and this created world. Only those who understand this can give people guidance in what to do and what not to do. It is not necessary to study great numbers of books or to rely upon those who speculate about the future of the silent masses. The essential requirement is to look upon humanity and the natural world with genuine empathy.

Let me give a simple example. Those who believe in taking control of resources by force, but do not believe in changing human hearts through spiritual practices and spiritual education, argue that the capitalists have plundered the wealth of society for so long that the question of compensation does not even arise. Those who do not believe in violence may argue that in the process of abolishing the capitalist system or socializing the means of production, society should pay adequate compensation for everything it acquires. Those who have genuine love for humanity, however, cannot accept either of those viewpoints.

If large enterprises are acquired by paying compensation, the owners will have to be paid in installments, which will take a long time. During this period the public will not get much benefit. Seen from this perspective, compensation is not a suitable solution. If, on the other hand, there is widespread confiscation of property, a large section of society may face economic deprivation, and as a result social balance may be lost. Not all those who own land or industry are healthy, able-bodied, or young. Many of them are sick, invalids, or elderly, and some are widows and minors. If all their property is forcibly confiscated without compensation, what situation will they be in? Furthermore, not all property owners are rich. Many are poor or belong to the lower middle class. Even if the policy of compensation or exchanging one property for another of similar value is not accepted, those who cherish genuine love for humanity will take a sympathetic view of the plight of those who find themselves in difficulty due to confiscation, and will act accordingly. A monthly pension or a lump sum should be allocated to old people, invalids, children, and destitute women. Opportunities to earn a decent livelihood must be provided to all those, young or middle-aged, who are able-bodied and healthy, if they have no alternative means of support. Such employment opportunities should take into consideration their abilities as well as their needs.

If a proprietor or an industrialist has invested all his capital in a large industrial enterprise, after his assets are confiscated, he or she should be employed in a similar profession according to his or her ability. If, out of vengeance, some local group or individual judges that such people should break rocks for the rest of their lives, these people may be compelled to resign themselves to a sad fate, but what is the benefit to society. Those who have assumed power may derive some sort of satisfaction from their suffering, but the people will not be served in any way by this vengeance. In fact, humanity will have been robbed of these people's skills, if they are willing to admit their injustice to society and make amends.

We must not forget for a moment that even those who have continuously exploited others and suppressed their legitimate rights are themselves human beings. The human family includes everyone. We have to move unitedly along the path of welfare by helping everyone to rectify themselves and adapt. If we try to progress while holding on to a feeling of revenge towards those who committed mistakes in the past, we will be following a suicidal path. Who among us has not been responsible for creating some injustice in our lives? [344] The balance on such a score card will be handled by the Divine. Karma will always be at work whether we can see it or not.

Having said this, the local socio-economies (samajs) begun by the local revolutionary leaders will always support the interests of the exploited community and will oppose all forms of exploitation. Insofar as poverty is the main problem of the masses of people today, local socio-economies (samaj movements) will give top priority to opposing economic exploitation, as this directly affects the livelihood and the very existence of the people. In addition to creating local socio-economies, the revolutionaries will also need to create anti-exploitation sentiments to unite the people against all forms of psycho-economic exploitation.

To accomplish their work at the local level, the revolutionaries can follow a seven-point strategy intended to isolate the economic exploiters within each samaj and mobilize the people to socialize local resources.

1: *The Spiritual Warriors Organize Themselves and Develop a Plan*

The first step for getting a nuclear revolution off the ground is for the local leadership to gather and develop a plan to guarantee the self-reliance of the local people. This plan must entail a program by which local people cooperate to meet their own basic needs independent of multinational corporations and non-local governments. The program must strive for self-reliance in the production of the goods and services required to meet everyone's need for food, clothing, shelter, health care, education, and physical security.

Of the basic necessities, food is the most vital because it must be consumed every day in order to live. Thus, it makes sense that the creation of a local self-reliant economy should entail food production. To increase the acreage available for the production of local food, the revolutionary leadership should divide local land into economic holdings and uneconomic holdings, according to current productivity. Economic holdings are those that produce agricultural wealth where the market price of the produce exceeds the costs of production (including capital, labor, and machinery).

Uneconomic holdings are those where the market price of the produce is less than the cost of production. As uneconomic holdings are not profitable, the landowners usually refrain from producing any crops on these lands. In the rural economy, there may be many plots of land which are not used for producing crops because they are uneconomical.

In the first phase of the socialization of land, the leaders will not raise the demand for a limit on land ownership, but the sale of agricultural land to outside interests will be prohibited and uneconomic landholdings will be brought under cooperative management. The responsibility for cultivating this land will not lie with the landowners but with locally formed cooperatives, preferably under the aegis and assistance of the local government.

The landowners of the uneconomic landholdings in each town will become members of the cooperatives in this phase. Thus, cooperatives will consist of those who merged their land together and those who work the land to make uneconomic landholdings economic. The landowners will provide their land, and in this phase, they will remain the owners of the land. As such, they will be compensated in commensurate value for their contribution. In no case, however, shall the value of the land contributed, secure more than fifty percent of the net profit. The remainder of the net profit shall be shared by the other members in the coop who work the land.

In this phase, the rivers and streams in the town should be harnessed for the collective welfare. For instance, by constructing embankments and small dams on the rivers, irrigation, electricity generation, and industries based on local needs can be established.

Steps must also be taken to alleviate the population pressure on land. An increasing percent of the rural population will have to be employed in industry by establishing agrico-industries and agro-industries. Agrico-industries are pre-harvest industries that produce farm inputs like seeds, soil, earth worms, farm equipment, fertilizer, etc. Agro-industries, on the other hand, comprise those post-harvesting industries which depend on agricultural products. These entail various mills and facilities for value-added production and distribution. There should also be provision for the preservation of crops by building stores and cold-stores under the control of local administrative boards. The cooperatives should be supplied with tools, manure, seeds, water pumps, and other farming equipment through producers' cooperatives. Consumers' cooperatives will purchase and distribute the commodities necessary for daily consumption to the population.

In the very first phase of establishing cooperatives, agricultural laborers, landless laborers, day laborers, and sharecroppers will come within the scope of cooperatives. From this phase, the education system in rural areas should be thoroughly reformed. To arouse the cooperative spirit among the people, there should be extensive training and education, and moral education must take precedence over everything else so that people do not give greater importance to individual interests at the expense of the collective interest.

In the second phase of implementing agricultural cooperatives, the economic holdings of the landowners should be brought under cooperative management. Only after all the uneconomic holdings in a town are brought within the scope of cooperatives should the economic holdings be brought under cooperative management. In this phase it will be easy to apply science and technology extensively in agriculture, increasing the amount of production.

In this second phase, all should be encouraged to join the cooperative system. The net profit will be increased in favor of the laborers working in the cooperatives so that no more than twenty-five percent of the net profit will go to the landowners and seventy-five percent to the laborers. Here laborers mean those who employ either their physical or psychic labor in the cooperative. The landowners will benefit in two ways. First, as landowners, they will get twenty-five percent of the net profit of the produce from the land, and secondly, if they are part of the cooperative labor force, they will be entitled to a portion of the seventy-five percent of the profit distributed among the cooperative members.

In this phase, there must also be a rapid and large-scale establishment of agrico-industries and agro-industries so that the rural population will gradually become more dependent on industry than on agriculture.

From this second phase forward, production for local consumption (not export or profit) will increase the standard of living of the rural population. They will experience real "social security" in that they will have the on-going security of having their minimum requirements of life met.

In the third phase, there should be rational distribution of land and redetermination of ownership. The rational distribution of land will depend on two factors—the minimum holding of land necessary to maintain a family, and the capacity of the farmer to utilize the land. In this phase, the landowners will not be able to employ individual laborers, landless laborers, or sharecroppers for the cultivation of land, so it will be more beneficial for them to participate fully in the cooperative system.

In this phase, it will be easy to establish larger cooperatives by the extensive application of science, but these cooperatives will not be anything like the huge collective farms of the Soviet Union or China. If cooperatives are allowed to become extremely large, it will be difficult to utilize natural resources efficiently and this will lead to complications in the sphere of production. One of the many defects of the collective farms in socialist countries is their unmanageable size.

With progressive socialism (PROUT), the farmers' cooperatives themselves will determine the size of the cooperatives. But while building up the cooperative system, two factors should be kept in mind: (1) the highest quantity and quality of production should be ensured through the application of science and technology while keeping production costs at a minimum, and (2) the cooperative members will be encouraged to achieve their highest level of psychic and spiritual development as well as commit to a minimum of physical labor.

In this third phase of implementing the cooperative system, one hundred percent of the net profit will be distributed among the cooperative members. The former landowners will identify fully with the cooperatives in this phase.

Through these three phases, it will be possible to reduce the excessive population pressure on land and to engage thirty to forty-five percent of the population in agriculture and ag-related industries. While, in the second phase, the problem of unemployment will be tackled through the establishment of agrico- and agro-industries, by the third phase there will be no unemployment problems for workers in agriculture. The rural area will be freed from the vexing problems of unemployment and lack of social security.

In the fourth phase of implementing the cooperative system, there will be no conflict over the ownership of land. The agrarian problems of every town will be solved. All the social security arrangements concerned with food, clothing, housing, education, and medical treatment will be provided to the people. In this phase, it will be possible to make the maximum utilization of the collective physical, psychic, and spiritual wealth of every town.

For the total implementation of the cooperative system, there must be proper psychic preparation through internal urge and external pressure, adjusting with the time factor, because people will never accept a system which is forcibly imposed on them. Such a change in the collective psychology will not occur overnight, but will depend on the sentiment of the people. The time period from the first phase to the fourth phase of the implementation of the cooperative system can be called the transitional period for the implementation of progressive socialism (PROUT).[345]

2: *100% Employment of the Local People*

It is the God-given right of all people to be guaranteed their minimum requirements of food, clothing, housing, health care, and education. This basic right should not be arranged through welfare or dole-outs by external forces but through guaranteed employment. Of course, those who are unable to perform any kind of work will be cared for by the local population.

Unemployment is the most critical economic problem in the world today and one hundred percent employment of the local people is aimed at solving this problem. By one hundred percent employment we mean the employment of those people who are ready, willing, and able to work and to put their effort into the creation of a local socio-economy. It stands to reason that we are not advocating employment within capitalist enterprises, but rather within those local businesses created by the local people with the explicit purpose of producing the goods and services required to meet the basic needs of the local people. This sector of the larger capitalist economy, we call the People's Economy. This local self-reliant economy is not based upon the profit motive, but on meeting basic needs. These needs will be met as local people develop a network of cooperatives which will interact within a system to meet each other's needs.

Local people are defined as those who have merged their individual socio-economic interests with the socio-economic unit in which they live. This is the primary consideration regardless of their age, gender, race, religion, or national origin. Even though people may live in a designated local area, they will not be considered members of the local samaj, nor reap its benefits, unless they make a commitment to join with the local people who are creating a samaj (local socio-economy).

Those who earn their livelihood in a particular socio-economic unit, but spend their earning in or send their earnings to another socio-economic unit are not considered under this definition to be local people. Those who hold franchises with multinational corporations, or administer outposts of multinational corporations, for example, will be considered non-local people because they regularly send capital out of the socio-economic unit from which they earned their income in order to benefit foreign stock holders and the parent companies. This action results in the drainage of capital necessary to stimulate the economic growth of the local socio-economy.

To create one hundred percent employment for all those ready, willing, and able to work within the local socio-economic unit will require that the capital, presently being drained from the local economy, remain in the local economy in order to provide the base for setting up the new economy. Capitalists derived their start-up capital by exploiting the wealth and labor of other people. The capitalists stole natural resources and enslaved people to provide free labor. Progressive socialism will derive its start-up capital by driving out the exploiters and by pooling the resources of the local people.

To build the local People's Economy, both short-term and long-term planning will be required. In the short term, labor intensive industries that produce the goods and services required to meet the basic necessities of life should be started immediately or further developed where they already exist. These industries should be based upon meeting local needs rather than upon export and personal profit. Surplus wealth must be generated internally to guarantee an adequate purchasing capacity for all those employed in these industries as well as ensure the continued existence and growth of these industries.

Of the five basic needs indicated above, food is the most critical, and securing food for the local people will require an agricultural revolution in which local farmers, working in a cooperative manner, are able to produce the food required by the local people in a sustainable manner.

In the medium to long-term, the local economy will require the creation of producer cooperatives to process the locally grown food into finished goods for the consumer. These may include different milling operations (flour, oil, textile, lumber, etc.), storage facilities, and parts production for different processing facilities, etc.

In addition, we will also require producer cooperatives to produce the seeds, equipment, tools, etc,. required by farmers to growth their crops.

In the long term, the local economy will be transformed into a three-tiered progressive socialist economy that will consist of: (1) small, private businesses that produce non-essential goods and services and amenities, (2) medium-scale producer and consumer cooperatives that produce essential and demi-essential goods and services, and (3) key industries that are run in partnership with the local government to produce energy, provide water, treat waste, provide public education, etc., at a "no-profit, no-loss" basis so that the cost of the infrastructure required by all local businesses and consumers is kept to a minimum. Such an economy will be based on the principles of economic decentralization, economic democracy, local self-reliance, maximum utilization and rational distribution of all resources, and be guided by a policy that continuously increases the standard of living of all local people.

Through the continuous process of creating new industries, new products, and new production techniques, which are based on the latest appropriate technology and scientific discoveries and which protect the local eco-systems, the economic vitality of the local economy will become sustainable and prosperous.

The local revolutionary leadership will have to closely study the vegetation in the local eco-systems to determine which plants may be maximally used to create more goods for the local people. The leaders should increase their plant and crop varieties by properly matching them with the soil, topography, and climate conditions of the local area. For example, reforestation can reclaim arid and semiarid land and some unique plants like ferns, which have the capacity to attract clouds, can transform the rainfall and weather patterns in an area. The creation of pre- and post-agricultural industries, based upon the productive potential of different plants, will help solve the unemployment problem by creating a range of new goods and services.

As surplus wealth is generated beyond the need of people to meet their basic needs and maintain their local businesses, it can be used to offer incentives to those individuals who most contribute to the collective welfare. These incentives can take the form of money, goods and services, free time, or public acknowledgement.

The creation of new cooperative businesses, which will meet the basic needs of the local people, may require job training. This training will have to be provided by the revolutionary organization.

3: *A Maximum Number of Industries Should Be Developed in the Local Area According to the Availability of Raw Materials and Local Consumption*

To fully empower the local people, all local resources should be controlled by the local people. This process will take some time but it must be done. The theft or monopolization of local resources by outside interests must be ended. This is the only way to ensure that the industries required to meet the local people's need for self-reliance can be materialized. Once the people have control over local resources, they can begin to develop local industries based upon these resources as the means to meet their basic needs.

This is an integral part of the polarization strategy. Its intention is to put economic wealth into the hands of the local people and divest outsiders of their economic power. In a Proutist economy most of the industries will be run by producer or consumer cooperatives. Only local people will work in and control these economic enterprises. As such, these local businesses will be established with the goal of meeting local needs and not export for personal profit. This can be accomplished when:

A. Local industries use available local raw materials and do not import raw materials from outside the socio-economic unit.

Raw materials are the basic ingredients or the resources necessary to make a finished product. For example, hemp is a plant that can be used to make thousands of different products. It can be used to make clothing, building materials, food, and medicine. These products can be made in small local factories set up by local cooperatives. Thus, if the topography of the local area favors the growth of hemp, then industries should be created around this raw material.

There are several reasons for this approach:

1. Industries based on locally available raw materials can produce commodities cheaper because they are located near ready supplies of the raw materials and this will ensure self-reliance. This advantage does not exist where there is dependence upon outside raw materials.
2. Raw material producers, that is, the farmers and/or producer cooperatives, will prosper as there will be ready markets for their products.
3. Farmers and producer cooperatives will feel secure that there will be sufficient raw materials to feed their businesses and they will be able to plan their future production efficiently. When local economies are dependent upon importing raw materials over which they have no control and the availability of those raw materials lessens or ceases, people face major economic crises.

4. Many multinational corporations deliberately sabotage the economic and political policies of a local area by preventing the growth of local industries based on local raw materials. They exploit the local market by selling them manufactured goods, often made from their own raw materials. Australia, for example, imports manufactured goods from Japan produced from Australian iron ore. Encouraging the growing of local industries based on local raw materials will destroy the stranglehold that multinational corporations have over local economies and end their drainage of capital necessary for the local areas economic growth.

B. Local raw materials are not exported and only manufactured goods may be exported, but only after the requirements of the local people for those goods are met first.

Under capitalism, local raw material prices in the export market are subject to manipulations and erratic fluctuations by capitalist vested interests through speculative commodity exchanges. Local economies have been devastated by such manipulations. For example, the Philippines, which has relied strongly on the export of sugar as a cash crop raw material, had to lay off more than two hundred and fifty thousand sugar workers when the price in the export world market dropped to one quarter of the cost of production. This situation impoverished large areas of the country and caused widespread malnutrition.

While manufactured goods are generally less subject to price manipulations and command better prices than raw materials, we must remember that within the local socio-economy (samaj), the purpose of production is to meet local needs not create products for export and personal profit. By manufacturing finished goods for local consumption, the local economy can keep capital from draining and improve the purchasing capacity of the local area people. This is the way to guarantee an improved standard of living for the local people.

4: *Do not use any outside finished products which can be locally produced or substituted for.*

Before considering the import of manufactured goods, the local planners should determine if, by the use of local raw materials, substitutes can be created for the manufactured goods that require importation.

Importation of manufactured goods means that local capital is being transferred to another socio-economic unit or to the multinational corporation that has produced the product. Drainage of capital is always detrimental to the economic growth of the socio-economy from which it is being taken away. Therefore, unnecessary importation should always be discouraged. In trade agreements between samajs, barter agreements should be arranged so that no net loss occurs to either of the trading partners.

This point requires that the local people support their local industries by purchasing their own finished products. They should buy the finished products of the local industries even if initially they may be of a lesser quality than finished goods manufactured outside the samaj. This is essential so as to ensure the continued economic survival and growth of the local economy. With continued local support, the local industries will eventually survive and progress to a stage when they will be able to produce goods which are of better quality. But, if people continue to purchase finished goods from outside rather than those made locally, then the local infant industries will struggle to remain operational. This will reintroduce unemployment and have negative impact on the local samaj.

To promote the local purchase of local goods is the task of the revolutionary forces. This approach is not new in history. For example, when Great Britain ruled India, India imported salt from Great Britain even though the potential for starting local salt industries existed in India. The Indian leadership then staged a civil disobedience movement and proceeded to make their own salt and boycott the salt that Britain was selling. This movement caught the imagination of the people and with their support, the Indian people became conscious of British exploitation. This movement brought down the price of salt and saved the country from the drainage of wealth into the coffers of Britain. More importantly, this movement helped to polarize the population into pro- and anti-British sentiments. It was not long after this that Britain left India.

Another example is the American trade union campaign to "Buy American." The unions started this campaign to keep American industries from dying and losing their jobs. This campaign was started because the multinationals were hiring foreign labor at a cheaper price and selling imported goods to Americans that American unions formerly produced. To do this, the multinationals used stores like Walmart to sell these foreign products to the American people. The unions needed the assistance of the American people to combat this assault on their lives by the multinational corporations. Unfortunately, it was not much of a success because the unions had not worked to convince the American public that the protection of unions was in the best interests of the American people as a whole. Consequently, many Americans opted to take the cheaper prices offered in Walmart even though the labels on the goods said "Made in China."

5: *Education from primary schools to universities should be in the local language.*

The sum total of human expression is culture, and language is the best medium to express human culture. While samaj policy is to encourage every language, each samaj should use the local language to inspire the feeling of self-respect among the local people. This will create positive feelings that will be projected in their life and work.

The use of a non-local language in the medium of instruction can result in the suppression and subjugation of the local language and inevitably the suppression of the local customs. This, in turn, can lead to an inferiority complex, defeatist mentality, and demoralization that makes the people easy prey to economic, political, and psycho-economic exploitation. This is the strategy employed by the French, Dutch, Spanish, German, English, and American colonizers and must be stopped. When the local people are able to throw off all psychic inferiority complexes, they will be able to become successful in the struggle for self-reliance.

The introduction of the local language as the medium of instruction will also put the local people in charge of their educational institutions, thus ridding them of the cultural prejudices, biases, and teachings that promote psycho-economic exploitation.

6: *The local language should be the medium of communication in governmental and non-governmental institutions and offices.*

By demanding that the local language be the medium of work, the non-locals who are controlling the local industries will be forced out of the local area or be forced to create more employment opportunities for the local people. Having said this, it will not hurt to teach English as a second language in the schools because it is the language that most closely resembles a world language.

7: *Meet the demands that are particular to the local area.*

This last point is to cover all such requirements needed to expedite the process of polarization. The local situation must be carefully studied and demands should be adopted as per the requirements of the particular locality. In one particular samaj, for example, the demand for the construction of bridges and roads to make raw materials accessible might be the first demand to tackle before any industry can be built or developed. In another samaj, the education of the people regarding the causes of their diet-related diseases may be the first demand.

This seven-point socio-economic strategy is a critical component of the grand strategy of the revolutionary leadership. It is, in essence, a means to create self-reliance among the people and weaken the capitalists financially. This strategy is intended to drive a wedge between the exploiters and the exploited people in the economic arena. For a more detailed presentation of this economic strategy, see the chapter on The Economics of Progressive Socialism in Book 2, *Universal Ideology: The Thought of P. R. Sarkar.*

The socio-economic campaign will always be under the control of the local people so as to prevent outside interests gaining control of the people's economic well-being.

Cultural Campaign

We have said before that there is only one human culture, although customs may differ from place to place. By the same reasoning, it is important for the revolutionary leadership to remember that human society is a singular entity. From a spiritual perspective, human society is indivisible. It has been kept divided to accommodate those with selfish agendas. Let us take the capitalists' agenda for example. It benefits them immensely to keep people divided and blaming each other for their problems. By their divide-and-conquer strategy, the capitalists maintain their domination.

Let us look at the divisive sentiments upon which capitalism has been build. Without these divisive sentiments, the exploitive capitalist system could never have been built, nor could it remain in force today. The primary sentiments are sexism, racism, classism, religious exclusivism, and national chauvinism. If you want to break down the reasons why you find ex-President Trump and many of his followers so offensive, it is because they embody all of these divisive sentiments. For example, one way capitalism exists is by exploiting women economically and keeping them in a socially inferior position. Trump reifies this sentiment in his behavior and in his comments about women.

In like manner, the capitalist economy in the United States is based on racism. First the European whites stole the land from the Native Americans and then they imported black slaves from Africa to work the land. Today, most non-whites continue to be exploited economically and kept in an inferior position socially. Trump has demonstrated himself to be a racist and a white supremacist. There would be no immigration crisis today if the immigrants were white. There would be no Muslim ban, if the Muslims were white. There would be no white supremacists in his administration if everyone in the US were white.

Trump is also a capitalist who looks after his own class interests first. Two of his greatest accomplishments to date have been giving the capitalists large tax breaks and cutting human services to the poor. While most Americans declare their income and pay taxes, Trump and other capitalists hide their income and do not pay taxes.

We previously addressed the difference between spirituality and religion. Spirituality seeks Oneness, while many religions are divisive. These divisions are based upon different definitions and symbols of God, which in and of themselves are inadequate and misleading because definitions and symbols of the Absolute will never contain the Absolute. Who cares whether the Hindus envision the Absolute as a union of male and female principles while the Christians envision the Absolute as a Father, Son, and Holy Ghost. Capitalism also utilizes religious divisiveness between Christians. The religions that supports Trump today are the Christian fundamentalists many of whose members were slave owners and oppressors of women. They maintain their same divisive sentiments today. To them, anyone who is not a white Protestant is inferior and not worthy of being an American.

Finally, capitalism utilizes the sentiment of nationalism, which is one of the greatest obstacles in today's world to a united humanity. In the US, Trump and his followers are white nationalists who oppose anyone who does not put America first.

By this simple accounting, we can see that Trump and many of his followers are among the most divisive people on earth. They are the counter-revolutionary die-hards for whom any change is unacceptable. In a nuclear revolution, a person's lack of humanity can be judged by the number of divisive sentiments they hold sacred. The more divisive sentiments that are held, the less of a human being they are. Capitalism is divisive and inhuman, as are its proponents.

We may understand spiritually that human society is one and indivisible, but we may remain forgetful of it. In most cases, forgetfulness is caused by falling under the spell of selfishness. It is evident that the behavioral patterns of selfish people are the same everywhere.

Today, people tend to classify human beings as rich or poor. From a spiritual perspective this is an artificial distinction. Those who are rich today may become poor tomorrow and conversely, those who are poor today may become rich tomorrow. To permanently divide humanity on the basis of poverty and affluence is a divisive tendency. People whom I dislike today as exploiters may deserve it. Under the influence of their karmic programming, however, the same exploiters may fall to the level of the exploited. Did the same people change overnight? No, that is not the case. It is only a reversal of roles. They changed from the role of the exploiters to the role of the exploited. Conversely, those who are exploited today may become the exploiters tomorrow. Therefore, to divide the human race on the basis of such a frail classification is not valuable. The exploited have the same value as the exploiters. The only line of demarcation between them is exploitation. If we remove exploitation from society there will be neither exploiters nor exploited. The fundamental disease is exploitation, and once it is removed, there will be no further demarcation on that basis. The opportunists who utilize a situation to create rifts in the society will not get any scope to do so when exploitation is removed. That is why intelligent humans who desire to promote the well-being of humanity will try to eradicate exploitation from the human society.

With this, the struggle between the exploiters and the exploited will eventally come to an end. Human beings will realize the supreme truth that all of humanity is bound together by common ties of family.[346]

This implies that those who are capitalists today may not be permanent enemies of society. If they adapt the principles of progressive socialism, they may find themselves counted among the revolution's greatest resources. The purpose of a nuclear revolution is not to destroy the capitalists, it is to destroy capitalism. A post-capitalist society will still require the services of those who possess a merchant psychology and are skilled in the workings of an economy. Having made this important qualification, it still may be necessary, after the revolution, to incarcerate those capitalists who fight the revolution by continuing their exploitation of the people. Such capitalists will have to be educated in the values and behaviors required to maintain a spiritually-based human society.

The same situation applies to those "policy-oriented intellectuals who have spent their lives protecting the capitalist system.

At various times, we come across ideological conflicts among human beings. What does this ideological conflict really mean? All human beings should be able to hold different ideas and views so long as their ideas and ideals are not detrimental to other human beings. Those ideologies, which are harmful for the human race, should never be tolerated. Notwithstanding this, different theories and doctrines, with their numerous interpretations and connotations, should be developed. No one should object to that, because intellectual clash and cohesion causes the human intellect to progress. Why should we keep all the doors of the human intellect closed? Let sufficient light and air enter the intellect for its overall enrichment.

Having said this, it is up to the revolutionary leadership to create a Cultural Campaign to dismantle the methods of psycho-economic exploitation used by the capitalists. These methods as described on page 282, include the promulgation of: (1) pseudo-culture, (2) the suppression of the indigenous language and customs; 3) the imposition of numerous restrictions on women; (4) an unpsychological educational system; (5) the suppression of true spiritual endeavor; (6) the division of the people into combative groups; (7) forcing women to use unnatural and harmful methods of birth control; and (8) placing the control over different mass media in the hands of the capitalists.

For such a campaign it is best to engage in methods of struggle that the revolutionary strategist Gene Sharp calls "nonviolent coercion" and "disintegration." *Nonviolent coercion* is the process by which the capitalists, even though they may initially remain in their positions and adhere to their original goals, are unable to act effectively because the people have usurped their political, social, and cultural dominance. *Disintegration* is the final stage of capitalism that occurs when the revolutionaries' self-direction, noncooperation, and defiance have become so complete that the capitalists lack any control over them. Under this condition, the capitalists' bureaucracy and corporations begin to break down.

Gene Sharp provides a list of tactics for Nonviolent Action in his book *The Politics of Nonviolent Action*. These tactics consist of specific actions that can be undertaken by the revolutionary organization to dismantle the power of the capitalist dictatorship and transfer that power to the local people.

The tactics are divided into three strategic areas: (1) Protest and Persuasion, (2) Noncooperation, and (3) Intervention. The purpose of these tactics is to create a social consciousness and move the people toward revolution.

See Appendices C, D, and E for a comprehensive list of the tactics developed by Gene Sharp.

Environmental Campaign

The protection of the environment, including plants, animals, and the life support systems of the planet, is a vital component of a nuclear revolution. Capitalists are destroying the natural world because they view the world as little more than a storehouse of raw materials by which to make their commodities. Furthermore, because the capitalist countries exploit the resources of other countries, they care little about the environmental damage that they impose on these foreign lands. Progressive socialism will remedy this situation because the resources of the natural world will be controlled by local people. This is to say that the resources on which the local people depend in order to survive will be cared for by the same local people. In such a situation, they abuse the natural environment at their own risk and they care for it for their own benefit.

Because the nuclear revolution will be primarily focused at the local area in rural areas, the farmers who are participants in the local economy will take steps to protect the environment. This means moving away from fossil fuels while using renewable energy. It means moving away from GMO seeds and using heirloom seeds instead. It means taking precautionary measures to protect the soil, water, air, and ozone of the planet.

The local community will also be benefitted by setting aside land for wildlife sanctuaries. Tree planting will be encouraged, not only to increase the use of a renewable resource to meet our basic needs, but also to enjoy their beauty and to combat climate change.

Farmers will be encouraged to cultivate perennial and bi-annual food crops to cut down on the energy expended per calorie of food production.

Farmers and all local people will be encouraged to eliminate harmful chemicals from their land, homes, and places of business. This will include chemicals in building materials, cleaners, clothing, household supplies, and food.

Our militias will also be trained to follow a policy of non-destruction of the natural environment should they be engaged in combat operations.

Our workers' cooperatives will conserve resources and energy for their own benefit. They will be motivated to build the best products that they can because they will not need to build depreciation into their products to continue their jobs. Without having to continually produce inferior-made products and unnecessary products, they will meet their basic needs, and also have leisure time to spend with their families or in personal development.

Respect for the natural environment will be taught in all schools and children will grow up with a love for life in all its forms.

Political Campaign

Last but not least, the Grand Strategy of nuclear revolution entails a political campaign. By this, we do not mean a campaign to elect politicians within the capitalist system. Rather, we mean a campaign to build an administrative structure for the revolutionary forces, as well as a governmental structure to replace the capitalist institutions after the call for revolution has been sounded and the revolution is successful.

The development of a formal political structure, as we have said, begins at the local level when the local revolutionary leadership begins the process of organizing the people to build a local economy and take up the struggle to disrupt the oppression of the capitalist system against the local people.

The local revolutionary leaders may hold local political offices or they may not. It does not matter. What matters is that their selfless service garners the attention and support of the local people who become willing to sit on organizing boards to transform the local community. It is from these boards that the revolution at the local level will gain its initial leadership.

The local boards will address such areas as food, housing, clothing, health care, education, social welfare, finance, energy, water, transportation, etc. Local boards will also evolve a local administrative structure that will include a Supreme Board, Executive Board, Legislative Board, and Judicial Board (See *Universal Ideology: The Thought of P. R. Sarkar* for a more detailed description of the function of these boards).

In the process of building local socio-economies, the time will come when adjacent samajs may want to join forces to work on common problems, improve infrastructure, or protect the natural environment. When two or more samajs link in this manner, they may form what

PROUT calls a federated state. In this instance, each samaj will send representatives to serve on similar boards at the federated state level. These boards will seek the welfare of all the people in the federated state.

In a similar manner, the representatives of the federated states will eventually send representatives to a World Government. The World Government will be a confederation or union of federated states. It will address issues that affect all of humanity. These issues will include the drafting of a World Constitution, laws, and regulations in support of the federated states. The federated states may choose not to enact the laws in their locality, but they will never violate them. The World Government may also address issues like defense, foreign relations, crime, environmental protection, etc. The World Government will not however concern itself with the People's Economies, nor with issues related to agriculture, industry, commerce, and trade other than to serve in a consultative capacity, or ensure that all local economies have the capability of meeting the basic necessities of their members.

See *Universal Ideology: The Thought of P. R. Sarkar* for a more in-depth presentation of the political structure of progressive socialism.

The Grand Strategy of a nuclear revolution will therefore consist of several campaigns, each of which contributes to meeting the overall goal of physical, mental, and spiritual balance of the individual and collective. Leaders may work on one or several of these campaigns in the overall pursuit of self-realization and service to suffering humanity and the planet.

Chapter Eleven: The Psychology of a Nuclear Revolutionary

Taking the Opposite Stance in Battle

BECAUSE THE STRATEGY OF nuclear revolution is polarizing by nature, wherein the people are engaged in a struggle to separate themselves from those that exploit them, the psychology of the revolutionary should be to take the opposite stance in battle against exploitive forces. P. R. Sarkar, a spiritual master and social revolutionary, tells us the meaning of "taking the opposite stance in battle."

This principle, he says, is applicable to all spheres of life—physical, mental, and spiritual. One should always remember this golden principle because life means struggle. Without struggle there can be no life. There can be no progress. Those revolutionaries who think they will advance with firm and rapid steps to revolution by avoiding struggle and emerge without even the slightest scratch on their bodies are surely mistaken. In fact, even in the smallest event, one will have to move ahead through struggle. Even during the act of walking, the ground tries to arrest our movement. As long as we are able to struggle against inertia, we will be able to move; but the moment we lose this capacity, we will immediately fall down. Life without struggle means death. Because we have to live through struggle, we must understand the principle of taking the opposite stance in battle.

In physical combat, this principle is quite obvious. When someone shoots a bullet at you from the north, you will certainly not shoot back from north to south. On the contrary, you will shoot from the opposite direction, from south to north. Similarly, if one thrusts a sword at you from north to south, you will certainly thrust back from south to north; this is the accepted principle of battle. When you must struggle against impenetrable darkness, you will certainly carry a torch or flashlight with you.

The Principle Applied in the Mental and Spiritual Realms

Now let us see how this golden principle of taking the opposite stance in battle is applicable in the psychic and spiritual spheres of the individual and also in the social, political, and economic life of the collective.

Regarding the psychic sphere, it is said, "Overcome anger by patience; overcome dishonesty by honesty; overcome greed by generosity; overcome falsehood by truth."[347]

If someone comes to you in anger and challenges you to a fight, your strategy should be to avoid anger because those who are angry have already lost control over their rational judgement. They also diminish their control over their hands and feet, and their entire body trembles with rage. If, at that time, you remain free from anger and keep your brain cool, you may more easily vanquish that angry person in battle with your calm judgement. But if you, too, become as angry as your opponent, this will be foolishness on your part. Hence you must never become angry at an angry person; this is the strategy of battle. It is not only the policy of a decent person, but also an effective strategy of war.

Similarly, the strategy of struggle against dishonest people is to be more honest. When people come to know about their dishonesty, they will support you, and victory will be yours.

If you want to struggle against a miser, start by doing charity for others, and that miser will surely be defeated. If someone will not even offer you a crust of bread, offer him a full meal, and he will be put to shame.

Suppose certain people are wicked; one should not hate them. Rather one should overlook their faults and encourage their humanity. However, one should not let the person's wickedness harm others. This principle of taking the opposite stance should be applied equally in all spheres of life.

When you are struggling against a liar you must be strictly truthful and you will be victorious. Hence the policy of taking the opposite stance in battle is the most effective strategy; the mind must be guided in the opposite direction of that which you are fighting against.

On the other hand, when someone is doing something positive, we should not take the opposite stance, but rather we should also take a positive stance. For example, when someone is happy, people often feel jealous. They do not try to let their minds flow in the same direction as that happy person; rather they feel psychic pain thinking of that person's happiness. The correct strategy, in this case, should be to arouse the feeling of friendliness in the mind towards that happy person. Instead of feeling jealous one should think, "So-and-so is my friend—how wonderful! She is so happy!" If the mind thinks thus, then all jealousy will be removed.

Suppose a person is very sad; seeing his miserable fate some people will say, "He deserved it!" But this is not the proper approach. The proper approach should be to show compassion, thinking, "What a pity he is in great trouble. Had I also been in such a difficult position, how badly I would feel." In this way the feeling of compassion should be aroused in one's mind.

Again, suppose someone is doing some good work, but others cannot tolerate him because they feel inferior to him. Therefore, they try to pull him down. This should not be done; rather we should encourage such people by saying, "Keep up the good work! I fully support you."

Now let us look at how this principle applies to our spiritual life. Spiritual masters have recognized three main paths to attain the Supreme Goal in life: by knowledge, action, and devotion. Of these three, the path of devotion is the highest. There is a particular reason behind this. The cultivation of too much knowledge can create two major defects: lethargy and vanity. You will notice that people of knowledge are often averse to physical labor. They sit idly and indulge in philosophical speculation for hours, but at the same time are dependent upon others for their sustenance. Yet, they remain vane in their self-assessment.

Armchair philosophers generally prefer to remain idle. Many do not even possess the desire to make their abstract theories stand up to reality—they have lost the link with reality. We have all seen this happen to intellectuals.

Rather, we should cultivate, not only our rationality, but also intuitional science (spirituality). This will inspire us with the greatest ideas and also motivate us to take action to serve others. The spiritual approach is to fight against divisiveness by seeking unity with all life and serving all life.

We should always remember that in every sphere of life—social, political, economic, psychic, and spiritual—when we meet any divisiveness or negativity, we should always follow the principle of "taking the opposite stance"; there is no exception to this rule. Victory will surely be ours.

Adopting a Universal Ideology

One should follow an ideology which encourages the development of a universal outlook. It should not be based on narrow ideas. In the past, ideologies were based on very limited ideas, which caused people endless troubles. Due to constant fighting between communities, the human society has suffered tremendously.[348]

According to a universal ideology, human society is one and indivisible. But today, due to superstitions, dogmatism, narrow-mindedness, group psychology, separatism, and all kinds of "isms," human society is split into numerous parts. None of the political, scientific, or religious isms of today are broad enough to bring humanity together and act for the welfare of the planet as a whole. In order to create this supreme unity, we need to adopt a path of synthesis that originates from the psychology of service and welfare. As such, a universal ideology should be propagated by local people wherever they are organizing to achieve nuclear revolution.[349]

Sociologists of the world are well aware that there are apparent differences amongst human beings due to environmental, physiological, geographical, and personal differences. But these differences are based on relative factors while essentially human society is one and indivisible.

Even so, there are those who take advantage of these apparent differences in order to promote their group at the expense of others. Human welfare can never be achieved by this path of self-serving psychology.

To establish a universal ideology in human society, the spiritual sentiment is indispensable. The unity that grows from collective efforts in local movements is the first step towards a greater unity. Every human being has a spiritual thirst. It is a thirst for limitlessness. Knowingly or unknowingly, human beings are searching for the Supreme Entity. Yet, ignorant of the right path, they remain confused. One of life's great tragedies is that so many people do not find the object of their search. Their entire life is spent searching everywhere, but in vain. If people are shown the right way, the entire humanity will converge on the same path. As fellow travelers on the same journey, we will move towards the same supreme goal with unison, with a single rhythm. So, for the unity of the entire humanity, the indispensable factor is spirituality. Spirituality teaches us that the Supreme Consciousness is our Father, the natural world is our Mother, and the entire universe is our homeland.[350]

As the local people cooperate to build their local socio-economic units, these local movements are bound to gain great popularity all over the world within a short time. While there may be diverse expressions of customs and socio-economic potentialities in different units, these differences should not be allowed to divide humanity. If the common sentiments of human beings are given prominence and the points of unity are made the basis of collective development, our apparent differences and diversity will enrich us all rather than tear us apart. If each socio-economic unit is inspired by a universal ideology and a universal outlook, human society will move ahead with accelerating speed towards a sublime ideal.

A universal ideology should be a prerequisite for local, socio-economic units. While a universal ideology will not be established on the earth overnight, it will come to fruition gradually, stage by stage. In doing so, it will include each and every person in the world, as well as animals, plants, and inanimate objects. If a single person remains outside the influence of universalism and becomes a victim of exploitation, then the foundation of our universal ideology will be undermined. Hence, PROUT has adopted a rational method to solve socio-economic problems that may be characterized as universal in spirit, but local in approach.

Once local socio-economic units are established throughout the world, how will exploitation be avoided in the future? Society will enjoy lasting protection from all types of exploitation only if a universal ideology, an empirical spirituality, spiritually-oriented cadres, and proper institutions are well established in social life.

A universal ideology should have several aspects. Primarily: (1) it must provide a rational analysis of socio-economic problems and appropriate, logical solutions; (2) secondly, it must not ignore the human need for psychic expansion and spiritual emancipation; and (3) it should be imbued with inherent dynamism and vitality so that it can guide humanity forward in its quest for all-round progress.

Such an ideology will be needed to protect society from all divisive tendencies and groupist sentiments which create narrow-mindedness. Spirituality does not recognize any unnatural distinctions between human beings. It stands for evolution and elevation and not for superstition or pessimism.

Spiritually-oriented cadres (spiritual warriors) will provide a moral check against all forms of exploitation and propagate moral and spiritual values throughout society according to the motto, "Self-realization and service to suffering humanity."

Finally, institutions that reflect the needs and aspirations of the people and work for the cause of human welfare and the welfare of the earth are also needed. These institutions will gain greater capacity as local socio-economic units join together to form greater networks.[351]

While revolutions are built upon anti-exploitation sentiments as a means to unify the people, if a nuclear revolution is successful, one day exploitation must cease. Even if it is not stopped completely, we can assume that its intensity will be much less than what it is today. In the absence of exploitation, the anti-exploitation sentiment will die out.

What will happen then? The sentiment of a universal ideology will keep people united. It will always maintain the capacity to unite the entire planet, and even the universe, into a nation. Then there will be only one nation—a universal nation.

Today revolutionaries, wherever they live, should, on one hand, propagate an anti-exploitation sentiment and, on the other hand, spread the theory of our common spiritual inheritance—that every living being is the child of the Supreme Consciousness, and that all the people of all nations belong to the same family. This will have to be explained to all because there will be a clash between different nations as long as national feelings exist. People may talk of disarmament, but military preparation will go on underground. If people dedicate themselves to the welfare of the entire human race, their respective nations will also benefit.

The theory of spiritual inheritance is based on the premise that the goal of all living beings is the Supreme Consciousness. This perspective has the power to keep human beings united for all time to come. No other theory can save the human race.[352]

For all the above reasons, a universal ideology should not remain a utopian dream. Rather, it should be the first expression of the practical wisdom of humanity.

The Need for Holistic Thinking

The current public education system prevalent in the West today was formulated to suit the needs of the capitalist class. The main purpose behind the current educational system is to create people with a compliant mentality to take orders in the work force. This type of education is fundamentally defective and creates obstacles in the evolution of humanity.

The system promotes analytical thinking as the only way of knowing and intentionally ignores holistically thinking. It is set up this way to protect the capitalist system which does not benefit from people thinking comprehensively.[353]

Therefore, in order to dispel the prejudices and divisive sentiments of the capitalists' educational system, the nuclear revolutionaries must be able to think comprehensively, synthetically, and universally. They must ensure that our strategies and tactics for creating self-reliant communities are integrated and comprehensive in their approach and shared with others in a systematic manner. To accomplish this, they must be able to think holistically and be grounded in the tenets of a universal ideology.

A spiritual warrior must be an intellectual without the vices of lethargy and vanity. Lethargy and vanity among the intellectuals and warriors serve the capitalist class in two ways. It keeps them dependent upon the existing system for their livelihood, not having the skills to stay alive by any other means. And secondly, it keeps them from changing the system to make it more useful in confronting the problems of the world and in their own lives. In this way, they remain distant from the pain and agony of the world's people and confused about their own identity as human beings.

The Position of Pure Devotion

Pure devotion is free from both lethargy and vanity. If the Lord comes to you and says, "What do you want?" what will you reply? You will probably ask for what you think you need. But because you do not know what you really need, you should say, "Oh Lord, you know best what my need is. You should give me what You feel is necessary for me." You must leave everything to Him. Why risk asking for anything? But what if He insists, saying, "No, you tell me what you want," then you should reply, "Give me absolutely pure devotion." Those who seek pure devotion will certainly acquire the things they need.

Do not be a devotee who says, "Oh Lord, I am your devotee, help me to pass my examination . . . Help me to find a good job . . . Help me get a new computer." This type of showy devotion that simply asks for mundane favors, is not devotion at all, because it demands everything except the Lord. Such devotees never say, "Oh Lord, be mine… I want You and only You." They always say, "I want this, I want that." When one asks only for the Supreme Divinity, this is the only true devotion.

So, the position of devotion is above action and knowledge.

There is a spiritual aphorism that states: "One should be as humble as the grass and as tolerant as the trees; one should give respect to those whom no one respects, and always give praise to the Lord."

As human beings we come to this earth only for a short while. After our stipulated period, we depart. All entities in this world merely come and go, nothing is fixed—everything is moving. When nothing ever ceases to move, can you remain permanently in this world? The effort to try and establish oneself permanently in this world is unnatural. Therefore, one should give up one's inflated ego, his or her desire for self-aggrandizement and prestige, and instead worship the Absolute Divinity.

Only the devotees can come in close contact with the Supreme Consciousness because the devotees' main goal of life is to serve the Lord. To serve the Lord, the devotees cannot remain away from Him. A philosopher or a person of action always maintains some distance from Him—they do not come close to Him because they are immersed in figuring Him out. Does he exist? Is he real? Why will not he give me what I want? Am I a fool for believing in a Supreme Consciousness?

When someone comes in the shelter of the Lord, He elevates them from all psychic distortions. There are certain religious who, at their place of worship, confess to the Lord, "I am a sinner . . . I am a sinner . . . ." This is a defective approach, an unpsychological approach. The Lord knows more than you whether you are a sinner or not, so what is the necessity of telling Him? What news can you report to a person who already knows all the intimate details of your life? Is it not ludicrous to retell the stories of your grandfather's house to your mother, for certainly she knows them better than you!

Then what should be the proper approach to the Divine. You should think within that you are ever pure and enlightened—you are a child of the Lord. You should think that you are like a newborn baby who remains untouched or unaffected by the dirt and dross of sin. The relationship between the Lord and human beings is one of a close family tie.

Therefore, devotion is the highest and best means to merge with the Divine Consciousness. Through devotion, the spiritual warriors will be guided in how to lead the revolution. P. R. Sarkar, as a yogic master, reminds us:

> "It is the frailty of human beings that when a few of them group together, they indulge in criticizing or mudslinging against others. This leads to psychic degradation. In fact, people should perform such actions which will lead to psychic elevation, actions which give the mind no scope for degeneration. The easiest way to do this is to sing the glories of the Supreme Entity, whose very thought automatically elevates the mind. Chanting the name of the Supreme Consciousness aloud is called kiirtan. The

> vibration created in the atmosphere by kiirtan will influence others also. Whenever you find time, do kiirtan loudly and you will never become degenerated. By chanting the Lord's holy name, the mind remains elevated. Mudslinging and criticizing are thereby stopped forever, and hence kiirtan averts the degradation of the human mind."[354]

To be victorious, our spiritual leaders, will have to practice tolerance. But this spirit of tolerance must be developed only in our personal lives. In our collective life, it must not be expressed toward the capitalists and the defenders of capitalism when their intent is to bring harm to others. If the capitalists illegally grab even one inch of our land, we must take one mile of the aggressor's land; no spirit of tolerance in this regard is permissible. If we show our tolerance to such aggression, then they will take one inch today and one mile tomorrow. This is the lesson of history. However, we can practice tolerance as much as possible in our individual lives as we work beside our neighbors to build a viable community. This will be considered a sign of greatness.

In individual life, we expect that people should respect and honor us, but we usually do not think about respecting and honoring others. We long for name and fame, and we burn with impatience because we are not receiving the honor we expect. Therefore, we must use the reverse strategy: give respect to those whom no one gives respect, and our minds will become balanced and at peace.

When the immigrants, refugees, and vagrants are stranded by the wayside, do you ever think to greet them? Are they not also expressions of the same One Supreme Consciousness? But when well-dressed, wealthy businessmen and leaders stand before us, we tend to bow humbly before them. Why this discrimination? All are manifestations of one and the same God: yet one person we greet warmly and another we utterly neglect. It shows that there is a hidden sense of flattery in our minds and we are secretly trying to gain honor from the elite. Rather, we must give respect to those who do not receive respect from anyone. In this way, our minds will become balanced and we will make progress as human beings.

"When trees are laden with fruit, they do not raise their heads high, rather they bend down low: this is the sign of greatness."[355]

Chapter Twelve: The Intellectual Revolution

"The character of good people is to fight against injustice, to firmly adhere to truth and righteousness, to protect the helpless and fight against all evil persons; those who are merely simple and are of a meek and goody-goody nature are not good people." P. R. Sarkar

Many revolutions in history and in the present are based upon defective approaches. Let us take the Robin Hood approach for example. This refers to those people who want to seize the wealth of the capitalists by violence and distribute it among those in need. In nearly every country of the world such Robin Hoods have emerged, but they have not been able to solve the problem of social injustice. The main reason is that no people can long survive on charity. Such an approach tends to create a society of beggars who believe that society owes them a living. Moreover, plundering the wealth of the rich does not destroy capitalism because, although the Robin Hoods may reduce the assets of the capitalists, they do not destroy the seed of capitalism. It may be possible to snatch away the wealth of the rich by violent means, but this will not permanently prevent them from again becoming rich. The capitalists who lose their wealth will plot greater conspiracies in the future.

Then where does the solution to this problem lie? A change of heart is absolutely necessary, but such a change will never be possible through violent means alone. If the capitalists' incessant hunger for wealth is held in check due to a fear of the collective force of society or if they pretend to be free from this hunger, it does not mean that they have acquired the peace of mind that comes with not being hungry or will not engage in ruthless acts of violence to appease their hunger if they get the opportunity.

On the other hand, some people are of the opinion that only humanistic appeals and no other approach can effect a change of heart among the capitalists. Although the principles of such people may be high, their appeals to do good will not find reception among the capitalists.

Such humanistic appeals do not hold much value for those with crude minds. To vibrate the minds of such people, it is, and will forever remain, necessary to influence their minds by giving them a rude shock. Otherwise, we will have to wait forever while the exploited people, on whose behalf these appeals are being made, will be long gone.

No matter how much importance was given to the benevolence of the human mind by the Gandhian non-violent movements, selfish and mean-minded people will never accept these principles. The bleeding and tear-gassed peaceful protesters will never soften the minds of ruthless exploiters. Gandhism may be an excellent utopian model, but in the harsh reality of the world it is unworkable and self-righteous.

While it is true that the human mind must be vibrated by humanistic values this alone is not enough. Rather, all necessary steps must be taken to create circumstantial pressure on the capitalists. This may mean violence in a totalitarian state and legal compulsion in a democratic one; it may require the use of one or the other depending upon social conditions, time, and place. But the primary objective is not simply to use any means available to create a vibration in the minds of exploiters. Proper moral training must also be imparted to keep the energy of that vibration alive, and there must be a ceaseless endeavor in individual and collective life to perpetuate the waves of that vibration. One must maintain one's zeal so that, with the passage of time, the momentum of the vibration does not slow down. Its vigor and vitality should not become stagnant. [356]

It must be remembered that the merchant psychology is one of the basic means by which people have power in this relative world and, as such, it will never be eliminated. The Social Cycle, by its natural movement will inevitably create the conditions in which that psychology will again hold sway over human society.

Therefore, those who depend solely on humanistic appeals or violence to overthrow capitalism without changing the hearts of the capitalists and of the people, without implementing local development programs, and without introducing moral and ideological education to address people's corrupt and divisive habits, are bound to fail.

That is why neither non-violent Gandhism nor communist brute force can bring about human welfare. People will have to adopt a strategy in which there is sufficient scope for both humanistic appeals, based upon a universal ideology, and for physical force, as well as the application of other types of force when necessary.

To build something on the basis of a universal ideology means to build something on the basis of real love for humanity and the natural world. It is not possible to build a genuine society and maintain a healthy planet if its most intelligent and active members, constantly evaluate their contribution to society in terms of profit and loss. When love for humanity is the primary concern, the question of individual profit and loss becomes secondary.

Having said this, we cannot entirely dismiss the question of personal rewards, because if people incur losses, if one's personal interest suffers seriously, or if, for some reason, one's survival is threatened, their love for humanity may be affected. In a society based on mutual love, everyone would have the right to redress their loss.

However, all crude objects of enjoyment are limited in nature, and thus their accumulation by any one individual is not desirable. They should be enjoyed by all, and no one should be deprived of their enjoyment due to another's excesses. Having said this, unless people's propensities are directed towards subtle spiritual forms of enjoyment, their minds will inevitably get enmeshed in thoughts of material enjoyment.

A person who runs after petty objects of enjoyment to fulfill his or her personal desires lacks the spirit needed to unite humanity. To unite humanity, one must love the divine element latent in every human being, which is the reflection of the Supreme Consciousness. This type of love does not arise in the hearts of those who are not inspired by or sympathetic to the lofty ideal of unqualified universalism. As P. R. Sarkar explains:

> It is completely meaningless for people to try to do something constructive if they lack Cosmic ideation—if they have not learned how to accept the most magnanimous, universal, omniscient Supreme Entity as the object of their devotion . . . .[357]

The human mind does not want to be suppressed. It wants to find ways to express itself. The endeavor to remove petty indulgences from the mind leads human beings slowly but surely to the Supreme Entity and establishes them in a universal ideology. Hence it will not do for those who are determined to solve the problems of humanity and the natural world to accept inferior ideals. Along with this, we must also acquire the strength we need to implement our ideals. The active endeavor to acquire this strength is called spiritual practice. It should be borne in mind that no theory will ever bring people salvation. In fact, it is the inner strength gained from spiritual practices that helps to expand the individual mind. The tremendous force acquired from spiritual practices helps to bridge the gap between the harsh realities of human existence and the supreme goal of human life. This is an eternal truth, applicable to all spheres of life—personal, social, economic, and spiritual.

This approach to revolution, therefore, assumes that we will need both violent and non-violent tactics to create a nuclear revolution. And beyond this, we will also need spiritual training and spiritual discipline to maintain a revolutionary spirit that will continually serve humanity and the planet. In our strategy, we should advance as far as we can using non-violent tactics and only use violent tactics when we are forced to do so because we are being attacked or physically oppressed by enemy forces and there are no other means to serve the people. Our militias are created for defense purposes and should only go on the offensive as part of a defensive strategy.

Thus, we can say that the spiritual warriors will capture power first by intellectual revolution, using militias to defend our gains, and, only when the time comes, by mobilizing the masses in a physical revolution, if necessary.

The Ideal of Cosmic Ownership

In a nuclear revolution, the primary intellectual component is based on the ideal of Cosmic Ownership. This is to say that the universe is a creation of the Supreme Entity and as such the ownership of the universe is with the Supreme Entity. We may enjoy and utilize the objects of this relative world, but we cannot declare that any item is mine. Everything is our common inheritance. We are to accept this factor with an appreciation of the requirements of time, place, and condition. Not only will this common spiritual factor unify the inhabitants of multilingual or multi-regional countries, the entire human society will come under the banner of this common spiritual ideal.

Insofar as every human being is the child of the Supreme and has the capacity to be one with Supreme Consciousness, the revolutionary leadership will need to make all people conscious of their rights to our common inheritance. No gender, class, race, generation, nation, or religion has the right to deprive others of their God-given right to our common inheritance. The best way to ensure this is to acknowledge that local resources are to be controlled by local people so as to facilitate their ability to meet their own basic needs—the needs for food, shelter, clothing, health care, and education.

To accomplish this task, the revolutionary leadership will have to create an intellectual revolution. In the first phase of this revolution, they will have to create events to popularize the tenets of the Universal Ideology among the disgruntled middle-class warriors and intellectuals. This is the first phase of intellectual education and will make possible the creation of revolutionary cadre.

Educating the Masses

In the second phase, the leaders and cadre will develop and spread the ideas of Universal Ideology to the general public. This work can involve public forums, mass rallies, distribution of popular literature, and all the other tactics indicated by Gene Sharp.

Gene Sharp, an Englishman, who is a student of revolutions, has compiled a list of political tactics for overthrowing oppressive regimes. In his book, From *Dictatorship to Democracy,* he categorizes tactics accordingly: Protest and Persuasion, Noncooperation, and Intervention. See Appendices B, C, and D for a complete list of tactics in these areas.

Strategies of Nonviolent Action

Protest and Persuasion

The strategy of Protest and Persuasion makes use of several tactics, which include: (1) Formal statements, (2) Mass communication, (3) Group activities, (4) Symbolic public acts, (4) Pressure on individuals, (5) Drama and music, (6) Processions, (7) Honoring the dead, (8) Public assemblies, and (9) Withdrawal and renunciation. See Appendix B for a list of specific tactics under each of these categories.

Noncooperation

The strategy of Noncooperation is divided into Social, Economic, and Political tactics. Social Noncooperation includes tactics like: (1) Ostracism of persons, (2) Noncooperation with social events, customs, and institutions, and (3) Withdrawal from the social system. These tactics can be performed by individuals or groups.

Economic Noncooperation within a nuclear revolution entails the creation of local, self-reliant economies. Please review the seven-point strategy above and also the book on *Universal Ideology: The Thought of P. R. Sarkar* for more details. In addition to this effort, Gene Sharp also proposes ways to weaken the capitalist economy by taking direct action against it. Tactics include different kinds of Economic Boycotts and Strikes. (See Appendix C).

Tactics of Political Noncooperation include: (1) Rejection of Authority, (2) Citizens' noncooperation with government, (3) Citizens' alternatives to obedience, (4) Actions by government personnel, (5) Domestic governmental action, and (6) International governmental action. See Appendix C for a list of tactics specific to social, economic, and political noncooperation.

Intervention

The tactics of Nonviolent Intervention are: (1) Psychological intervention, (2) Physical intervention, (3) Social intervention, (4) Economic intervention, and (5) Political intervention. Interventions differ from noncooperation in that intervention tactics entail those actions specifically taken to stop or hinder the working of the capitalist social, economic, and political systems and to end their control over the minds of the people. See Appendix D for a breakdown of tactics specific to psychological, physical, social, economic, and political interventions.

In the process of utilizing these strategies and tactics for mass mobilization, the people will need to develop an ideological consciousness even though they may not be immediately interested in developing an ideological education.

The intellectual revolution stands for the propagation of ideals. While these ideals are essential for individual and social progress, to materialize them will take a long time. Suffering humanity cannot afford to wait for this. Therefore, a nuclear revolution can never be simply an intellectual revolution; nor can the running of progressive socialists for government offices within a capitalist democracy immediately benefit the people of the world or protect the planetary life support systems, in a timely manner. These goals can only be met by the elimination of the capitalist system.

When the hopes and aspirations of people cannot be fulfilled in a democratic framework, a physical revolution is likely to take place. Such a revolution, although it may be undesirable to many in the moment, is irresistible.

Under such conditions, the revolutionary leadership will create a revolutionary consciousness among the people in anticipation of a call for physical revolution. A physical revolution entails a fight against all factors which go against the principle of public well-being. The revolutionaries will inaugurate a new era of revolution against all sorts of divisive tendencies and social evils including racism, sexism, classism, religious fundamentalism, and national chauvinism. If a country's laws are not strong enough to rectify the conduct of immoralists, the revolutionaries will do something concrete to stop them.

Unless a benevolent, Neo-Humanistic social system, under the guidance of spiritual warriors can be established, the maximum utilization and rational distribution of our mutual God-given resources will not be possible. Progressive socialization will not thrive within a capitalist democratic framework.[358]

Under such conditions it will be the duty of intellectuals and warriors to help the revolutionary leadership to educate and eventually mobilize the masses.

The leaders are those who follow a Universal Ideology and perform spiritual practices based on the principles of self-realization and selfless service to humanity and the planet. Such people will be the guiding personalities of the society of tomorrow. It is the duty of the conscious people to snatch away the physical and intellectual leadership from the hands of the capitalists and the hands of politicians who are engaged in nothing but mudslinging and in deflecting the people's attention from the real cause of their misery. If the spiritual warriors get the active support of the people, revolution is bound to come.

In the case where a government adopts the ideals of progressive socialism, the rule of spiritual warriors will prevail. If the same is not adopted by a government, a physical revolution is sure to come and ultimately the power will be transferred to the spiritual revolutionaries.[359]

Spiritual revolutionaries, therefore, are not inactive witnesses. They are active participants to see that no person or class exploits the rest. For this, they may need to resort to physical violence in order to strike at the source of the power of the capitalists or whichever class is tending to become the exploiter. In case the warrior class is becoming exploitive, the spiritual warriors may have to resort to physical force, and in an age where the intellectual class is dominating, they will have to bring about a revolution in the intellectual field.[360]

Chapter Thirteen: Next Steps

The Process of Self-Realization

For those who are sick and tired of being exploited and seeing their fellow human beings exploited, the first step in beginning a nuclear revolution is to start the process of self-realization. This entails learning the principles and methods of a universal ideology. One can receive an overview of a universal ideology by reading *Universal Ideology: The Thought of P.R. Sarkar* available on Amazon.com and Barnes and Noble as well as at many book stores.

The process of self-actualization entails performing spiritual practices. These practices should include: (1) acting according to the principles of a universal morality, (2) eating nutritious food, (3) getting proper rest, (4) exercising (especially yoga that keeps the spine and glandular system in good operating order), (5) practicing acts of selfless service to others and the environment, and (6) learning how to meditate.

By meditation, we do not mean concentrating on a candle or practicing mindfulness. We mean ideating on the Absolute Consciousness with a feeling of love. We mean adapting a daily practice to reduce the distance between one's individual consciousness and the Absolute Consciousness.

From this description of the process of self-realization, it should be clear that it is unrelated to religion, although some elements of religions may be helpful to those revolutionaries who find them helpful. They must, however, be aware of the corrupting influence of dogma which tends to be divisive.

Environmental Protection

All efforts to protect the natural environment must be continued and expanded. By environmental protection, we mean the practice of protecting the natural environment by individuals, organizations, and governments. Its objectives are to conserve natural resources and the existing natural environment and, where possible, to repair damage and reverse trends. Environmental protection includes the conservation of minerals, soil, water, air, ozone, forests, all species of insects, plants, animals, sea creatures, etc. It includes protection of the climate by reducing and ultimately eliminating the burning of fossil fuels.

Sustainable Energy

We should continue to promote and expand the use of sustainable energy. Sustainable energy is a form of energy that is safe for all life forms and that can meet today's demand for energy without being in danger of expiring or being depleted, and can be used over and over again. Sustainable energy should be widely encouraged as it does not cause harm to the environment and is available in every locality.

Sources of renewable energy include solar, wind, geothermal, hydropower and ocean energy. It does not include any sources that are derived from fossil fuels. Fossil fuels are not considered to be sustainable energy because they are limited in supply and cause immense pollution by releasing harmful gases. The burning of fossil fuels also heats up the atmosphere and destroys the earth's benevolent climate. Fossil fuels normally include coal, oil, and natural gas.

Continue the Cultural War against Consumerism, Sexism, Racism, Capitalist Dominance, National Chauvinism, and Religious Bigotry.

Our culture war is expressed in the conflict between the values, morality, and lifestyle of the capitalist exploiters and their proponents versus the the exploited population. We need to express those values that oppose the exploitative values of capitalism so that the people begin to see the source of their oppression and are persuaded to stand with the revolutionaries in our quest for a society that supports the people and the natural world.

By cultural war, in this context, we do not mean the cultural war between the Left and the Right, or between the conservatives and liberals that is promoted by the capitalist system. This is, by and large, a contrived cultural war to keep the people from focusing on the real cause of their suffering, which is capitalist oppression.

Build Community and Local Political Power

Start local movements, with or without the support of local governments, to meet the people's basic needs. In these movements, develop a local leadership that is able to address the various projects that are required to establish an autonomous basic needs infrastructure. For example, these projects could entail the creation of an autonomous local food system, a project to put green energy on the ground, a project to fund and develop local workers cooperatives, etc. See the discussion of Carbondale Spring below for a model of such a movement. In these movements local planning bodies will emerge and leaders will be identified.

Training Courses

Once a handful of spiritual warriors begins to form at the local level by virtue of their own spiritual intention and self-discipline and are able to create a local self-reliance movement, it will be helpful to have training courses available to help them with their task of building a local People's Economy. There are many organizations that work at the local level in the fields of permaculture, intentional communities, alternative currencies, etc., who have valuable information that can be used in the development of such training courses. Of particular help is the "Tools to Change the World," a Study Guide based on the Progressive Utilization Theory (PROUT) written by Dada Maheshvarananda and Mirra Price. It consists of learning modules that introduce key concepts that a new leadership will require to understand the physical, mental, and spiritual transformations required of the individual and social organizations. The Study Guide is available at www.prout.info.

Socio-Economic Models (Samajs)

Carbondale Spring: A Local Socio-Economic Model

As stated many times, a way for people to free themselves from capitalist exploitation is to meet our own basic needs by cooperatively building our own local economies using our own labor and the local natural resources to do so.

This will require a democratic planning process to build a coordinated production and distribution network of worker's cooperatives that are based on the shared principles of economic democracy and local self-reliance.

Decentralized economic planning is performed with the intention of producing those goods and services that local people require to meet their basic needs. It stands in opposition to the centralized planning of capitalist multinational corporations and the centralized planning of communist governments. It is the primary solution to overcoming capitalist exploitation and to provide a base for the people to launch a nuclear revolution.

All local activists, regardless of their primary issue, should unite with other local activists to create the building blocks of a local, basic needs economy.

Carbondale Spring is a movement started by residents of Carbondale, Illinois to address the economic and environmental challenges that face the local population. These economic problems included:

1. The loss of local small businesses;
2. The dependence on a food system that is undependable and the purveyor of toxic chemicals. The majority of the food supplied by this system not only lacks nutrition, it creates sickness and disease. The top four diseases in the United States are all diet-related. These include heart disease, diabetes, obesity, and a range of cancers caused by agricultural chemicals and the so-called food we ingest;
3. Our energy supply is based upon fossil fuels which are non-renewable, poisonous to the environment, and are responsible for heating up the atmosphere beyond the capacity of human beings and the natural world to survive and prosper; and
4. The dependence upon a police force that does not prevent crime, but rather perpetrates a psychology that maintains that the only way to solve human conflict is with armed force. The maintenance of this police force costs the people of Carbondale millions of dollars each year that can be better used to make Carbondale a model, self-reliant, city of the future.

To begin to address these problems, Carbondale Spring is developing six projects:

1. The creation of a Chamber of Cooperatives that contains a revolving loan fund to provide administrative assistance and funding to people who want to create

a worker's coop whose purpose is to meet the community's needs instead of siphoning profit from the community.

2. A food autonomy project with the aim of creating a local food system that provides safe, nutritious food at affordable prices and keeps money circulating within the local economy.
3. A group-buy solar energy project to put solar panels on residential properties and use economies of scale to assist low-income home owners to install solar as well. This will contribute to a safe, healthy environment, create more local jobs and keep money circulating in the local community.
4. The creation of a team of care workers to assist people in need of help in non-violent circumstances when guns are not required and only escalate tensions. This team of care workers will react to 911 calls and be used to answer calls for help when guns are not necessary.
5. A media team to continually promote the values, ideas, and the activities of Carbondale Spring to the public.
6. A campaign to shrink the police department budget and use freed up city funds to advance programs like those mentioned above.

Additional Models

According to the Prout Institute there are four important questions to ask about any economic development program: (1) Does it help meet people's essential needs? (2) Does it provide opportunities for meaningful work and increasing purchasing capacity? (3) Do local people control and manage it—and do so for their common welfare? and (4) Does it help keep resources, productive capacity, and wealth working regionally for local development—and not drained away to serve outside interests?

Two other models should also be looked at based upon these four criteria: The Emilia Romagna Model in Italy and the Timbaktu Collective in India.

The Emilia Romagna Model

Emilia Romagna is a region located in northern Italy. It contains 7.3% of the population of Italy, has 9% of its manufacturing firms, and produces 9.3% of its GDP. Other things to note about this model is that the region has the second highest rate of employment and the highest per capita income. It is tenth in performance among the European Union's one hundred and twenty-two economic regions.

The region is home to fifteen thousand coops that produce about 40% of the region's GDP. Fifty-seven percent of the population participates in a coop and 10% of the population is employed in a coop. A second feature of Emilia Romagna is that its production is highly diversified with 97.5% of its firms having fifty employees or less.

During WWII when the Nazis invaded Italy, the people of Emilian Romagna, which was a socialist stronghold, hid their machines and tools so they could not be stolen or taken over by the Nazis. After the war as Europe struggled to rebuild, the people here were able to jumpstart their local economy. This gave them the initiative to decentralize their economy and to put it in the hands of local people.

Growth was generated from within the region and profits were reinvested back into the local economy. The first generation of coops focused on meeting people's basic needs. In time, these created subsidiary industries that gave rise to more coops and small businesses. They also gave rise to secondary cooperatives in finance, business development, education/training, and research/development. The coops clustered by sector around cities.

Another concomitant development was the creation of a partnership between consortia of business sectors with the local government to create a service sector that was funded by the local businesses, labor unions, and universities to provide strategic assistance to start ups. By law, three percent of the profits of coops have to go to assist new firms. In addition the law requires that commercial coops are exempt from seventy percent of taxes if they invested in reserves to grow the local industries. Credit coops also help finance start ups. Policy coops have the right to hold shares in private companies, who can also then control those firms, to eliminate competition locally.

Some of the key lessons derived from Emilia Romagna include:

- Strong relationships and focus on local opportunities;
- Networks of small enterprises, often in clusters by sector and geographic region;
- Key role for cooperatives;
- Public-private agency for regional development;
- Local service centers for applied assistance;
- Diversified financial mechanisms; and
- Supportive legislation[373]

Timbaktu Collective - India

The Timbaktu Collective is located in the southeast area of India in the state of Andhra Pradesh, Anantapur District.

The vision of the Timbaktu Collective is that rural communities should govern themselves, live in social and gender harmony, and create a sustainable lifestyle.

They are fighting against the legacy of British colonialism and the imposition of the so-called Green Revolution based on monocropping, fossil fuel chemicals, and production for export, which has undercut the autonomy of local villages.

Timbaktu Collective is organized by concentrating on complimentary functions. These are: (1) organic farming and marketing, (2) rural enterprise and management, (3) animal husbandry, (4) alternative banking and credit, (5) legal aid and counseling, (6) ecological restoration and biodiversity, (7) people's education, 8) cultural activities and spaces, and 9) a focus on people with disabilities and social handicaps.

Their strategies include: (1) local self-governance, (2) village-based organizations, (3) producer-owned businesses that pool their resources, (4) cooperatives, (5) model projects, and (6) education and cultural activities.

When the collective originated, the area was mostly desert. They planted a huge variety of trees to reestablish biodiversity and turned the landscape green.

One of the keys to accomplishing this amazing task was to focus on adult education. Today, the people have created schools across the region to educate children about holistic farming and the regional culture.

The collective links the farmers with local and regional markets and helps to manage that process. They offer training classes, farm visits, peer reviews. etc., that help the farmers work cooperatively instead of competitively.

Women run credit cooperatives that pool local capital and then loan it out to other local enterprises and new businesses. A loan is guaranteed by having four members of the network back that loan. This builds strong personal relationships between people and provides support for each new business or coop. The women have united their cooperatives into a federation that allows them to maintain their local autonomy, but also allows them to coordinate their activities and strengthen the local economy as a whole

Once crops are grown, they are processed thru locally created production facilities that turn the raw foods into marketable products. The products are marketed throughout India and can also be ordered on-line.

Cultural celebrations also help bring the people together.

The success of the Timbaktu Collective vision can be witnessed by the participation of local people in its work.

- Over 21,000 women are members in 4 credit coops;
- Over 2,000 shareholder members are in the farmer coop that produce the Timbaktu organic products;
- The tree-growers coop has regenerated over 7,500 acres which has supported the livelihoods of over 1,000 familes;
- Residential schools now serve children from 48 villages, including 31 community vegetable gardens; and
- A livestock coop supports 625 landless laborers who tend sheep and goats;

The key lessons learned from the Timbuktu project include:

- Collective vision;
- Integrated farming operations;
- Key role for cooperatives;
- Business networks and clusters;
- Empowering governance structure;
- Key role for schools; and
- Regeneration and sustainability of resource base.[374]

Lessons from Cuba and Kurdistan

Although we covered these areas in earlier chapters, it is worth repeating our analysis of the lessons we have learned from their efforts and the struggles they continue to face.

Cuba

It can be argued that the Cuban people are the most community-minded people on the planet. They have a deep community spirit that allows them to work together to meet their own basic needs. As such, they can cooperatively adapt to new challenges without the fear of facing danger alone.

This community spirit derives from their having worked together to overthrow the US-based multinational capitalists that were exploiting their resources and keeping them impoverished. Once they drove the capitalists from their land they cooperatively set up their own local economy.

They have also learned that it is not feasible to depend upon Russia or the other so-called Communist elites for help.

They learned this lesson when the Soviet Union collapsed and Russia no longer supplied Cuba with oil. The consequence was that Cuba's agricultural system collapsed. There was no longer any oil and gas for tractors, transportation, chemical fertilizers, herbicides, etc. But, instead of searching for more oil to keep their traditional fossil fuel based agricultural system in operation, the Cuban people unitedly created an alternative system based upon organic production. The Castro government hired Permaculture experts from Australia who came to Cuba and taught the Cuban people how to grow food organically in their backyards, on their roof tops, and in vacant lots throughout the city of Havana. The system spread from there. After two very lean years of food rationing, the people had their new system and were self-reliant in food production.

Revolutionary Cuba has always focused on meeting the people's basic needs. Aside from food, they have also tackled education and health care. For years, they have sent teachers and doctors to other exploited countries around the world to offer assistance to the local people there.

The people of the world have much to learn from this little country and it will behoove revolutionaries everywhere to take a close look at the Cuban people and their accomplishments.

Kurdistan

The Kurds in Turkey and Syria have developed a progressive socialist political-economy that is unrivaled on the planet today. Its features include local economic self-reliance, a holistic ideology, equal opportunities for women, cultural pluralism, religious tolerance, universalism (instead of nationalism), environmental protection, local workers cooperatives, and a political process based upon neighborhood decision-making.

Perhaps the most amazing thing about this extraordinary accomplishment is that it was created in the face of tremendous opposition from the Syrian government, ISIS, Turkey, Iran, and Iraq, the betrayal by the US and Russian governments, and internal conflicts between Kurdish militias.

Rojava, the center of this great accomplishment has recently been overrun by Turkish forces and by the time this book is published, Rojava may no longer exist. If so, the world will have lost a great treasure.

Common Denominators in the Above Models

The local economy development models discussed above share the same following features:

- A strong vision to catalyze local potential and inspire the people;
- Regenerative ecology to restore and maintain the local resource base;
- A balanced mix of small businesses and cooperatives to foster innovation and build resilience;
- Diversified capitalization to empower entrepreneurs and facilitate community supported investment;
- Integration of Production and Distribution Networks;
- Engaged learning;
- Participatory governance and decision making;
- Formal authority to establish autonomy, secure legal standing, and facilitate reciprocal partnerships; and
- Celebration of Culture and Identity to promote common causes and unity in diversity.[375]

Chapter Fourteen: Defending Nuclear Revolution

We do not mean to harm anyone. But when unrighteous people use force against you, you must act in self-defence. This is not violence, it is self-defence. P. R Sarkar

ATTACKING THE EXPLOITATION OF capitalism can take many forms. Earlier we discussed the non-violent approaches of Gene Sharp and also the creating an economic alternative that focuses on a PROUT model where everyone is able to meet their basic necessities. Both of these approaches will result in lessening the domination of the capitalist exploiters and reducing their ability to make exhorbitant profits. As history has demonstrated, once something poses a real threat to their status quo, exploitive capitalists will stop at nothing to prevent their loss of profits. Whether they manipulate a country's military against its own people, hire mercenaries, or exploit the anger and militaristic spirit of right wing groups, they will no doubt attempt to prevent alternatives to capitalism from gaining a foothold. Consequently, to create a nuclear revolution, we may have to be prepared to defend what is rightfully ours. At times, violence may be the only alternative.

Violence vs Non-Violence

Many people may feel it is wrong to advocate violence even in defense of one's own people and property. However, this is a false notion. P. R. Sakar makes a clear distinction between non-harm and violence. One of the main moral principles of PROUT is ahimsá, which Sarkar explains as follows:

> According to its (ahimsa) meaning, one will have to guide one's conduct carefully to ensure that one's thoughts or actions cause pain to nobody and are unjust to none. Any thought or action with the intention of causing harm to someone else amounts to himsá. The existence of life implies destruction of certain lower forms, no matter whether there is intention of doing harm or not. The process of respiration kills thousands of millions of protoplasmic cells. Wherether one knows it or not, in every action such living cells are dying and being destroyed. The use of prophylactics means destruction of millions of disease-carrying germs. The crop-eating insects, parasites, mosquitoes, bugs, spiders, etc., are being killed in innumerable ways. This is necessary to maintain one's livelihood; it is not with the intention of causing pain to them. Such acts also, therefore, cannot be classed as himsá; they are to be done for self-defense.[361]

He continues:

> The champions of non-violence (so-called ahimsa) have, therefore, to adopt hypocrisy and falsehood whenever they seek to use this so-called ahimsá for their purposes. If the people of one country conquer another country by brute force, the people of the defeated nation must use force to regain their freedom. Such a use of force may be crude or subtle and, as a result, both the body and mind of the conquerors may be hurt. Where there is an application of force, it cannot be called non-violence.[362]

Creating a revolution, whether intentionally violent or non-violent is by its very nature violent. Revolution is the application of tremendous force in order to move the society forward, accelerate the speed of the social cycle and end exploitation. In the wake of a true revolution, radical change will occur in individual and collective life and far reaching changes will take place in the collective psychology of the people. What PROUT advocates is nuclear revolution where every aspect of collective life—social, economic, political, cultural, psychic, and spiritual—is completely transformed. New moral and spiritual values will arise creating greater impetus for accelerated progress. The collective psychology of the old era will be completely replaced by a new one. Only a violent force can create such a revolution. Hence we can only expect that such a force will cause an equal and opposite reaction from those who see a threat to their exploitative activities.

Although the main strategy of our revolutionary leadership is to weaken the capitalists by encouraging noncooperation and political defiance among the people, the day may come when the local people will be required to defend themselves from armed attack. There will be little room for negotiation. As P. R. Sarkar describes it: "Those at the helm of society are absorbed in strife and suspicion, or in reckoning their loss and profit They have no desire to think of humanity. In order to gratify their self-interest they are ready to chew the humans' bones. There is no room for mercy, sympathy, or fellow-feeling."[363]

Currently, around the world there is an increasing awareness of capitalist's economic exploitation. This is apparent in the number of protests throughout the world and the devastation being created by the wars and climatic upheaval leaving millions homeless and without the means to meet their basic needs. And yet the capitalists turn a deaf ear to the cries of the hungry and homeless, looking only for more profit and more control. They prefer that the common people remain ignorant and blame their destitution on women, minorities, immigrants, etc. They especially do not want people to gain spiritual knowledge or have equal opportunites to attain any self-realization. Therefore, The only way to end the suffering of the common people is to end exploitative capitalism. Those who dedicate themselves to this endeavor have to accept that they may one day be called upon to forcefully defend themselves and others. Fighting against an oppressor is bravery; refusing to use of force to do so is a cowardly position. According to P. R. Sarkar:

> Those who are not adequately equipped to oppose an evil-doer should make every endeavour to gain power and then make the proper use of this power. In the absence of ability to resist evil, and in the absence of even an effort to acquire such ability, declaring oneself to be non-violent in order to hide one's weaknesses before the oppontent may serve a political end, but it will not protect the sanctity of righteousness.[364]

Only by understanding that the evil of exploitation will not be solved by complaining about it, but by having the spirit to fight against all odds to establish a society free from exploitation, will we be able to solve the problems facing humanity today. Revolutionaries will need to adopt a path that has sufficient scope for humanism, but allows for the use of force if it becomes necessary to protect any progress made in an attempt to move humanity forward.

However, before attempting to mount a defense against an aggressor, no matter what form it takes, it is important to be prepared for it in advance. To be successful, you must have an assessment of your own strength, a strong organization, a preplanned strategy, and sound leadership. As P. R. Sarkar states:

> Revolution and war are almost the same. Revolution is a type of war. The difference between the two is that in war force is applied on behalf of a person or state, but in revolution, force is applied by a group of people to establish a society free from exploitation. A revolutionary organization is esssential for revolutionary war. During the psychic preparation for revolution, the disgruntled sections of society will have to build a revolutionary organziation which will create a congenial environment for revolution.[365]

It is hoped that local activists, who see the need for revolution, understand the need for strategic planning. Many activists become so caught up in their day to day struggle that they do not want to dedicate time to strategic planning. This is a major flaw in their work to serve the people.

Developing the appropriate strategy necessary to defend one's community, will depend on time, place, and circumstance. If one is in a situation where there is minimal chance of attack by an enemy, then merely maintaining a defensive contingency in one's overall stategic planning may suffice. As the probability of attack increases, a community may have to begin training its members in self-defense and developing a strategic defense plan. In the extreme case, a community may need to establish a local militia and begin implementing its strategic defense plan. Also as smaller communities begin to grow, they may pose a greater threat to the capitalist system and therefore, their need for implementing a defensive strategy may increase.

Even in a defensive struggle, it will be necessary to have a strong revolutionary organization to organize and mount a defense to protect an exploitive free society. Without it, the chance of success is minimal. To ensure preparation of a winnable defensive strategy, it must become part of the overall thinking of any revolutionary organization or community. Obviously, the amount of preparation and the resources applied to a defensive strategy will depend on the location, the strength, and the level of oppressiveness of the exploiters. It will also depend on the size and strength of the revolutionary force and the level of threat it poses to the exploiters. For example, a small community or even group of communities developing a socio-economic unit or units may pose minimal threat, but as the communities grow larger and begin to create greater networks, the threat to exploiters will increase and the need to prepare a defensive strategy and apply resources to defend the community/organization will increase proportionately. Finally, it will also depend on how much experience and training community members already have in opposing exploitation.

Strategy for Mounting a Defensive Struggle

In order to help communities begin to think about establishing a defensive strategy, it may be helpful at this point to provide some pointers in mounting a defense of a community. The most orderly, coherent, and comprehensive analysis of the rules of combat was written by the Chinese general Sun Tzu in his *Art of War*.[366] Although his book was written as a strategy for war, many of his strategies can be also be applied to a defensive struggle

According to Sun Tzu, the keys to success are careful planning and accurate information about the opposition. In possessing these two keys, a community's defensive troop can engage in battle without fear and trepidation because they know what they are doing and they can choose to fight only when they know they will be successful. What follows is a summary of Sun Tzu's thinking. It will give the reader a look at the thoughts and feelings of one of the greatest generals in all of human history regarding armed conflict. It should be known that Sun Tzu's *Art of War* is still being studied in military academics around the world, including West Point.

According to Sun Tzu, combat should be structured according to five factors. The first is termed the Tao. The Tao, (right way) causes the people to be fully in accord with the revolutionary leadership. The second is Heaven which encompasses sun and shade, cold and heat, and the constraints of the seasons (i.e., ability to adapt to unpredictability). The third, Earth, encompasses near and far, difficult or easy, expansive or confined, fatal or tenable terrain (i.e., choosing the right terrain for a battle). The fourth, is the General. This factor encompasses wisdom, credibility, benevolence, courage, and strictness. The fifth factor is the laws for military organization and discipline. It encompasses organization and regulations, the right way of command, and the management of logistics.

Therefore, when the leaders of a local defensive troop estimate comparative strength, they must ask the following questions:

- Which ruler has the Tao? (Who do the people follow because they believe in his/her cause?)
- Which general has greater ability?
- Who has gained the advantages of Heaven and Earth?
- Whose laws and orders are more thoroughly implemented?
- Whose forces are stronger?

- Whose officers and troops are better trained?
- Whose rewards and punishments are clearer?

By knowing the answers to these questions, one will know the chance of victory and defeat. Let us take a closer look at these five factors.

The Tao (The Right Way)

As an all encompassing philosophy, the Tao allows one to understand the interconnected relationship between all living things and their beginning and ending cycles. It is understanding that everything has an order and harmony and that everyone is part of the whole.

Understanding Taoist principles requires one to have a deep connection with the earth and all of its elements and creatures and to understand that all things are interdependent.

Sally Painter in her article, "Taoism Principles You Can Understand," reviews a number of Taoist principles that can readily be applied to defensive combat:[367]

- Learn to be an observer of the life around you.
- Bring all life into harmonious and balanced energy.
- Being balanced means you are just as aggressive as you are passive.
- Having compassion is necessary.
- You recognize that all people have value and are worthy.
- The path leads you to a life of moderation.
- Refilling the well of spiritual energy is a must for healthy life.

Approaching combat with these principle helps one maintain a spiritual approach, while continuing to maintain balance and reverence for all life. It also helps one keep in mind that combat, while violent, is still done without an intent to harm even the enemy.

According to Sun Tzu, one who knows the enemy and knows himself will not be endangered in a hundred engagements. One who does not know the enemy, but knows himself will sometimes be victorious, sometimes meet with defeat. One who knows neither the enemy nor himself will invariably be defeated in every engagement."

Ability to Adapt to Unpredictability (Tactics)

According to Sun Tzu all warfare is the Tao of deception. "Therefore, although you are capable, display incapability to the enemy. When committed to employ your forces, feign inactivity. When your objective is nearby, make it appear as if distant; when far away, create the illusion of being nearby."

What enables the army to withstand the enemy without being defeated are the *unorthodox* and *orthodox*. In general, in battle one engages with the orthodox and gains victory through the unorthodox. "Thus one who excels at sending forth the unorthodox is as inexhaustible as Heaven, as unlimited as the Yangtze and Yellow rivers. In warfare the strategic configurations of power do not exceed the unorthodox and orthodox, but the changes of the unorthodox and orthodox can never be completely exhausted. The unorthodox and orthodox mutually produce each other, just like an endless cycle."

"Intermixed and turbulent, the fighting appears chaotic, but they cannot be made disordered. In turmoil and confusion, their deployment is circular, and they cannot be defeated."

"Simulated chaos is given birth from control; the illusion of fear is given birth from courage; feigned weakness is given birth from strength. Order and disorder are a question of numbers; courage and fear are a question of the strategic configuration of power; strength and weakness are a question of the deployment of forces.

The art of deception is a practical application of the general principle of *taking the opposite stance in battle* that we discussed above.

Regarding combat tactics, Sun Tzu says, one who excels at warfare first establishes himself in a position where he cannot be defeated while not losing any opportunity to defeat the enemy."

"For this reason, the victorious army first realizes the conditions for victory, and then seeks to engage in battle. The vanquished army fights first, and then seeks victory."

In preparing for combat, it is necessary to assess the strength of the opposing forces. If a defensive troop must be used in battle, Sun Tzu gives the following advice:

> "If your strength is ten times theirs, surround them; if five, then attack them; if double, then divide your forces. If you are equal in strength to the enemy, you can engage him. If fewer, you can circumvent him. If outmatched, you can avoid him. A small enemy that acts inflexibly will become the captives of a large enemy."

"Thus if I determine the enemy's disposition of forces while I have no perceptible form, I can concentrate my forces while the enemy is fragmented. If we are concentrated into a single force while he is fragmented into ten, then we attack him with ten times his strength. Thus we are many and the enemy is few. If we can attack his few with our many, those who we engage in battle will be severely constrained."

In regard to the act of commanding the army, Sun Tzu says that commanding a large number is like commanding a few. It is a question of dividing up the numbers. Fighting with a large number is like fighting with a few. It is a question of configuration and designation."

He continues: "After estimating the advantages in accord with what you have heard, put it into effect with strategic power, supplemented by field tactics that respond to external factors." Strategic power is "controlling the tactical imbalance of power in accord with the gains to be realized."

"One who employs strategic power commands men in battle as if he were rolling logs and stones. The nature of wood and stone is to be quiet when stable but to move on precipitous ground. If they are square, they stop, if round, they tend to move. Thus the strategic power of one who excels at employing men in warfare is comparable to rolling round boulders down a thousand-fathom mountain. Such is the strategic configuration of power."

Regarding the strategic movement of troops, Sun Tzu says: "Thus one who excels at moving the enemy deploys in a configuration to which the enemy must respond. He offers something that the enemy must seize. With profit he moves them, with the foundation he awaits them."

"Thus one who excels at warfare seeks victory through the strategic configuration of power, not from reliance on men. Thus he is able to select men and employ strategic power."

Regarding field tactics, Sun Tzu says: "In general, whoever occupies the battleground first and awaits the enemy will be at ease; whoever occupies the battleground afterward and must race to the conflict will be fatigued. Thus one who excels at warfare compels men and is not compelled by other men."

To compel the enemy to come of their own volition, extend some apparent profit. In order to prevent the enemy from coming, show them a potential harm.

"Thus if the enemy is rested you can tire him; if he is well fed you can make him hungry; if he is at rest you can move him. Go forth to positions to which he must race. Race forth where he does not expect it."

"To ensure taking the objective in an attack, strike positions that are undefended. To be certain of an impregnable defense, secure positions that the enemy will not attack." Thus when someone excels in attacking, the enemy does not know where to mount his defense; when someone excels at defense, the enemy does not know where to attack. Subtle! Subtle! It approaches the formless. Spiritual! Spiritual! It attains the soundless. Thus he can be the enemy's Master of Fate."

"To effect an unhampered advance, strike their vacuities. To effect a retreat that cannot be overtaken, employ unmatchable speed. Thus if I want to engage in combat, even though the enemy has high ramparts and deep moats, he cannot avoid doing battle because I attack objectives he must rescue."

"If I do not want to engage in combat, even though I merely draw a line on the ground and defend it, he will not be able to engage me in battle because we thwart his movements."

"The location where we will engage the enemy must not become known to them. If it is not known, then the positions they must prepare to defend with be numerous. If the positions the enemy prepares to defend are numerous, then the forces we will engage will be few.

"Thus if they prepare to defend the front, to the rear there will be few men . . . . If there is no position left undefended, then there will not be any place with more than a few. The few are the ones who prepare against others; the many are the ones who make others prepare against them."

Thus if one knows the field of battle and knows the day of battle, he can traverse a thousand miles and assemble to engage in combat. If he does not know the field of battle nor know the day for battle, then the left flank cannot aid the right nor the right flank aid the left; the front cannot aid the rear nor the rear aid the front. . . . Even though the enemy is more numerous, they can be forced not to fight."

"Thus critically analyze them to know the estimations for gain and loss. Stimulate them to know the patterns of their movement and stopping. Determine their disposition of force to know the tenable and fatal terrain. Probe them to know where they have an excess, where an insufficiency.

"Thus the pinnacle of military deployment approaches the formless. If it is formless, then even the deepest spy cannot discern it or the wise make plans against it."

"Men all know the disposition by which we attain victory, but no one knows the configuration through which we control the victory. Thus a victorious battle strategy is not repeated, the configurations of response to the enemy are inexhaustible."

"The army's disposition of force is like water. Water's configuration avoids heights and races downward. The army's disposition of force avoids the substantial and strikes the vacuous. Water configures its flow in accord with the terrain; the army controls its victory in accord with the enemy. Thus the army does not maintain any constant strategic configuration of power, water has no constant shape. One who is able to change and transform in accord with the enemy and wrest victory is termed spiritual.

"If the enemy is numerous, disciplined, and about to advance, how should we respond to them? First seize something that they love, for then they will listen to you."

It is the nature of the army to stress speed; to take advantage of the enemy's absence; to travel unanticipated roads, and to attack when the enemy is not alert."

"When you mobilize the army and form strategic plans you must be unfathomable."

"When the soldiers and officers have penetrated deeply into enemy territory, they will cling together. When there is no alternative, they will fight."

In defensive situations, "Do not rely on the enemy not coming, but depend on us having the means to await them. Do not rely on their not attacking, but depend on us having an unassailable position."

Regarding, "Military Disposition," we learn that "One who cannot be victorious assumes a defensive posture; one who can be victorious attacks. In these circumstances, by assuming a defensive posture, strength will be more than adequate, whereas in offensive actions, it would be inadequate."

"Those who excel at defense bury themselves away below the lowest depths of Earth. Thus they are able to preserve themselves and attain complete victory."

Sun Tzu also cautions against "lengthy campaigns." A victory that is long in coming will blunt the fighters' weapons and dampen their spirit as well as make the revolution's resources inadequate. This will exhaust our capacity to fight.

Choosing the Right Terrain

"The strategy for employing the military in the field depends upon the terrain on which a battle is to be fought. There is dispersive terrain, light terrain, contentious terrain, traversable terrain, focal terrain, heavy terrain, entrapping terrain, encircled terrain, and fatal terrain." And depending on the terrain, one should take different strategic approaches:

> When the revolutionaries fight in their own territory, it is 'dispersive terrain.' On dispersive terrain do not engage the enemy.
>
> When they enter someone else's territory, but not deeply, it is 'light terrain.' On light terrain do not stop.
>
> If when we occupy it, it will be advantageous to us, while if they occupy it, it will be advantageous to them, it is 'contentious terrain.' On contentious terrain do not attack.

> When we can go and they can also come, it is 'traversable terrain.' On traversable terrain do not allow your forces to become isolated.
>
> Land of the counter-revolutionaries surrounded on three sides such that whoever arrives first will gain the masses of All under Heaven is 'focal terrain.' On focal terrain unite and form alliances with nearby revolutionaries.
>
> When one penetrates deeply into enemy territory, bypassing numerous cities, it is 'heavy terrain.' On heavy terrain plunder for provisions.
>
> Where there are mountains and forests, ravines and defiles, wetlands and marshes, wherever the road is difficult to negotiate, it is 'entrapping terrain.' On entrapping terrain move through quickly.
>
> Where the entrance is constricted, the return is circuitous, and with a small number they can strike our masses, it is 'encircled terrain.' On encircled terrain use strategy.
>
> Where if one fights with intensity he will survive but if he does not fight with intensity he will perish, it is 'fatal terrain.' On fatal terrain engage in battle.

Leadership

Regarding leadership, Sun Tzu has this to say: "It is essential for a general to be tranquil and obscure, upright and self-disciplined, and able to stupefy the eyes and ears of the officers and troops, keeping them ignorant. He alters his management of affairs and changes his strategies to keep other people from recognizing them. He shifts his position and traverses indirect routes to keep other people from being able to anticipate him."

"At the moment the general has designated his troops, it will be as if they ascended a height and abandoned their ladders. The general advances with them deep into the territory of the enemy and then releases the trigger. He commands them as if racing a herd of sheep—they are driven away, driven back, but no one knows where they are going."

In regard to giving orders: "If orders are consistently implemented to instruct the people, then the people will submit. If orders are not consistently implemented to instruct the people, then the people will not submit. One whose orders are consistently carried out has established a mutual relationship with the people."

A wise general will concentrate on securing provisions from the enemy. "One bushel of the enemy's foodstuff is worth twenty of ours." He will also reward soldiers who steal weapons from the enemy. He also treats the captured soldiers well.

Military Organization and Discipline

Sun Tzu also warns about the dangers of the civilian leadership interfering in the militias actions by: (1) not knowing that the militia should not advance, but instructs them to advance or does not know that the militia should not withdraw and orders a retreat; this is termed "entangling the army"; (2) not understanding the militia's military affairs but directs it in the same way as his civilian administration confuses the officers; and (3) not understanding the militia's tactical balance of power, but undertakes responsibility for command will make the officers will be doubtful.

Such cautions speak to the need for a high degree of cooperation and communication between the leaders of the militia and the civilian forces. It speaks to the need to let the warriors be warriors and intellectuals be intellectuals.

Use of Spies

Sun Tsu also talks about the beneficial use of spies. He says, "The means by which enlightened rulers and sagacious generals moved and conquered others, so that their achievements surpassed the masses, was advanced knowledge.

"Advanced knowledge cannot be gained from ghosts and spirits, inferred from phenomena, or projected from the measures of Heaven, but must be gained from men for it is the knowledge of the enemy's true situation."

"Thus there are five types of spies to be employed: local spy, internal spy, turned spy [double agent], dead [expendable spy], and the living spy. When all five are employed together and no one knows their Tao, this is termed 'spiritual methodology.' They are a leader's treasures."

- "Local spies – employ people from the local district.
- "Internal spies – employ their people who hold government positions.
- "Double agents – employ the enemy's spies.
- "Expendable spies – are employed to spread disinformation outside the state. Provide our expendable spies with false information and have them leak it to the enemy agents.
- "Living spies – return with their reports.

"Thus no relationship is closer than with spies; no rewards are more generous than those given to spies; no affairs are more secret than those pertaining to spies."

"Unless someone has the wisdom of a Sage, he cannot use spies; unless he is benevolent and righteous, he cannot employ spies; unless he is subtle and has a ready insight into things, he cannot perceive the substance in intelligence reports. It is subtle, subtle! There are no areas in which one does not employ spies."

"In general, as for the armies you want to strike . . . you must first know the names of the defensive commander, his assistants, staff, door guards, and attendants. You must have our spies search out and learn them all."

"You must search for enemy agents who have come to spy on us. Tempt them with profits, instruct and retain them. Thus double agents can be obtained and employed. Through knowledge gained from them, you can recruit both local and internal spies. Through knowledge gained from them, the expendable spy can spread his falsehoods, can be used to misinform the enemy. Through knowledge gained from them, our living spies can be employed as times require."

"The ruler must know these five aspects of espionage work. This knowledge inevitably depends on turned spies; therefore, one must be generous to double agents."

"Enlightened rulers and sagacious generals who are able to get intelligent spies will invariably attain great achievements."[368]

Finally, Sun Tzu had these points to offer.

"Those that the ancients referred to as excelling at warfare conquered those who were easy to conquer. Thus the victories of those that excelled in warfare were not marked by fame for wisdom or courageous achievement. Thus their victories were free from errors. One who is free from errors directs his measures toward certain victory, conquering those who are already defeated."

He reminds us that, "attaining one hundred victories in one hundred battles is not the pinnacle of excellence. Subjugation of the enemy's army without fighting is the true pinnacle of excellence."

"Thus, the highest realization of warfare is to attack the enemy's plans; next is to attack their alliances; next to attack their army; and the lowest is to attack their fortified cities. By this analysis, we can see that following the strategy of noncooperation and nonviolent intervention is paramount. It is by these means that we will preserve the people while destroying the plans of the capitalists.

Considering that in many cases, the enemy could be more experienced, more trained, and with greater access to more sophisticated weapons, the five points of Sun Tsu are worth including in any defensive strategy: (1) always maintain one's balance and centeredness, (2) control the tactics of the battle, while keeping one's strategy unpredictable to the enemy, (3) choosing the best terrain for the battle, (4) choosing the best leadership, and (5) maintaining both internal and external discipline and open communications with civilian leadership.

Mao TseTung, who produced the strategy and tactics for modern guerilla warfare, acknowledged that it was Sun Tzu'sArt of War that gave him practically all of the ideals he incorporated into his theory.

The primary aim of Mao's system of guerrilla war was to direct small but frequent violent attacks against enemy bases, depots, and lines of commiication. His attacks occurred in many different places and at unexpected times. The goal was to force the enemy to disperse his forces widely and in small detachments to protect vital points. Even guarded, these positions remained vulnerable, because the enemy could not predict which positions Mao's forces would attack. The purpose was to confuse the enemy, make him uneasy and insecure and, therefore, demoralize him. This was the process of nullifying the enemy's strength.

Although, Mao believed that a guerrilla war was an offensive war, there is much that can be learned from him for a defensive ooperation.

Role of Policy, Strategy, and Tactics

It is critical that in defending their communities, our local leaders must also understand the role of policy, strategy, and tactics. The policy that governs the grand strategy will be determined by specified governing boards of the local revolutionary leadership. The strategy and tactics for specific defensive engagement, however, will fall to the local militia in-charge.

According to B. H. Liddell Hart, a contemporary British military strategist, the responsibility of the military in-charge is to "apply most profitably, in the interest of the higher war policy, the force allotted to him within the theatre of operations assigned to him. If he considers that the force allotted is inadequate for the task indicated he is justified in pointing this out, and if his opinion is overruled he can refuse or resign the command; but he exceeds his rightful sphere if he attempts to dictate to the government what measure of force should be placed at his disposal."[369]

Hart goes on to say that the government, which formulates the war policy, and has to adapt it to conditions which often change as an operation progresses, can rightly intervene in the strategy of a military campaign, not merely by replacing a commander in whom it has lost confidence, but by modifying his object according to the needs of a defensive policy. While it should not interfere with him in the handling of his tools, it should indicate clearly the nature of his task. Thus, grand strategy has more than the simple objective of mounting a defensive strategy. When a government appreciates that the enemy has the military superiority, either in general terms or in a particular theater, it may wisely enjoin a strategy of limited aim.[370]

Applying Hart's strategy to the defense of a local community, enjoining a strategy of limited aim will occur if the revolutionary leadership believes it can achieve its desired aim through other than military means. Another reason for adopting a strategy of limited aim is to await a change in the balance of force. This change can be sought and achieved by draining the enemy's force, weakening him by pricks instead of risking blows. The essential condition of such a strategy is that the drain on the enemy should be disproportionately greater than on oneself.

At this point the reader should realize that policy is created by the revolutionary leadership as the means to articulate the goals of a nuclear revolution. The grand strategy is the plan to put this policy into execution. The grand strategy covers both combat and non-violent operations. It calculates and develops the physical and human resources of the people and it regulates the distribution of power between the different forces and campaigns.

The local revolutionary leadership will, therefore, determine the purpose of the local defense and employ it to meet the objectives of its grand strategy. The militia in-charge will then take his or her orders and create a strategy and tactics to fulfill these orders. Given the fact that a local militia has neither the resources nor the skills of a standing army, it is most likely that its strategy will be a strategy of limited aim. At minimum, this limited aim would be to defend the local people on their farms, in their households, at their places of work, and during events that are created to forward the aims of the nuclear revolution.

As far as possible, the revolutionary leadership should initially separate non-violent from violent struggle by geography, groups, timing, issues, etc., so as to protect the majority of the people from violence whenever the capitalist representatives happen to be present.

Needless to say, the highest standard of moral behavior by our leaders and cadre is required in order to gain the long-term support of the people. The revolutionary leadership must be fearless and be able to maintain discipline under the most adverse circumstances.

Maintaining One's Balance

As expressed repeatedly, violence in the act of self-defense does not go against any moral or spiritual principles. In fact, it is cowardice to refrain from use of force when it is necessary to protect one's self and and others from those wishing to commit harm. One cannot expect to tear down an old corrupt system and build an alternative that benefits all without expecting to encounter significant resistance, including violent attacks. However, one must never cross the line where violence exceeds the need for self-defense.

In the heat of combat, it is easy to want to seek revenge on an enemy, especially one that may have committed atrocities. However, it must be remembered that violence, with the intention of harming someone, does not solve anything. P. R. Sarkar cautions us that:

> Because whatever poisonous tendencies of the individual and collective minds may be destroyed by violence, the seeds of those tendencies remain embedded in the mind itself. When the pressure of circumstances is relaxed they may again sprout forth, creating ever greater evil.[371]

And if this type of violence is not checked, he warns:

> If any society overemphasizes brute force, the members of that society will start fighting each other and eventually become independent, self-styled autocarts. In such a society the last vestiges of morality will disappear and chaos will prevail.[372]

In creating an alternative system that benefits all people, we must create a change of heart within the people; one that follows the principles of Neohumanism. This change will never be possible through the use of violence.

Benefitting all means everyone, even those that may want to do harm to us. We must keep in mind that those who participate in exploitation and intend us harm today are also human beings and participants in the human family. Instead of seeking revenge on them, we must strive to rectify their offenses and bring them around to becoming productive human beings. If we try to make progress in creating a new system, our attempt will be fruitless if we hold on to feelings of revenge for those who committed crimes in the past. Everyone has the potentiality to be good productive citizens and it is more important to tap that potentiality then to punish them for past mistakes.

Chapter Fifteen: One Human Society and One World Government

THIS FINAL CHAPTER IS intended to help us envision the consequences for humanity if we are able to successfully create a nuclear revolution. To begin with, we would all be members of one human society; citizens of the Earth sharing a common governmental structure that would seek the benefit of every human being and every other life form on the planet.

Our one world government would be composed of representatives of all the Earth's people and they would have been chosen because they understand the people's problems and their needs and have dedicated their lives to helping to resolve those problems and meet those needs. We would have a government that would be composed of those with the highest moral integrity, whose only mission is to create greater peace and unity among us and to not promote their own selfish interests for-personal wealth and self-aggrandizement.

We would live in a society, where every human being, in every locality of the world, would have their basic needs met and would know that they and their children will always have their basic needs met because that is their goal as well as that of all of their neighbors. This constitutes a form of social security that can never be taken away or controlled by forces from afar who have their own agenda of how to use local resources.

We will live in a world where the common value of the people is to support one another and where good-will between people is the norm rather than the exception. We will live in a world where every life is held sacred and people are encouraged to act with the best of intentions.

We will live in a world where our customs reflect an appreciation for diversity and the value of human unity.

We will live in a world where the existence of economic decentralization and political unity will be buttressed by: (1) a world government whose role is to promote a common philosophy of life that seeks the well-being of all, (2) a constitutional structure that puts this value into law, (3) a common penal code that provides justice and safety for all, and (4) a world government that promotes the availability of the minimum requirements of life for one and all.

A Common Philosophy of Life

A common philosophy of life requires a clear understanding that the development of every person requires growth in all the three spheres—physical, mental/emotional, and spiritual. A common philosophy moves beyond the opinion that spirituality is a utopian philosophy bereft of practicalities pertaining to actual problems of life, or that it is a device to befool the people. Rather, with spirituality as a base, we will be able to develop a rational philosophy that is common to humanity in general and that is revolutionary and ever-progressing.[376]

Same Constitutional Structure

Despite the reactionary forces operating at the behest of the ruling classes of the global economy, a social blending of humanity is already in process and needs a common constitutional structure to cement the solidarity of the world.

Points of a world constitution will address political issues like the relationsips of the leaders who preside over the different branches of the world government; the moral character of leaders, the equal status of all languages, the equal rights for all people, the establishment of review boards to assure economic and political equality, the right of self-determination, the right to education free from political interference, and the identity between law and constitution.

There will also be a Bill of Rights which will ensure that: (1) complete security be guaranteed to all the plants and animals on the planet; (2) each samaj will guarantee purchasing power to all its citizens to meet basic their needs; (3) the four fundamental rights of spiritual practice, cultural legacy, education, and indigenous linguistic expression are guaranteed; and that (4) if the practice of any of these rights conflicts with cardinal human values, such practices should be immediately curtailed. Cardinal human values must take precedence over all other rights.[377]

Common Penal Code

The idea of crime is presently based on different concepts of virtue and vice. To simplify matters, our common penal code will view all those actions which help in the growth of the spiritual, mental, and physical aspects of human beings as virtuous deeds, and those actions which go against humanity's spiritual, mental, and physical development as "vices." This conception of virtue and vice will apply commonly to humanity in general.

To ensure that our laws are continually able to contribute to the development of human beings, our legislation must be progressive and capable of gradual adjustment with the prevalent conditions. Because conditions are always changing there should be a never-ending effort to amend the penal code, and its goal should be rectification, not punishment.[378]

The primary purpose of the penal code will not be focused on punishing people but on rectifying the negative behavior, where love is the driving force and not revenge. See Chapter 4 of *Universal Ideology: The Thought of P. R. Sarkar* for a more thorough review of the common penal code.

Minimum Essentials of Life

Progressive Socialism (PROUT) has a rich body of information upon which to build a post-capitalist economy that ensures that local people are able to provide for their own basic needs in a sustainable manner. Proutists put a great deal of emphasis on this point because we acknowledge that human beings cannot pursue subtle goals if their energy is spent only trying to stay alive. Mr. Sarkar, the founder of PROUT speaks directly to this point:

> The availability of the minimum essentials of life plays a vital part not only in achieving world brotherhood but also in the development of human personality. This should be tackled on a world footing, and should be based on certain fundamental presumptions. Every human being has certain minimum requirements which he or she must be guaranteed. Guaranteed availability of foodstuff, clothing, medical assistance, and housing should be arranged so that human beings may be able to utilize their surplus energy (energy up till now engaged in procuring the essentialities of life) in subtler pursuits. Side by side, there should be sufficient scope for providing other amenities in a progressive manner. To fulfill the above responsibilities, enough purchasing capacity should be created.
>
> If the supply of requirements be guaranteed without any conditions of personal skill . . . the individual may develop the psychology of idleness. The minimum requirements of every person are the same, but diversity is also the nature of creation. Special amenities should, therefore, be provided so that the diversity in skill and intelligence is fully utilized, and talent is encouraged to contribute its best towards human development. It will, therefore, be necessary to make provision for special emoluments which can [serve as] special amenities of

> life according to the age and time. But, at the same time, there should be a constant effort to reduce the gap between the amount of special emoluments and the bare minimum requirements of the average individual. The guaranteed supply of minimum requirements must be liberalized by increasing the provision of special amenities pertaining to the age and also, simultaneously, by bringing about a decrease in the provision of special emoluments given to the few. This never-ending effort of proper economic adjustment must ceaselessly continue at all times with a view to assisting the spiritual, mental, and physical evolution of human beings, and to let humanity develop a Cosmic sentiment for a Cosmic ideal and world fraternity.[379]

In conclusion, a nuclear revolution is based upon a universal spiritual sentiment. Its leaders are spiritual warriors whose dual mission is devotion to divine consciousness and the welfare of all. To achieve a society that acts for the benefits of all will require a struggle against those forces who would destroy people and the planet for their own selfish means or who, out of ignorance, will stand in the way of human development, because they see a personal benefit in keeping others subservient. This struggle is a total struggle between the forces of progress and the forces of decay. The battle exists on the physical, mental and spiritual levels, within the individual as well as within the collective.

A nuclear revolution represents the most viable opportunity to unite humanity against all divisive behaviors and protect our planet before it is too late. As such, we encourage every human being on the planet to reject any divisive sentiments and to immediately undertake the performance of self-purification and selfless service to those most in need. By the grace of the Divine may we be successful.

Appendix A: The Eight Occult Powers

The 8 occult powers are:

Aṅimá
The ability to "convert one's psychic existence into a small point and then transform it into a minimum entity." One may understand anything by entering into each and every physical particle and becoming one with the different waves of expression and emanation, by assuming the wavelengths of objects or ideas. Unless and until one's very psychic existence is converted into a point, one cannot move or dance according to the wavelengths of objects or ideas.

Laghimá
This ability makes the mind light, free from the bondages of so many liabilities. A carefree mind, freed from the fetters of materialistic bondages, can understand and think properly. By this occult power, one may understand any idea, subtle or crude, in the abstract realm. Through laghimá, you can acquire the power to study others' minds and alleviate their sorrows and sufferings.

Mahima
The third occult power brings expansion of mind. The radius of the mind may cover the entire universe, and we may have ideas about different subjects even without reading books and literature. And, in this way, too, we may feel our oneness with the varied entities of this universe —unity in variety and unity in diversity. By associating our benevolent thoughts with each and every entity, we will contribute to universal progress and prosperity.

Iishitva

This power enables the spiritual aspirant to guide the minds of others who suffer from different psychic ailments. So many people in this world are crying in pain and agony. So many miseries and afflictions have paralyzed human beings physically and mentally. This ability helps to lead afflicted humanity to physical progress and psychic well-being.

Vashitva

Vashitva helps a person to bring defective ideas, or people goaded by defective ideas, unto the path of supreme greatness. If people are to be guided towards progress, they cannot be expected to reach the state of welfare if they work haphazardly and do not follow the right path. So, to really help people, we will have to direct them along the right path to their goal.

Prápti

This power allows one to help himself or herself and others to acquire and benefit by the grace of the Supreme.

Prakámya

This power allows one to have the right mode of thinking aimed at promoting universal welfare, and bringing light to the entire universe. Through it, spiritual aspirants acquire what is needed to serve the entire world.

Antaryámitva

With this power one is able to enter within the ectoplasmic or endoplasmic structures of others, and thereby to know the pains and pleasures, the hopes, aspirations and longings of others to guide them properly. Regarding this eighth occult power, it cannot be attained by spiritual practice alone. It requires the special grace of the Supreme Consciousness.[380]

By doing spiritual practices, an ordinary person can become extraordinary and possess the eight occult powers. Over the course of human history, many persons have risen to this rank.

Having said this, once one obtains these powers, it is better for the spiritual warrior to surrender these powers back to the Supreme Consciousness. If he or she fails to do this, what will happen? Well, after acquiring various occult powers, he or she will develop some ego. One will think that he or she alone is great and unknowingly will gradually degenerate. Therefore, after attaining internal powers, one should surrender one's soul, endowed with the acquired occult powers, to the Supreme Consciousness. What will be the result? One will become one with the Supreme Consciousness; one will become the Supreme Consciousness and, in doing so, will experience pure love unadulterated by separation.

Appendix B: Guidelines for Planning a Socio-Economic Unit (Samaj)

Initially, local planners will have to consider the following four factors: (1) the Five Fundamental Principles of PROUT,[381] (2) the physical demand at present and the physical demand in the foreseeable future, (3) the physical supply at present and the physical supply in the foreseeable future, and (4) the maximum utilization of land.

To address these factors, the local revolutionary leaders will need to develop a Master Plan for their area. A Master Plan may include the following information:

1. A study of the local demographics, including migration patterns;
2. A detailed study of the surplus and deficit of labor trends within the Socio-Economic Unit (SEU). If there is a surplus of manual labor, jobs should be created that make use of manual labor. The goal of an SEU is one hundred percent full employment. Labor can be paid for in local currency that can be exchanged for goods and services produced within the SEU;
3. A study of local natural resources, including rocks, clay, sand, minerals, fuels, forests, plants, animals, etc., and a determination of how to optimize their use in a sustainable manner to meet the basic needs. The study should include a plan to expand local utility of these resources;

4. A definition of the specific goods and services required to meet the basic needs of the local population, as well as the processes and components required to produce them. Seeds, fertilizers, hand tools, mills, etc., can be set up as local businesses and supported by SEU participants;
5. A study of the current land used by farms, industry, and trade, as well as the variety, size, and distribution of the available goods and services required to meet basic needs within the SEU;
6. A specific plan for food production. A food plan, for example, might define the calories required in the local diet and the composition of those calories in terms of proteins, carbohydrates, fats, vitamins, and minerals and then define the foods required that can be grown locally to meet those needs. Depending upon the number of participants in the SEU, planners can then calculate the quantity of foods required; who among the participants will grow the food and handle their processing, storage, and distribution; how much land is required to produce the food, etc. In this manner, it will also be able to define the unit cost per food item;
7. A study of the housing stock and the materials and costs associated with retrofitting them to use sustainable energy;
8. A study of the health services within the locality and a plan to shift services to local organizations. The study should focus on prevention, nutrition, and plant-based medicines;
9. A study of how to improve the capacity of the education system to address local needs;
10. A plan to conserve local raw materials and eco-systems by reducing the outflow of raw materials from the local economy;
11. A determination of a minimum and maximum wage that meets basic needs and rewards exceptional contributions;

12. A model calculation of factors required to set up an agriculture or pre- and post-agriculture production system. Factors to consider are: the labor requirements, the cost of production, availability of local raw materials, the productive capacity, the purchasing capacity of the local people, the collective necessity, and the environmental sustainability of the endeavor;
13. The creation of training materials to teach people how to set up and manage cooperatives;
14. A survey to identify the current local market for locally produced goods. All goods and services produced to meet basic needs should be sold only in local markets. If the local goods do not meet the needs or aspirations of the people, immediate steps should be taken to increase quality, reduce prices, and increase the supply of local goods to discourage imports;
15. A plan to meet the emergency food, clothing, and shelter needs of local people;
16. A marketing strategy to educate the local people about the benefits of buying local products (psycho-economy);
17. A plan to create a local currency to facilitate the exchange of goods and services. This may include a Time Banking system,[382] LETS system,[383] ROCS,[384] barter,[385] a local currency,[386] or combination of these methods. Each economic exchange that does not require dollars will add that amount to the local money supply and allow dollars to be used for exchange within the larger existing economy. Local money should be kept in constant circulation and not hoarded;
18. An exploratory strategy to involve local government in the development of the SEU;
19. A strategy to conserve SEU resources through nonuse, reuse, repurposing, recycling, and retrofitting;
20. A strategy for creating events which allow people to

partake in recreation and higher personal pursuits and to build trust in community. Holding public award events for those who have made extraordinary contributions to the welfare of the people in the SEU may be one kind of event. Events to share ideology, to educate the community, to increase culture, to have fun, and to worship together are all positive events to bring people together;

21. A plan to ensure the security of the local people if a local police force is unable to do so. Local people have the God-given right to meet their own basic needs and any forces that intend to deny people this right threaten their survival and are to be considered enemies of the people and the people will have to defend against such enemies;
22. A short-term (6 months) and long-term (3 years) plan of action; and
23. An initial financial plan and budget based upon the needs of local labor and resource utilization. A system such as SHARE,[387] a revolving loan fund,[388] or a credit union[389] will eventually have to be established in order to allow local people to invest in the local unit.

Appendix C: The Tactics of Nonviolent Protest and Persuasion[390]

Formal statements
Public speeches
Letters of opposition or support
Declarations by organizations and institutions
Signed public statements
Declarations of indictment and intention
Group or mass petitions

Communications with a wider audience
Slogans, caricatures, and symbols
Banners, posters, and displayed communications
Leaflets, pamphlets, and books
Newspapers and journals
Records, radio, and television
Skywriting and earthwriting

Group representations
Deputations
Mock awards
Group Lobbying
Picketing
Mock elections

Symbolic public acts
Display of flags and symbolic colors
Wearing of symbols
Prayer and Worship

Delivering symbolic objects
Protest disrobing
Destruction of own property
Symbolic lights
Displays of portraits
Paint as protest
New signs and names
Symbolic sounds
Symbolic reclamations
Rude gestures

Pressures on individuals
"Haunting" officials
Taunting officials
Fraternization
Vigils

Drama and music
Humorous skits and pranks
Performance of plays and music
Singing

Processions
Marches
Parades
Religious processions
Pilgrimages
Motorcades

Honoring the dead
Political mourning
Mock funerals
Demonstrative funerals
Homage at burial places

Public assemblies
Assemblies of protest or support
Protest meetings

Camouflaged meetings of protest
Teach-ins

Withdrawal and renunciation
Walk-outs
Silence
Renouncing honors
Turning one's back

Appendix D: The Tactics of Noncooperation[391]

Social Noncooperation

Ostracism of persons
Social boycott
Selective social boycott
Lysistratic nonaction (sex strikes)
Excommunication
Interdict

Noncooperation with social events, customs, and institutions
Suspension of social and sports activities
Boycott of social affairs
Student strike
Social disobedience
Withdrawal from social institutions

Withdrawal from the social system
Stay-at-home
Total personal noncooperation
Flight of workers
Sanctuary
Collective disappearance
Protest emigration (hijrat)

The Tactics of Economic Noncooperation
Economic Boycotts

Action by consumers
Consumers' boycott
Nonconsumption of boycotted goods
Policy of austerity
Rent withholding
Refusal to rent
National consumers' boycott
International consumers' boycott

Action by workers and producers
Workmen's boycott
Producers' boycott

Action by middlemen
Suppliers' and handlers' boycott

Action by owners and management
Traders' boycott
Refusal to let or sell property
Lockout
Refusal of industrial assistance
Merchants' "general strike"

Actions by holders of financial resources
Withdrawal of bank deposits
Refusal to pay fees, dues, and assessments
Refusal to pay debts or interest
Severance of funds and credit
Revenue refusal
Refusal of a government's money

Actions by governments
Domestic embargo
Blacklisting of traders
International sellers' embargo
International buyers' embargo
International trade embargo

The Tactics of Economic Noncooperation

The Strike

Symbolic strikes
Protest strike
Quickie walkout (lightning strike)

Agricultural strikes
Peasant strike
Farm workers' strike

Strikes by special groups
Refusal of impressed labor
Prisoners' strike
Craft strike
Professional strike

Ordinary industrial strikes
Establishment strike
Industry strike
Sympathetic strike

Restricted strikes
Detailed strike
Bumper strike
Slowdown strike
Working-to-rule strike
Reporting "sick" (sick-in)
Strike by resignation
Limited strike
Selective strike

Multi-industry strikes
Generalized strike
General strike

Combination of strikes and economic closures
Hartal (closing of shops and businesses)
Economic shutdown

The Tactics of Political Noncooperation

Rejection of authority
Withholding or withdrawal of allegiance
Refusal of public support
Literature and speeches advocating resistance

Citizens' noncooperation with government
Boycott of legislative bodies
Boycott of elections
Boycott of government employment and positions
Boycott of government departments, agencies, and other bodies
Withdrawal from government educational institutions
Boycott of government-supported organizations
Refusal of assistance to enforcement agents
Removal of own signs and placemarks
Refusal to accept appointed officials
Refusal to dissolve existing institutions

Citizen's alternatives to obedience
Reluctant and slow compliance
Nonobedience in absence of direct supervision
Popular nonobedience
Disguised disobedience
Refusal of an assemblage or meeting to disperse
Sit-down
Noncooperation with conscription and deportation
Hiding, escape, and false identities
Civil disobedience of "illegitimate" laws

Action by government personnel
Selective refusal of assistance by government aides
Blocking of lines of command and information
Stalling and obstruction

General administrative noncooperation
Judicial noncooperation
Deliberate inefficiency and selective noncooperation by enforcement agents
Mutiny

Domestic governmental action
Quasi-legal evasions and delays
Noncooperation by constituent governmental units

International governmental action
Changes in diplomatic and other representation
Delay and cancellation of diplomatic events
Withholding of diplomatic recognition
Severance of diplomatic relations
Withdrawal from international organizations
Refusal of membership in international bodies
Expulsion from international organization

Appendix E: The Tactics of Nonviolent Intervention[392]

Psychological Intervention
Self-exposure to the elements
The fast
Fast of moral pressure
Hunger strike
Satyagrahic fast (non-violent resistance)
Reverse trial
Nonviolent harassment

Physical Intervention
Sit-in
Stand-in
Ride-in
Wade-in
Mill-in
Pray-in
Nonviolent raids
Nonviolent air raids
Nonviolent invasion
Nonviolent interjection
Nonviolent obstruction
Nonviolent occupation

Social intervention
Establishing new social patterns

Overloading of facilities
Stall-in
Speak-in
Guerrilla theater
Alternative social institutions
Alternative communication system

Economic intervention
Reverse strike
Stay-in-strike
Nonviolent land seizure
Defiance of blockades
Politically motivated counterfeiting
Preclusive purchasing
Seizure of assets
Dumping
Selective patronage
Alternative markets
Alternative transportation systems
Alternative economic institutions

Political intervention
Overloading of administrative systems
Disclosing identities of secret agents
Seeking imprisonment
Civil disobedience of "neutral" laws
Work-on without collaboration
Dual sovereignty and parallel government

As a final thought, P.R. Sarkar reminds us that:

A careful study of the social history of the world will reveal that until now every attempt at counter-revolution has not only caused enormous psychic and financial suffering and plunged humanity into the mire of gloom and despair, but has also lengthened the period of social contraction. This in turn, in the next phase, has helped to accelerate the speed of the period of social expansion—has inspired the chariot of revolution to advance towards victory with greater momentum. P. R. Sarkar

Endnotes

1 See Chuck and Tom Paprocki, *The Untold Story of Western Civilization*, Volume 5, for an in-depth analysis of the US capitalists' global strategy and its impact on the world's people.
2 See Alfred North Whitehead, *Process and Reality*, Max Planck, *The Nature of Matter*, Stephen Hawking, *The Great Design*, Peter Baksa, *The Point of Power*, Robert Lanza, *Biocentrism: How Life Creates the Universe*, Amit Goswami et al, *The Self-Aware Universe: How Consciousness Creates the Material World.*
3 Constitution of the United States of America 1789 (rev. 1992), Constitute, https://www.constituteproject.org/constitution/United_States_of_America_1992.
4 Anna Pigott, "Capitalism is killing the world's wildlife populations, not humanity," Phys.org. November 2, 2018, https://phys.org/news/2018-11-capitalism-world-wildlife-populations-humanity.html.
5 Laura Shin, "The 85 Richest People In The World Have As Much Wealth As the 3.5 Billion oorest," Forbes, January 23, 2014, https://www.forbes.com/sites/laurashin/2014/01/23/the-85-richest-people-in-the-world-have-as-much-wealth-as-the-3-5-billion-poorest#8f67fd017531.
6 Marilyn Geewax, "The Tipping Point: Most Americans No Longer Are Middle Class. NPR, December 9, 2015, https://www.npr.org/sections/thetwo-way/2015/12/09/459087477/the-tipping-point-most-americans-no-longer-are-middle-class.
7 Ben Harack, "How much would it cost to end extreme poverty in the world?" Vision of Earth, August 26, 2011, https://www.visionofearth.org/economics/ending-poverty/how-much-would-it-cost-to-end-extreme-poverty-in-the-world/.
8 Amanda Dixon, "How Many Americans Are Underemployed?" SmartAsset, May 15, 2019, https://smartasset.com/retirement/how-many-americans-ore-underemployed.
9 Amanda Dixon, "How Many Americans Are Underemployed?" SmartAsset, May 15, 2019.
10 Bouree Lam, "The Surging Cost of Basic Needs," The Atlantic, June 2, 2016, https://www.theatlantic.com/business/archive/2016/

household-basic-spending/485330/.

11 Brianna Vannozzi, "Study: 37 Percent of NJ Households Can't Afford Basic Needs," NJTV News, January 17, 2017, http://www.njtvonline.org/news/video/study-37-percent-nj-households-cant-afford-basic-needs/.

12 "The Effects of Ozone Depletion On Your Skin And Eyes," Sunsafe RX, July 20, 2019, https://www.sunsaferx.com/health-and-wellness/effects-ozone-depletion-skin-eyes/.

13 Wikipedia, "Ozone depletion," last modified May 25, 2019, https://en.wikipedia.org/wiki/Ozone_depletion.

14 Michael Casey, "Millions dying around the world due to air pollution," CBS News, September 16, 2015, https://www.cbsnews.com/news/millions-dying-around-the-world-due-to-air-pollutio/.

15 "Water Pollution," Everything Connects, accesssed July 10, 2020, https://www.everythingconnects.org/water-pollution.html#:~:text=According%20to%20UN%2DWater%2C%2075,%2C%20lakes%2C%20ponds%20and%20groundwater.

16 Josh Fox, "Fracking Facts," Welcome to Gasland, Gasland Part II, https//gaslandthemovie.com/whatsfracking/faq/water-used.

17 Mariah Blake, "Halliburton Fracking Spill Mystery: What Chemicals Polluted an Ohio Waterway?" Mother Joes, July 14, 2014, http://www.motherjones.com/politics/2014/07/halliburton-ohio-river-spill-fracking/.

18 "Water Pollution," Water Pollution, November 20, 2013, http://www.everythingconnects.org/water-pollution.html.

19 "Threats: Soil Erosion andDegradation," World Wildlife Fund, accessed June 24, 2020, http://www.worldwildlife.org/threats/soil-erosion-and-degradation.

20 Chris Arsenault, "Only 60 Years of Farming Left If Soil Degradation Continues," Scientific American, December 5, 2014, https://www.scientificamerican.com/article/only-60-years-of-farming-left-if-soil-degradation-continues/.

21 "Life on Land," UN Sustainable Development Goals, accessed June 24, 2020, http://www.un.org/sustainabledevelopment/biodiversity/.

22 Eva Botkin-Kowacki, "So 1million species are at risk of extinction. Now What?" The Christian Science Monitor, May 14, 2019, https://www.csmonitor/Environment/2019/0514/So-1-million-species-are-at-risk-of-extinction.-Now-what.

23 Gayathri Vaidyanathan, "How to Determine the Scientific Consensus of Global Warming," Scientific American, July 24, 2014, https://www.scientificamerican.com/article/how-to-determine-the-scientific-consensus-on-global-warming/.
24 "Scientific Consensus: Earth's Climate is Warming," NASA, May 23, 2019, https://climate.nasa.gov/scientific-consensus.
25 David Roberts, "Exxon researched climate science. Understood it. And misled the public," Vox, August 23, 2107, https://www.vox.com/energy-and-environment,2017/8/23/16188422/exxon-climate-change.
26 Peter Coy and Christopher Flavelle, "Harvey Wasn't Just Bad Weather. It Was Bad City Planning," Bloomberg Business Week, August 32, 2017, https://www.bloomberg.com/news/features/2017-08-31/a-hard-rain-and-a-hard-lesson-for-houston.
27 Aamer Madhani and Roger Yu, "Some Harvey-battered homeowners thought skipping flood insurance was low-risk gamble," USA Today, August 31, 2017, https://usatoday.com/story/news/2017/08/31/some-harvey-battered-homeowners-thought-skipping-flood-insurance-low-risk-gamble/619593001/.
28 Alexis C. Madrigal, "The Houston Flooding Pushed the Earth's Crust Down 2 Centimeters," The Atlantic, September 5, 2017, https://www.theatlantic.com/technology/archive/2017/09/hurricane-harvey-deformed-the-earths-crust-around-houston/538866/.
29 Ian Bremmer, "Houston, Mumbai, Ontario, Macau, Niger, Hong Kong, Istanbul, Ireland, Sierra Leone . . . All Flooded," Daily Kos, August 30, 2017, https://m.dailykos.com/stories/2017/8/30/1694839/-Houston-Mumbai-Ontario-Macau-Niger-Hong-Kong-Istanbul-Mainland-China-Sierra-Leone?detail-emaildkre.
30 Zbigniew Brzezinski, The Grand Chessboard: American Primcy and its Geostrategic Imperatives (New York: Basic Books, 1997).
31 Global Research, "Numbers of Casualties Resulting from America's Post 9/11 Wars of Aggression," April 29,2018, https://www.globalresearch.ca/numbers-of-casualties-resulting-from-americas-post-911-wars-of-aggression/5638317.32 Katie Bardaro and all, "CEO Pay: How Much Do CEOs Make Compared to Their Employees?" PayScale, accesssed June 24, 2020, http://www.payscale.com/data-packages/ceo-pay/full-list.
33 Stan Choe, "CEOs get $800,000 pay raise, leaving workers fur-

ther behind," Associated Press, May 24, 2019, https://www.apnews.com/1a68c31f2cb54b15af536cbe168442bf.

34 Tae Kim, "Airline shares decline as Wall Street worries the stocks will become a poor investment again," CNBC, Apr. 27, 2017, https://www.cnbc.com/2017/04/27/airline-shares-decline-amid-worries-the-stocks-will-become-a-poor-investment-again.html.

35 Vijay Prashad, "The human carnage from billionaires trying to carve up the planet to build their empires is astounding," Salon, Aug. 25, 2017, http://www.salon.com/2017/08/25/the-human-carnage-from-billionaires-trying-to-carve-up-the-planet-to-build-their-empires-is-astounding_partner/.

36 J. Gerassi (ed.), *VENCEREMOS! THE SPEECHES AND WRITINGS OF ERNESTO CHE GUEVARA* (New York: Macmillan, 1968), p. 398. from Che Guevara, "Man and Socialism in Cuba."

37 Erich Fromm, *The Art of Loving*, (New York: Harper Row, 1956).

38 Karl Marx and Frederick Engels, Manifesto of the Communist Party, MIA: Marxists: Marx and Engels: Library: 1848: Manifesto of the Communist Party: Chapter 1, https://www.marxists.org/archive/marx/works/1848/communist-manifesto/ch01.htm.

39 P. R. Sarkar, *Problems of the Day* (Trimohan Bhagalpur: Ananda Marga Publications, 1958), The Electronic Editions of the Works of P. R. Sarkar, Version 1.6.0.4, Ananda Marga Publications, 2009.

40 P. R. Sarkar, "Shudra Revolution and Sadvipra Society," *Human Society, Part 2* (Kolkata: Ananda Marga Publications, 1967), The Electronic Editions of the Works of P. R. Sarkar, Version 1.6.0.4, Ananda Marga Publications, 2009.

41 P. R. Sarkar, "Nuclear Revolution," *Prout in a Nutshell Volume 4 Part 21 [a compilation]*.

42 P. R. Sarkar, "Nuclear Revolution," *Prout in a Nutshell Volume 4 Part 21 [a compilation]* (Ranchi, Ananda Marga Publications, 1969), The Electronic Editions of the Works of P. R. Sarkar, Version 1.6.0.4, Ananda Marga Publications, 2009.

43 Although P. R. Sarkar never refers to his methodology as wave theory, we use this term because it best describes his theory of motion and station. Sarkar himself offers no name for his methodology

44 See 1) Vladimir Lenin, "Summary of Dialectics," *Collected Works* (Moscow: Progress Publishers, 1914), 2) J.V. Stalin, *Dialectical*

and Historical Materialism, 1938, 3) Mao Tse-tung, "On Contradiction," *Selected Works*, 1937.

45 Wikipedia, "Materialism," last modified May 9, 2019, https://en.wikipedia.org/wiki/Materialism.

46 P. R. Sarkar, *Subhasita Samgraha: A Treatise on Intuitional Science* (Kolkata: Ananda Marga Publications,) 33.

47 P. R. Sarkar, *The Human Society Part II* (Kolkata, Ananda Marga Publications, 1967), 1.

48 P. R. Sarkar, *The Human Society Part II* (Kolkata, Ananda Marga Publications, 1967), 5.

49 P. R. Sarkar, *The Human Society Part II* (Kolkata, Ananda Marga Publications, 1967), 1-2.

50 P. R. Sarkar, *The Human Society Part II* (Kolkata, Ananda Marga Publications, 1967), 7.

51 P. R. Sarkar, *The Human Society Part II* (Kolkata, Ananda Marga Publications, 1967), 7.

52 P. R. Sarkar, *The Human Society Part II* (Kolkata, Ananda Marga Publications, 1967), 5.

53 P. R. Sarkar, *Subhasita Samgraha: A Treatise on Intuitional Science, Part IV* (Kolkata, Ananda Marga Publications, 1968), 33.

54 P. R. Sarkar, *Ananda Marga Elementary Philosophy* (Kolkata, Ananda Marga Publications, 1998), 14.

55 See Julian Jaynes, *The Origins of Consciousness in the Breakdown of the Bicameral Mind* (Boston: Houghton Mifflin Co., 1976) for an anthropological analysis of how and when human beings developd a distinct individual consciousness.

56 P. R Sarkar, "The Human Search for Real Progress," *Prout in a Nutshell*, Volume 2, Part 6 (Kolkata: Ananda Marga Publications, 1967), The Electronic Edition of the Works of P. R. Sarkar, Version 1.4.0.6, Ananda Marga Publications, 2009.

57 P. R. Sarkar, "Fighting at Each and Every Step," *Ánanda Vacanámrtam Part 34* (Kolkata: Ananda Marga Publications, date unknown), The Electronic Edition of the Works of P. R. Sarkar, Version 1.4.0.6, Ananda Marga Publications, 2009.

58 See P. R. Sarkar, *PROUTist Economics: Discourses on Economic Liberation* (Kolkata: Ananda Marga Publications, 1987). This is a compilation of all of Sarkar's discourses and writings on his Progressive Utilization Theory (PROUT). The material is taken from the original

compilation entitled PROUT in a Nutshell (1987). In October 1991 the last part of the first edition of the series, Part 21, was published in English.

59 P. R. Sarkar, *The Human Society, Part II* (Kolkata, Ananda Marga Publications, 1967), 5.

60 P. R. Sarkar, *The Human Society, Part II* (Kolkata, Ananda Marga Publications, 1967), 6.

61 P. R. Sarkar, *The Human Society, Part II* (Kolkata, Ananda Marga Publications, 1967), 6.

62 P. R. Sarkar, "The Cosmic Brotherhood," *Idea and Ideology* (Kolkata, Ananda Marga Publications, 1959), The Electronic Edition of the Works of P. R. Sarkar, Version 1.4.0.6, Ananda Marga Publications, 2009.

63 P. R. Sarkar, "The Cosmic Brotherhood," *Idea and Ideology, The Electronic Edition of the Works of P. R. Sarkar, Version 1.4.0.6* Ananda Marga Publications, 2009.

64 P. R. Sarkar, "The Cosmic Brotherhood," *Idea and Ideology.*

65 P. R. Sarkar, "There Should Be Subjective Approach through Objective Adjustment," *Ánanda Vacanámrtam Part 31* (Kolkata, Ananda Marga Publications, date unknown), The Electronic Edition of the Works of P. R. Sarkar, Version 1.4.0.6, Ananda Marga Publications, 2009.

66 P. R. Sarkar, "Minimum Requirements and Maximum Amenities," *Prout in a Nutshell Volume 4 Part 17 [a compilation]* (Kolkata: Ananda Marga Publications, 1989), The Electronic Edition of the Works of P. R. Sarkar, Version 1.4.0.6, Ananda Marga Publications, 2009.

67 Aristotle, "Politics," Translated by Benjamin Jowett, http://classics.mit.edu/Aristotle/politics.5.five.html.

68 Aristotle, "Politics," Translated by Benjamin Jowett.

69 Barry Strauss, "Ancient Uprisings That Changed the World," History Net, May 3, 2011, http://www.historynet.com/ancient-uprisings-that-changed-the-world.htm.

70 Wikipedia, "Revolution," last modified May 1, 2019, https://en.wikipedia.org/wiki/Revolution.

71 Jeff Goodwin, *No Other Way Out: States and Revolutionary Movements, 1945-1991* (Cambridge, Eng., Cambridge University Press, 2001), 5.

72 History.com editors, "Soviet Union" History.com, last edited May 16, 2019, https://www.history.com/topics/russia/history-of-the-soviet-union.
73 Norwich University Online, Consequences of the Collapse of the Soviet Union, Norwich University, October 2, 2017, https://online.norwich.edu/academic-programs/resources/consequences-of-the-collapse-of-the-soviet-union.
74 Wikipedia, "Yulia Tymoshenko," last edited July 3, 2019, https://en.wikipedia.org/wiki/Yulia_Tymoshenko.
75 Pavel Polityuk and Richard Balmforth, "Yanukovich declared winner in Ukraine poll," The Independent, Monday, 15 February 2010, https://www.independent.co.uk/news/world/europe/yanukovich-declared-winner-in-ukraine-poll-1899552.html.
76 "Belarus: Events of 2017," World Report, 2019, Human Rights Watch, https://www.hrw.org/world-report/2018/country-chapters/belarus.
77 "Belarus: Events of 2018," World Report 2019, Human Rights Watch, https://www.hrw.org/world-report/2019/country-chapters/belarus.
78 Wikipedia, "Delyan Peevski," last edited July 15, 2019, https://en.wikipedia.org/wiki/Delyan_Peevski.
79 "An article about the Bulgarian mafia," StreetGangs.Com, Mar. 25, 2009, http://www.streetgangs.com/billboard/viewtopic.php?t=43324.
80 Wikipedia, "Bulgarian Mafia," last edited June 3, 2019, https://en.wikipedia.org/wiki/Bulgarian_mafia.
81 "Film: Once Upon a Time in the East, Transcript," Journeyman.TV, https://www.journeyman.tv/film_documents/5424/transcript/.
82 Wikipedia, "Politics of Bulgaria," last edited March 11, 2019, https://en.wikipedia.org/wiki/Politics_of_Bulgaria.
83 Susanna Kim, "Hong Kong Student Leader 'Optimistic' Before Government Deadline to Clear Protesters, ABC News, October 5, 2014, https://abcnews.go.com/International/hong-kong-student-leader-optimistic-government-deadline-clear/story?id=25978457.
84 "Transcript of remarks by CY Leung at media session," Press Releases, April 22, 2015, https://www.info.gov.hk/general/201504/22/P201504220388.htm.
85 Wikipedia, 2014 Hong Kong protests," last edited September 3,

2019, https://en.wikipedia.org/wiki/2014_Hong_Kong_protests.

86 Wikipedia, "2017 Hong Kong Chief Executive election," last edited September 2, 2019, https://en.wikipedia.org/wiki/2017_Hong_Kong_Chief_Executive_election

87 Nicola Slawson, "Hong Kong: Riot Police Crackdown On Protesters Occupy Parliament On Anniversary Of Handover to China," Huffington Post, 1/7/19, https://www.huffingtonpost.co.uk/entry/hong-kong-protesters-occupy-parliament_uk_5d1a1a11e4b082e5536d3740?guccounter=1.

88 Verna Yu, "We have no other choice': as China erodes democracy Hong Kong citizens prepare to leave," The Guardian, June 24, 2019, https://www.theguardian.com/world/2019/jun/25/we-have-no-other-choice-as-china-erodes-democracy-hong-kong-citizens-prepare-to-leave

89 Hong Kong: Extradition Changes Threaten Civil Society, Human Rights Watch, June 6, 2019, https://www.hrw.org/news/2019/06/06/hong-kong-extradition-changes-threaten-civil-society#:~:-%E2%80%9CNo%20one%20will%20be%20safe,journalists%2C%20and%20social%20workers.%E2%80%9D&text=The%20Chinese%20government%20regularly%20brings,oppose%20government%20policy%20in%20China.

90 Hong Kong-China extradition plans explained, BBC, August 22, 2019, https://www.bbc.com/news/world-asia-china-47810723.

91 Shibana Mahatani, "Masks, cash and apps: How Hong Kong's protesters find ways to outwit the surveillance state," The Washington Post, June 15, 2019, https://www.washingtonpost.com/world/asia_pacific/masks-cash-and-apps-how-hong-kongs-protesters-find-ways-to-outwit-the-surveillance-state/2019/06/15/8229169c-8ea0-11e9-b6f4-033356502dce_story.html.

92 Shibana Mahatani, "Masks, cash and apps: How Hong Kong's protesters find ways to outwit the surveillance state," The Washington Post, June 15, 2019, https://www.washingtonpost.com/world/asia_pacific/masks-cash-and-apps-how-hong-kongs-protesters-find-ways-to-outwit-the-surveillance-state/2019/06/15/8229169c-8ea0-11e9-b6f4-033356502dce_story.html.

93 Shibani Mahtani, "Masks, cash and apps: How Hong Kong's protesters find ways to outwit the surveillance state," The Washington Post, June 15, 2019.

94 Verna Yu, "Hong Kong's Carrie Lam defiant in face of extradition climbdown," The Guardian, June 15, 2019, https://www.theguardian.com/world/2019/jun/15/hong-kong-carrie-lam-defiant-extradition-climbdown.
95 Lily Kuo, "Hong Kong 'umbrella movement': nine convicted over protests," The Guardian, April 9, 2019, https://www.theguardian.com/world/2019/apr/09/hong-kong-umbrella-movement-protesters-guilty-over-pro-democracy-rallies-jail.
96 Louisa Lim, "Hong Kong is not China yet, but that feared day is coming ever nearer," The Guardian, June 16, 2019, https://www.theguardian.com/commentisfree/2019/jun/16/hong-kong-not-china-yet-but-feared-grim-day-coming-ever-nearer.
97 Louisa Lim, "Hong Kong is not China yet, but that feared day is coming ever nearer," The Guardian, June 16, 2019.
98 Yonden Lhatoo, "There will be blood if Hong Kong protesters continue to blur the line between people power and mob rule," South China Morning Post, June 22, 2019, https://www.scmp.com/news/hong-kong/politics/article/3015667/there-will-be-blood-if-hong-kong-protesters-continue-blur.
99 Tony Cheung, "Hong Kong's largest pro-establishment part 'would not oppose' city leader Carrie Lam announcing full withdrawal of extradition bill to heal rifts in society," South China Morning Post, June 23, 2019, https://www.scmp.com/news/hong-kong/politics/article/3015722/hong-kongs-largest-pro-establishment-party-urges-city.
100 Tony Cheung, "Hong Kong's largest pro-establishment part 'would not oppose' city leader Carrie Lam announcing full withdrawal of extradition bill to heal rifts in society," South China Morning Post, June 23, 2019.
101 Tony Cheung, "Hong Kong's largest pro-establishment party 'would not oppose' city leader Carrie Lam announcing full withdrawal of extradition bill to heal rifts in society," South China Morning Post, June 23, 2019, https://sg.news.yahoo.com/hong-kong-largest-pro-establishment-094132117.html.
102 Verna Yu, "'We have no other choice': as China erodes democracy Hong Kong citizens prepare to leave," The Guardian, June 24, 2019, https://www.theguardian.com/world/2019/jun/25/we-have-no-other-choice-as-china-erodes-democracy-hong-kong-citizens-prepare-to-leave#img-1.

103 Verna Yu, "'We have no other choice': as China erodes democracy Hong Kong citizens prepare to leave," The Guardian, June 24, 2019.
104 Verna Yu, "'We have no other choice': as China erodes democracy Hong Kong citizens prepare to leave," The Guardian, June 24, 2019.
105 Verna Yu, "We have no other choice': as China erodes democracy Hong Kong citizens prepare to leave," The Guardian, June 24, 2019.
106 Kelly Olsen, "Hong Kong activist Wong: World leaders 'must' raise our eroding freedoms at G-20," CNBC, June28, 2019, https://www.cnbc.com/2019/06/28/world-leaders-must-raise-hong-kong-at-g-20-activist.html.
107 Kelly Olsen, "Hong Kong activist Wong: World leaders 'must' raise our eroding freedoms at G-20," CNBC, June 28, 2019, https://www.cnbc.com/2019/06/28/world-leaders-must-raise-hong-kong-at-g-20-activist.html.
108 Anne Marie Roantree and Noah Sin, "Hong Kong prepares for pro-democracy march amid anger over extradition bill," Business Day, June 30, 2019, https://www.businesslive.co.za/bd/world/asia/2019-06-30-hong-kong-prepares-for-pro-democracy-march-amid-anger-over-extradition-bill/.
109 Jasmine Andersson, "Hong Kong leader Carrie Lam condems 'extreme use of violence' after protest over China extradition law," INEWS, July 2, 2019, https://inews.co.uk/news/world/hong-kong-leader-carrie-lam-protest-china-extradition-law-handover-anniversary-308961.
110 Shibani Mahtani, "Why Hong Kong protesters rage on, even though they cannot win," The Washington Post, July 3, 2019, https://www.washingtonpost.com.
111 "Hong Kong protesters storm parliament on anniversary of return to China," The Philadelphia Inquirer, July 3, 2019, https://www.inquirer.com/news/nation-world/hong-kong-protesters-storm-parliament-anniversary-return-china-20190701.html.
112 James Griffiths, Hong Kong is protesting a China extradition bill, but violence could open divisions," CNN, July 2, 2019, https://www.cnn.com/2019/07/01/asia/hong-kong-protests-divide-intl-hnk/.
113 Shibani Mahtani and Timothy McLaughlin, "Hong Kong police move to forcibly clear protesters occupying legislature complex," Washington Post, July 1, 2019, https://www.washingtonpost.com/

world/hong-kong-police-clash-with-protesters-on-anniversary-of-return-to-china/2019/06/30/7f0a05ae-9b3c-11e9-83e3-45fded8e8d2e_story.html

114 Alvin Lum, "It wasn't violence for violence's sake': the only unmasked protester at storming of Hong Kong's legislature gives his account of the day's drama," South China Morning Post, July 5, 2019, https://www.scmp.com/news/hong-kong/politics/article/3017327/it-wasnt-violence-violences-sake-only-unmasked-protester.

115 Sophia Yan, "Chinese tourists look on in amazement as Hong Kong street protesters march through city," The Telegraph, July 7, 2019, https://www.telegraph.co.uk/news/2019/07/07/chinese-tourists-look-amazement-hong-kong-street-protesters/.

116 Sophia Yan, "Chinese tourists look on in amazement as Hong Kong street protesters march through city," The Telegraph, July 7, 2019.

117 Sophia Yan, "Chinese tourists look on in amazement as Hong Kong street protesters march through city," The Telegraph, July 7, 2019.

118 Jean-Francois Dupre, "The bill is dead? Chief Exec. Carrie Lam is throwing dust in Hongkongers' eyes – democracy is the only solution," Hong Kong Free Press, July 9, 2019, https://www.hongkongfp.com/2019/07/09/bill-dead-chief-exec-carrie-lam-throwing-dust-hongkongers-eyes-democracy-solution/.

119 Michael Chugani, Hong Kong's protest stalemate can only be broken by giving the city the democracy it deserves," South China Morning Post, July 25, 2019, https://www.scmp.com/comment/opinion/article/3019865/hong-kongs-protest-stalemate-can-only-be-broken-giving-city.

120 Alec Larson, "Anti-Extradition Bill Protesters Occupy Hong Kong Airport," Newsy, July 26, 2019, https://www.newsy.com/stories/anti-extradition-bill-protesters-occupy-hong-kong-airport/.

121 Nectar Gan and Kristin Huang, "Will China send in troops to stamp out protests in Hong Kong?" South China Morning Post, July 24, 2019, https://www.politico.com/story/2019/07/24/china-troops-hong-kong-protest-1432265.

122 Antony Dapiran, "China is looking for a quick economic fix in Hong Kong. It's missing the point," The Guardian, July 30, 2019, https://www.theguardian.com/commentisfree/2019/jul/30/china-economic-hong-kong-carrie-lam.

123 James A. Dorn, "If protesters want to protect Hong Kong's

way of life, they must win the war of ideas," South China Morning Post, August 2, 2019, https://www.scmp.com/comment/opinion/article/3020897/if-protesters-want-protect-hong-kongs-way-life-they-must-win-war.

124 "Hong Kong: Timeline of extradition protests," BBC, September 4, 2019, https://www.bbc.com/news/world-asia-china-49340717.

125 "Hong Kong activists arrested: Joshua Wong and others charged," BBC, August 30, 2019, https://www.bbc.com/news/world-asia-china-49520025.

126 Hong Kong police storm metro system after protests, BBC, September 1, 2019, https://www.bbc.com/news/world-asia-china-49540751.

127 "Hong Kong: Timeline of extradition protests," BBC, September 4, 2019, https://www.bbc.com/news/world-asia-china-49340717.

128 Shelly Banjo, Natalie Lung, Annie Lee and Hannah Dormido, "How Hong Kong's Leaderless Protest Army Gets Things Done," Bloomberg, August 22, 2019, https://www.bloomberg.com/graphics/2019-hong-kong-airport-protests/.

129 Shelly Banjo, Natalie Lung, Annie Lee and Hannah Dormido, "How Hong Kong's Leaderless Protest Army Gets Things Done," Bloomberg, August 22, 2019.

130 Shelly Banjo, Natalie Lung, Annie Lee and Hannah Dormido, "How Hong Kong's Leaderless Protest Army Gets Things Done," Bloomberg, Aug. 22, 2019.

131 Zeeshan Aleem, "Hong Kong demonstrators defy lockdowns to protest a sweeping new national security law," Vox, May 24, 2020, https://www.vox.com/world/2020/5/24/21268893/hong-kong-protest-china-coronavirus-law.

132 "Population of Cities in Hong Kong (2019), World Population Review, accessed September 4, 2019, http://worldpopulationreview.com/countries/hong-kong-population/cities/.

133 "Offshore Tax Havens," TaxHavens.biz, accessed September 2, 2019, http://www.taxhavens.biz/other_tax_havens/tax_haven_hong_kong/.

134 Wikipedia, "Tax haven," last edited August 31, 2019, https://en.wikipedia.org/wiki/Tax_haven.

135 Tsering Namgyal, "Hong Kong Protests May Give Marx the Last Laugh," Asia Sentinel, July 31, 2019, https://www.asiasentinel.

com/politics/hongkong-protests-marx-last-laugh/.
136 Cannix Yau, Viola Zhou, "What hope for the poorest? Hong Kong wealth gap hits record high," Report finds richest 10 per cent in city earn 44 times that of the poorest, South China Morning Post, June 9, 2019, https://www.scmp.com/news/hong-kong/economy/article/2097715/what-hope-poorest-hong-kong-wealth-gap-hits-record-high.
137 Serge Berthier, "The Real Social Structure of Hong Kong," Asian Affairs Analysis, May 2003, http://asian-affairs.com/HongKong/structureofhongk.html.
138 Serge Berthier, "The Real Social Structure of Hong Kong," Asian Affairs Analysis, May 2003.
139 Serge Berthier, "The Real Social Structure of Hong Kong, Asian Affairs, Issue 19, May, 2003.
140 Leia Wang, "The Other Side of Hong Kong: A Dire Housing and Class Division Crisis," Glimpse from the Globe, July 14, 2018, http://www.glimpsefromtheglobe.com/regions/asia-and-the-pacific/the-other-side-of-hong-kong-a-dire-housing-and-class-division-crisis/.
141 Leia Wang, "The Other Side of Hong Kong: A Dire Housing and Class Division Crisis," Development and Human Rights, East Asia and the Pacific, Economics, July 14, 2018.
142 Oswald Chan, "Fixing social ladder for Hong Kong youth," ChinaDaily, August 27, 2018, https://www.chinadailyhk.com/articles/131/242/53/1535340475392.html.
143 "Letters | Hong Kong people, young or old, have it rough: no wonder they have taken to the streets," South China Morning Post, July 8, 2019, https://www.scmp.com/comment/letters/article/3017421/hong-kong-people-young-or-old-have-it-rough-no-wonder-they-have.
144 "Letters | Hong Kong people, young or old, have it rough: no wonder they have taken to the streets," South China Morning Post, July 8, 2019, https://www.scmp.com/comment/letters/article/3017421/hong-kong-people-young-or-old-have-it-rough-no-wonder-they-have.
145 Kanis Leung, "US-China trade war brings Hong Kong's economic expansion to near standstill," South China Morning Post, July 31, 2019, https://www.scmp.com/news/hong-kong/hong-kong-economy/article/3020808/us-china-trade-war-brings-hong-kongs-economic.

146 Josephine Ma, Cahterine Wong, Echo Xie, "Why China is looking to Shenzhen-and not Hong Kong-to reinvent its economic future," South China Morning Post, July 31, 2019, https://www.scmp.com/news/china/politics/article/3020710/why-china-looking-shenzhen-and-not-hong-kong-reinvent-its.
147 Josephine Ma, Cahterine Wong, Echo Xie, "Why China is looking to Shenzhen-and not Hong Kong-to reinvent its economic future," South China Morning Post, July 31, 2019.
148 Josephine Ma, Cahterine Wong, Echo Xie, "Why China is looking to Shenzhen-and not Hong Kong-to reinvent its economic future," South China Morning Post, July 31, 2019.
149 Josephine Ma, Catherine Wong, Echo Xie, "Why China is looking to Shenzhen-and not Hong Kong-to reinvent its economic future," South China Morning Post, July 31, 2019.
150 Ambrose Evans-Prichard, "Washington can pull the plug on Hong Kong's economic model at any moment," The Telegraph, July 16, 2019, https://www.telegraph.co.uk/business/2019/06/16/washington-can-pull-plug-hong-kongs-economic-model-moment/.
151 Ambrose Evans-Pritchard, "Washington can pull the plug on Hong Kong's economic model at any moment," The Telegraph, June 16, 2019.
152 Dan Cohen, "Behind a made-for-TV Hong Kong protest narrative, Washington is backing nativism and mob violence," The GrayZone, August 17, 2019, https://thegrayzone.com/2019/08/17/hong-kong-protest-washington-nativism-violence/.
153 John V. Walsh, "'Hands Off Hong Kong' – The Cry that Seldom is Heard," Global Research, September 10, 2019, https://www.globalresearch.ca/hands-off-hong-kong/5688696.
154 Jonathan Allen, "U.S. should side with Hong Kong protesters, says ex-Pentagon chief Mattis," Reuters, September 9, 2019, https://www.reuters.com/article/us-usa-trump-mattis/u-s-should-side-with-hong-kong-protesters-says-ex-pentagon-chief-mattis-idUSKCN1VU1YF?utm_source=applenews.
155 Twitter Safety, "Information operations directed at Hong Kong, August 19, 2019, https://blog.twitter.com/en_us/topics/company/2019/information_operations_directed_at_Hong_Kong.html.
156 Sonali Kolhatkar, "Understanding Hong Kong's Mass Movement," Truthdig, September 5, 2019, https://www.truthdig.com/arti-

cles/hong-kongs-mass-movement-isnt-controlled-by-the-u-s/.
157 Daniella Genovese, "London Stock Exchange rejects Hong Kong takeover offer," Fox Business, September 13, 2019, https://www.foxbusiness.com/markets/london-stock-exchange-rejects-hong-kong-takeover-offer.
.158 Wikipedia, "Sahel," last edited July 20, 2019, https://en.wikipedia.org/wiki/Sahel.
159 Wikipedia, "Ansar (Sudan), last edited December 18, 2018, https://en.wikipedia.org/wiki/Ansar_(Sudan).
160 Wikipedia, "Ansar (Sudan), last edited December 18, 2018, https://en.wikipedia.org/wiki/Ansar_(Sudan).
161 Dr. Riek Machar Teny-Dhurgon, Chairman and Commander-in-Chief, SSIM/A, "South Sudan: A History of Political Domination - A Case of Self-Determination, (Riek Machar)," accessed July 26, 2019, http://www.africa.upenn.edu/Hornet/sd_machar.html.
162 Wikipedia, "History of Sudan," last edited July 13, 2019, https://en.wikipedia.org/wiki/History_of_Sudan.
163 Wikipedia, "History of Sudan," last edited July 13, 2019.
164 Wikipedia, "War in Darfur," last edited June 30, 2019, https://en.wikipedia.org/wiki/War_in_Darfur.
165 Wikipedia, "National Islamic Front," last edited July 25, 2019, https://en.wikipedia.org/wiki/National_Islamic_Front.
166 Wikipedia, "History of Sudan," last edited July 13, 2019.
167 "A New Nation Torn Apart: Families Flee to Escape Conflict," June 24, 2019, Mercy Corps, https://www.mercycorps.org/articles/south-sudan/south-sudan-crisis.
168 Megan Specia, 383,000: Estimated Death Toll in South Sudan's War," September 26 2018, New York Times, https://www.nytimes.com/2018/09/26/world/africa/south-sudan-civil-war-deaths.html
169 "Sudan crisis: The ruthless mercenaries who run the country for gold," BBC, July 20, 2019, https://www.bbc.com/news/world-africa-48987901.
170 Wikipedia, "Rapid Support Forces," last edited July 21, 2019, https://en.wikipedia.org/wiki/Rapid_Support_Forces.
171 Zeinab Mohammed Salih and Jason Burke, "Sudanese doctors say dozens of people raped during sit-in attack," June 11, 2019, The Guardian, https://www.theguardian.com/world/2019/jun/11/sudan-troops-protesters-attack-sit-in-rape-khartoum-doctors-report.

172 Lydia Wang, "As Sudan Faces Political Upheaval & a Food Crisis, Here's How to Help," June 15, 2019, Refinery 29, https://www.refinery29.com/en-us/2019/06/235515/sudan-political-food-crisis-how-to-help.

173 Khalid Abdelaziz, "Update 2: Sudan military says it thwarts coup attempt, arrests senior officers," Reuters, July 24, 2019, https://af.reuters.com/article/sudanNews/idAFL8N24P74E.

174 What's at stake in next round of Sudan negotiations? Al-Jazeera, July 19, 2019, https://www.aljazeera.com/news/2019/07/stake-sudan-negotiations-190719072738212.html.

175 What's at stake in next round of Sudan negotiations? Al-Jazeera, July 19, 2019.

176 Samy Magdy, "Sudan's army chief among several arrested in new coup plot," Associated Press, July 24, 2019, https://apnews.com/ffe2b9e1b37747d7b09760d3d12c33a0.

177 "Sudan's Protesters Say 'Revolution Has Won,' Military Rulers Hail Power-sharing Deal Top of Form Bottom of Form," The Associated Press, July 05, 2019, Haaretz, https://www.haaretz.com/world-news/sudan-s-protesters-claim-victory-for-their-revolution-1.7450404.

178 Ali A. Ali, "Sudan's path to democracy is blocked by a modern slavery pact," Al-Jazeera, July 1, 2019, https://www.aljazeera.com/indepth/opinion/sudan-path-democracy-blocked-modern-slavery-pact-190701071559618.html.

179 Daniel P. Owen, Anna Autio, Olivier Mukarji, "Preventing conflict through human rights-based early warning in Africa," World Bank, February 4, 2019, http://blogs.worldbank.org/dev4peace/preventing-conflict-through-human-rights-based-early-warning-africa?CID=SURR_TT_WBGCities_EN_EXT.

180 Mohamed Suliman, "Civil War in Sudan: The Impact of Ecological Degradation," Institute For African Alternatives UK (IFAA) ENCOP - Sudan Programme, accessed July 27, 2019, http://www.africa.upenn.edu/Articles_Gen/cvlw_env_sdn.html.

181 Mohamed Suliman, "Civil War in Sudan: The Impact of Ecological Degradation," Institute for African Alternatives UK (IFAA) ENCOP - Sudan Programme, accessed July 27, 2019.

182 Mohamed Suliman, "Civil War in Sudan: The Impact of Ecological Degradation," Institute For African Alternatives UK (IFAA) ENCOP - Sudan Programme, accessed July 27, 2019.

183 Mohamed Suliman, "Civil War in Sudan: The Impact of Ecological Degradation," Institute For African Alternatives UK (IFAA) ENCOP - Sudan Programme, accessed July 27, 2019.
184 Mohamed Suliman, "Civil War in Sudan: The Impact of Ecological Degradation," Institute For African Alternatives UK (IFAA) ENCOP - Sudan Programme, accessed July 27, 2019.
185 United Nations Environmental Programme, "Agriculture and the Environment," Sudan Post-Conflict Environmental Assessment, https://postconflict.unep.ch/publications/sudan/08_agriculture.pdf.
186 "Sudan: Plan of Action (2015-2019), Summary," Food and Agriculture Organization of the United Nations, accessed July 29, 2019, http://www.fao.org/3/a-i4786e.pdf.
187 Mohamed Suliman, "Civil War in Sudan: The Impact of Ecological Degradation," Institute For African Alternatives UK (IFAA) ENCOP - Sudan Programme, accessed July 27, 2019.
188 Geoffrey York, "After defying violence for months, Sudan's fearless protesters are skeptical of latest agreement," The Globe and Mail, July 5, 2019, https://www.theglobeandmail.com/world/article-after-defying-violence-for-months-sudans-fearless-protesters-are/.
189 Geoffrey York, "After defying violence for months, Sudan's fearless protesters are skeptical of latest agreement," The Globe and Mail, July 5, 2019, https://www.theglobeandmail.com/world/article-after-defying-violence-for-months-sudans-fearless-protesters-are/.
190 Samia Al-Nagar and Liv Tønnessen, "Sudan: 'I'm Against All of the Laws of This Regime' - What Sudan's Women Want," AllAfrica, July 2, 2019, https://allafrica.com/stories/201907040417.html.
191 "Sudan after Bashir: 'The Revolution is on the Curriculum,'" BBC News, December 28, 2019, https://www.bbc.com/news/world-africa-50835344.
192 Wikipedia, "Kurdistan," last edited July 29, 2019, https://en.wikipedia.org/wiki/Kurdistan
193 Wikipedia, "Treaty of Lausanne," last edited July 19, 2019, https://en.wikipedia.org/wiki/Treaty_of_Lausanne.
194 Norvell B. DeAtkine, The Fears of Iran and its Forgotten Kurds, accessed August 5, 2019, https://limacharlienews.com/mena/fears-of-iran-forgotten-kurds/.
195 "Mahabad," The Kurdish Project, accessed August 3, 2019, https://thekurdishproject.org/kurdistan-map/iranian-kurdistan/ma-

habad/.
196 "Iranian Kurdistan," Unrepresented Nations and Peoples Organization, accessed August 5, 2019, https://unpo.org/members/7882.
197 "Iranian Sunnis: Religious Freedom," Dream of Iran, accessed August 3, 2019, http://dreamofiran.com/dossiers/iranian-sunnis-religious-freedom/.
198 Wikipedia, "Iranian Kurdistan," last edited July 26, 2019, https://en.wikipedia.org/wiki/Iranian_Kurdistan.
199 Chris Zambelis, "The Factors Behind Rebellion in Iranian Kurdistan," March 2011, Vol 4, Issue 3, Combating Terrorism Center Sentinel, West Point, accessed August 3, 2019, https://ctc.usma.edu/the-factors-behind-rebellion-in-iranian-kurdistan/.
200 Iran (Rojhelat or Eastern Kurdistan), The Kurdish Project, accessed August 2, 2019, https://thekurdishproject.org/kurdistan-map/iranian-kurdistan/.
201 Wikipedia, "Kurds in Iran," last edited July 19, 2019, https://en.wikipedia.org/wiki/Kurds_in_Iran.
202 Chris Zambelis, "The Factors Behind Rebellion in Iranian Kurdistan," March 2011, Volume 4, Issue 3, Combating Terrorism Center Sentinel, West Point, accessed Aug. 3, 2019.
203 The Kurdish Project, May 29, 2020, https://mail.google.com/mail/u/0/#label/BOOK%2FBook+3%2FKurds/FMfcgxwHNgWwltFlFrxlFkhjqGLPzVzj.
204 Wikipedia, "Iraqi Kurdistan," last edited August 2, 2019, https://en.wikipedia.org/wiki/Iraqi_Kurdistan.
205 Wikipedia, "Ansar al-Islam," last edited June 30, 2019, https://en.wikipedia.org/wiki/Ansar_al-Islam.
206 See *The Untold Story of Western Civilization*, Volume 5, Chapter 3 for an analysis of the rise of ISIS and US policy objectives in the Middle East.
207 Aaron Mak, "Hundreds of Thousands of Iraqis Turn Out to Protest US Military Presence," Slate, January 26, 2020, https://slate.com/news-and-politics/2020/01/iraqis-protest-u-s-occupation-following-the-killing-of-iranian-general-qassem-soleimani.html.
208 Mariya Petkova, "Why are Iraqi Kurds not taking part in protests?" Al-Jazeera, November 11, 2019, https://www.aljazeera.com/news/2019/11/iraqi-kurds-part-protests-191111125744569.html
209 Paul J. White, "Economic marginalization of Turkey's Kurds:

The failed promise of modernization and reform," Journal of Muslim Minority Affairs, April 1998, Volume 18, Issue 1, 139-158, accessed August 8, 2019, https://www.academia.edu/6977296/Economic_marginalization_of_Turkey_s_Kurds_The_failed_promise_of_modernization_and_reform.

210 G. Chaliand, A.R. Ghassemlou, M. Pallis, *A People Without A Country* (London: Zed Books, 1992), 58.

211 Paul J. White, "Economic marginalization of Turkey's Kurds: The failed promise of modernization and reform," Journal of Muslim Minority Affairs, April 1998, Volume 18, Issue 1, 139-158.

212 Wikipedia, "Abdullah Ocalan," last edited July 31, 2019, https://en.wikipedia.org/wiki/Abdullah_%C3%96calan#Democratic_confederalism.

213 Paul White, "Democratic Confederalism and the PKK's Feminist Transformation," *The PKK: Coming Down from the Mountains* (London, Zed Books, 20015, 126-149.

214 Paul White, "Democratic Confederalism and the PKK's Feminist Transformation," The PKK: Coming Down from the Mountains.

215 Peter Beaumont, "Erdoğan issues stark 'final warning' to Turkey's Gezi Park protesters," June, 14, 2013, https://www.theguardian.com/world/2013/jun/13/turkey-gezi-park-protesters.

216 Constanze Letsch, Nadia Khomami, "Turkey terror attack: mourning after scores killed in Ankara blasts," The Guardian, October 11, 2015, https://www.theguardian.com/world/2015/oct/10/turkey-suicide-bomb-killed-in-ankara.

217 Wikipedia, "Turkey-ISIL conflict," last modified August 9, 2019, https://en.wikipedia.org/wiki/Turkey%E2%80%93ISIL_conflict.

218 Liz Sly, Turkey strikes Kurdish militants in Iraq, ends truce of more than 2 years, July 25, 2015, https://www.washingtonpost.com/world/turkey-strikes-kurdish-militants-in-iraq-ending-truce/2015/07/25/1b7405b8-32b7-11e5-97ae-30a30cca95d7_story.html?noredirect=on.

219 Constanze Letsch, Nadia Khomami, "Turkey terror attack: mourning after scores killed in Ankara blasts, The Guardian,, October 11, 2015 https://www.theguardian.com/world/2015/oct/10/turkey-suicide-bomb-killed-in-ankara.

220 Ezgi Erkoyun, "Turkey detains four more Kurdish mayors on alleged terror links," Reuters, May 15, 2020, https://www.reuters.com/

article/us-turkey-security-kurds-idUSKBN22R0YG.

221 "Turkish army forces kill senior PKK terrorist in northern Iraq," TRT World, July 7, 2019, https://www.trtworld.com/turkey/turkish-army-forces-kill-senior-pkk-terrorist-in-northern-iraq-28058

222 "Turkish air strikes kill at least three PKK terrorists in northern Iraq," TRT World, August 4, 2019, https://www.trtworld.com/turkey/turkish-air-strikes-kill-at-least-three-pkk-terrorists-in-northern-iraq-28749.

223 "Erdogan Warns NATO Allies to Back Its Fight Against Terror," US News and World Report, December 5, 2029, https://www.usnews.com/news/world/articles/2019-12-05/erdogan-warns-nato-allies-to-back-its-fight-against-terror

224 Tim Cushing, "Turkish Court Says Government's Two-Year Ban of Wikipedia Violates Freedom of Expression Rights," TechDirt, December 27, 2019, https://www.techdirt.com/articles/20191226/09093643633/turkish-court-says-governments-two-year-ban-wikipedia-violates-freedom-expression-rights.shtml.

225 Human Rights Watch, "Turkey: Events of 2019," World Report 2020, hrw.org/world-report/2020/country-chapters/turkey#.

226 Wikipedia, "Syrian Civil War," last edited August 13, 2019, https://en.wikipedia.org/wiki/Syrian_Civil_War.

227 Wikipedia, "Syrian Civil War," last edited August 13, 2019.

228 Spyridon A. Plakoudas, "The PKK and the Guerrilla Tradition of Turkey's Kurds," February 22, 2015, https://www.academia.edu/15541931/The_PKK_and_the_Guerrilla_Tradition_of_Turkeys_Kurds.

229 Wikipedia, "Manbij offensive," last edited November 14, 2019, https://en.wikipedia.org/wiki/Manbij_offensive.

230 Conflict Between Turkey and Armed Turkish Groups, Council on Foreign Relations last edited August 6, 2019, https://www.cfr.org/interactive/global-conflict-tracker/conflict/conflict-between-turkey-and-armed-kurdish-groups

231 Wikipedia, "Rojava," last edited August 13, 2019, https://en.wikipedia.org/wiki/Rojava.

232 Wikipedia, "Rojava," last edited August 13, 2019.

233 "Persecution and Discrimination against Kurdish Citizens in Syria, Report for the 12th session of the UN Human Rights Council," Office of the United Nations High Commissioner for Human Rights. 2009, https://lib.ohchr.org/HRBodies/UPR/Documents/ses-

sion12/SY/KIS-KurdsinSyria-eng.pdf.
234 Wikipedia, "Regions of North and East Syria," last edited July 16, 2019, https://en.wikipedia.org/wiki/Regions_of_North_and_East_Syria.
235 "The experiment of West Kurdistan (Syrian Kurdistan) has proved that people can make changes," Tahrir-ICN, August 27, 2019, https://tahriricn.wordpress.com/2014/08/27/the-experiment-of-west-kurdistan-syrian-kurdistan-has-proved-that-people-can-make-changes/.
236 Ahmad Samir, "Rojava's Jazira Canton Now Has 87 Cooperatives With 30,000 Members," Hawar News, April 2, 2018, https://mesopotamia.coop/rojavas-jazira-canton-now-has-87-co-operatives-with-30000-members/.
237 View The Constitution of the Rojava Cantons at https://civiroglu.net/the-constitution-of-the-rojava-cantons/.
238 View The Constitution of the Rojava Cantons at https://civiroglu.net/the-constitution-of-the-rojava-cantons/.
239 "Building Autonomy Through Ecology in Rojava," Co-operation in Mesopotamia, February 28, 2018, https://mesopotamia.coop/building-autonomy-through-ecology-in-rojava/.
240 Mireille Court, Chris Den Hond, "Rojava occupied," Le Monde diplomatique, February 2020, https://mondediplo.com/2020/02/04rojava.
241 Matthew Petti, "A Kurdish Experiment in Decentralized Government," Reason, April, 2020, https://reason.com/2020/03/07/a-kurdish-experiment-in-decentralized-governance/
242 "Make Rojava Green Again," Internationalist Commune, June, 2018, https://internationalistcommune.com/make-rojava-green-again-picture-gallery-june-2018/.
243 "Kurdish Women in Culture," The Kurdish Project, accessed July 11, 2020, https://thekurdishproject.org/history-and-culture/kurdish-women/kurdish-women-in-culture/.
244 Zaher Baher, "The Experiment of West Kurdistan (Syrian Kurdistan) has proved that people can make changes," Libcom.org, August 26, 2014, https://libcom.org/news/experiment-west-kurdistan-syrian-kurdistan-has-proved-people-can-make-changes-zaher-baher-2.
245 R. Lowe, "The Emergence of Western Kurdistan and the Future of Syria," in Romano D., Gurses M. (eds) *Conflict, Democratization, and the Kurds in the Middle East* (New York, Palgrave Macmillan, 2014).

246 "The Kurds in Turkey," Federation of American Scientists, accessed July 11, 2020, https://fas.org/asmp/profiles/turkey_background_kurds.htm.
247 Joris Leverink, "Murray Bookchin and the Kurdish resistance," Roar, August 9, 2015, https://roarmag.org/essays/bookchin-kurdish-struggle-ocalan-rojava/.
248 "The Aims of the Mesopotamian Ecology Movement," Mesopotamian Ecology Movement, September 5, 2016, http://new-compass.net/articles/mesopotamian-ecology-movement-presents-its-aims.
249 Pierre Bance, Making Rojava green again," openDemocracy, June 21, 2019, https://www.opendemocracy.net/en/north-africa-west-asia/making-rojava-green-again/.
250 Rosa Burc, Fouad Oveisy, "Rojava Is Under Existential Threat," Jacobin, February 22, 2019, https://www.jacobinmag.com/2019/02/rojava-united-states-withdrawal-syria-erdogan.
251 See Book 2 on Ideology for a discussion about sentiments and their impact on human society
252 Chris Hedges, "The Capitalists Are Afraid," Truthdig, September 9, 2019, https://www.commondreams.org/views/2019/09/09/capitalists-are-afraid.
253 K. K. Tse, "Hong Kong's protests are just the tip of the iceberg: capitalism is in crisis across the globe," South China Morning Post, September 13, 2019, https://www.scmp.com/comment/opinion/article/3026675/hong-kongs-protests-are-just-tip-iceberg-capitalism-crisis-across.
254 Richard J. Barnet and Ronald E. Muller, *Global Reach: The Power of the Multinational Corporations* (New York: Simon and Schuster, 1974), 133-134.
255 P. R. Sarkar, "Nuclear Revolution," *Prout in a Nutshell Volume 4 Part 21 [a compilation]* (Kolkata: Ananda Marga Publications, date unknown), The Electronic Edition of the Works of P. R. Sarkar, Version 1.4.0.6, Ananda Marga Publications, 2009.
256 John Bloom, "Bloom: One Hundred Years: In Recognition of Rudolf Steiner's Threefold Commonwealth," RSF, August 22, 2017, http://rsfsocialfinance.org/2017/08/22/bloom-one-hundred-years-recognition-rudolf-steiners-threefold-commonwealth/.
257 Alvin Lum, "'They want freedom the same way we wanted it':Cheers and support in Lithuania as Hong Kong embraces the 'Baltic

Way'" South China Morning Post, August 24, 2019, https://www.scmp.com/news/hong-kong/politics/article/3024175/they-want-freedom-same-way-we-wanted-it-cheers-and-support.
258 Alvin Lum, "'They want freedom the same way we wanted it': Cheers and support in Lithuania as Hong Kong embraces the 'Baltic Way'" South China Morning Post, Aug. 24, 2019.
259 Wikipedia, "History of Cuba," last edited July 12, 2019, https://en.wikipedia.org/wiki/History_of_Cuba.
260 Wikipedia, "Cuba," last edited July 13, 2019, https://en.wikipedia.org/wiki/Cuba.
261 Wikipedia, "Guantanamo Bay Naval Base," last edited July 17, 2019.
262 Wikipedia, "Granma (Yacht)," last edited February 24, 2019, https://en.wikipedia.org/wiki/Granma_(yacht)
263 Wikipedia, "Operation Verano," last edited April 20, 2019, https://en.wikipedia.org/wiki/Operation_Verano.
264 Wikipedia, "Operation Verano," last edited April 20, 2019.
265 Wikipedia, "Bay of Pigs Invasion," last edited July 24, 2019, https://en.wikipedia.org/wiki/Bay_of_Pigs_Invasion.
266 Wikipedia, "Cuba," last edited July 13, 2019.
267 "Cuba," Human Development Indices and Indicators: 2018 Statistical Update Briefing note for countries on the 2018 Statistical Update, http://hdr.undp.org/sites/all/themes/hdr_theme/country-notes/CUB.pdf.
268 Juan José Cabello, Dunia Garcia, Alexis Sagastume, "An approach to sustainable development: the case of Cuba," Environment, Development and Sustainability, August 2012, Vol. 14, Issue 4, 573-591, https://link.springer.com/article/10.1007%2Fs10668-012-9338-8.
269 "Permaculture Design in Cuba," Urban Farmer, accessed July 25, 2019, http://www.theurbanfarmer.ca/cuba-pdc.
270 Carey Clouse, "Cuba's Urban Farming Revolution: How to Create Self-Sufficient Cities," The Architectural Review, March 17, 2019, https://www.architectural-review.com/essays/cubas-urban-farming-revolution-how-to-create-self-sufficient-cities/8660204.article.
271 Patrick Oppmann, "Fidel Castro survived 600 assassination attempts, officials say," CNN, November 26, 2016, https://www.cnn.com/2016/08/12/americas/cuba-fidel-castro-at-90-after-assassination-plots/index.html.

272 Wikipedia, "1973 Chilean coup d'etat," last edited January 9, 2019, https://en.wikipedia.org/wiki/1973_Chilean_coup_d%27%C3%A9tat.
273 The Church Committee (formally the United States Senate Select Committee to Study Governmental Operations with Respect to Intelligence Activities) was a U.S. Senate select committee in 1975 that investigated abuses by the Central Intelligence Agency (CIA), National Security Agency (NSA), Federal Bureau of Investigation (FBI), and the Internal Revenue Service (IRS). It was chaired by Idaho Senator Frank Church (D-ID).
274 The "40 Committee," was a semi-clandestine body of which Henry Kissinger was the chairman between 1969 and 1976. The committee maintained ultimate supervision over United States covert actions overseas (and possibly, at home) during this period. See David Wise, "The Secret Committee Called '40'" New York Times, Jan. 19, 1975, https://www.nytimes.com/1975/01/19/archives/the-secret-committee-called-40-at-least-in-theory-it-controls-the.html.
275 "Foreign Relations of the United States, 1969-1976, Volume XXI, Chile, 1969-1973," Office of the Historian, Department of State, https://history.state.gov/historicaldocuments/frus1969-76v21/d93.
276 Wikipedia, "Rene Schneider," last edited July 14 2019, https://en.wikipedia.org/wiki/Ren%C3%A9_Schneider.
277 Tim Weiner, *Legacy of Ashes: The History of the CIA* (New York: Anchor Books, 2008), 361.
278 Wikipedia, "United States intervention in Chile," last edited July 18, 2019, https://en.wikipedia.org/wiki/United_States_intervention_in_Chile.
279 Wikipedia, "United States intervention in Chile," last edited July 18, 2019.
280 Wikipedia, "Alejandrina Cox incident," last edited November 22, 2016, https://en.wikipedia.org/wiki/Alejandrina_Cox_incident.
281 "CIA Activities in Chile," Central Intelligence Agency, Sept. ember 18, 2000, https://www.cia.gov/library/reports/general-reports-1/chile/index.html.
282 Wikipedia, Human rights violations in Pinochet's Chile," last edited December 25, 2018, https://en.wikipedia.org/wiki/Human_rights_violations_in_Pinochet%27s_Chile.
283 Maolis Castro, "Chile's Government Pledges New Constitution in Bid to Quell Protests," The Wall Street Journal, November 11, 2019,

https://www.wsj.com/articles/chiles-government-pledges-new-constitution-in-bid-to-quell-protests-11573510278
284 Rodrigo Espinoza-Troncoso & Michael Wilson-Becerril, "What is Behind State Violence in Chile?" Aljazerra, December 4, 2019. https://www.aljazeera.com/indepth/opinion/state-violence-chile-191202110752549.html.
285 "Chile: Police Reforms Needed in the Wake of Protests," Human Rights Watch, November 26, 2019, https://www.hrw.org/news/2019/11/26/chile-police-reforms-needed-wake-protests#.
286 Richard J. Barnet, Ronald E. Muller, *Global Reach: The Power of the Multinational Corporations* (New York: Simon and Schuster, 1974), 87.
287 Wikipedia, "Sandanista National Liberation Front," last edited July 21, 2019, https://en.wikipedia.org/wiki/Sandinista_National_Liberation_Front.
288 "Sandinista Revolution," accessed July 21, 2019, https://vianica.com/go/specials/15-sandinista-revolution-in-nicaragua.html.
289 Wikipedia, "Nicaraguan Socialist Party, last edited July 3, 2019, https://en.wikipedia.org/wiki/Nicaraguan_Socialist_Party.
290 Wikipedia, "Sandinista National Liberation Front," last edited July 21, 2019.
291 "Revolutionary Forces," NACLA, September 25, 2007, https://nacla.org/article/iii-revolutionary-forces.
292 Wikipedia, "Sandinista National Liberation Front," last edited July 21, 2019.
293 "Revolutionary Forces," NACLA, September 25, 2007.
294 "Revolutionary Forces," NACLA, September 25, 2007.
295 Wikipedia, "Los Doce," last edited October 15, 2016, https://en.wikipedia.org/wiki/Los_Doce.
296 Wikipedia, "Pedro Joaquin Chamorro," last edited May 7, 2019, https://en.wikipedia.org/wiki/Pedro_Joaqu%C3%ADn_Chamorro_Cardenal.
297 Karen DeYoung, Nicaragua from Carter to Reagan," The Washington Post, November 8, 1987, https://www.washingtonpost.com/archive/entertainment/books/1987/11/08/nicaragua-from-carter-to-reagan/0a63330a-6035-4093-9887-1ff9ebfe8c20/.
298 Wikipedia, "Nicaraguan Revolution," last edited July 9, 2019, https://en.wikipedia.org/wiki/Nicaraguan_Revolution.

299 Wikipedia, "Nicaraguan Revolution," last edited July 9, 2019.
300 Roland Matthews, "Iran-Contra Affair," accessed July 21, 2019, https://www.britannica.com/event/Iran-Contra-Affair.
301 "Nicaragua and the US Media," Fair, October1, 1987, accessed December 6, 2019, https://fair.org/extra/nicaragua-and-the-us-media/.
302 Robert Parry and Peter Kornbluh, "Reagan's Pro-Contra Propaganda Machine," The Washington Post, September 4, 1988, https://www.washingtonpost.com/archive/opinions/1988/09/04/reagans-pro-contra-propaganda-machine/42d256fc-9d93-4174-b629-06d91f9124c6/?noredirect=on&utm_term=.9ccc3d6f00d2.
303 History.com Editors, "Sandinistas are defeated in Nicaraguan elections," last edited July 17, 2019, https://www.history.com/this-day-in-history/sandinistas-are-defeated-in-nicaraguan-elections.
304 "Sandinista Revolution," Vianica.com, accessed July 21, 2019.
305 Richard E. Feinberg, "Nicaragua: Revolution and restoration," November 2018, https://www.brookings.edu/research/nicaragua-revolution-and-restoration/.
306 Wikipedia, "Nicaraguan Social Security Institute," last edited June 24, 2019, https://en.wikipedia.org/wiki/Nicaraguan_Social_Security_Institute.
307 Wikipedia, "Nicaragua Canal," last edited July 14, 2019, https://en.wikipedia.org/wiki/Nicaragua_Canal.
308 Wikipedia, "Nicaraguan Social Security Institute," last edited June 24, 2019, https://en.wikipedia.org/wiki/Nicaraguan_Social_Security_Institute.
309 Tanya Paz, "The Flame that ignited Nicaragua's protests," AIDA, May 30, 2018, https://aida-americas.org/en/blog/the-flame-that-ignited-nicaragua-s-protests.
310 Carl David Goette-Luciak, "How a journalist's death live on air became a symbol of Nicaragua's crisis," The Guardian, May, 29, 2018, https://www.theguardian.com/world/2018/may/29/nicaragua-journalist-killed-live-on-air-angel-gahona.
311 Wikipedia, "Timeline of the 2018 Nicaraguan protests," last edited June 24, 2019, https://en.wikipedia.org/wiki/Timeline_of_the_2018_Nicaraguan_protests.
312 Wikipedia, "2018-2019 Nicaraguan protests," last edited July 19, 2019, https://en.wikipedia.org/wiki/2018%E2%80%932019_Nica-

raguan_protests.
313 Wikipedia, "2018-2019 Nicaraguan protests," last edited July 19, 2019.
314 John Holman, "Nicaragua revolution: Divisions continue after 40 years," Aljazeera, July 20, 2019, https://www.aljazeera.com/news/2019/07/nicaragua-revolution-divisions-continue-40-years-190720160626454.html.
315 "US cracks down on Cuba, Nicaragua and Venezuela," Aljazeera, April 17, 2019,
https://www.aljazeera.com/news/2019/04/cracks-cuba-nicaragua-venezuela-190417192724288.html.
316 "'El Covid-19 Zekeda' The Dictator's Resposnde To The Unknown Deaths," Today Nicaragua, May 27, 2020, https://todaynicaragua.com/el-covid-19-zekeda-the-dictators-reposnde-to-the-unknown-deaths/.
317 For example, see "Special interests shovel cash to democrats and republicans alike," San Francisco Chronicle, September 24, 2017, http://www.sfchronicle.com/politics/article/Special-interests-shovel-cash-to-California-12170800.php.
318 Egberto Willies, "How do we avert the inevitable Plutocrat driven civil war?" Daily Kos, May 30, 2019, https://egbertowillies.com/2019/05/30/civil-war-plutocrat/.
319 See Chuck and Tom Paprocki, The Untold Story of Western Civilization, Volume 5 which accounts the behavior of the capitalist class in America since WWII.
320 P. R. Sarkar, "The Excellence of God-Centred Philosophy," *Prout in a Nutshell Volume 4 Part 18* [a compilation] (Kolkata: Ananda Marga Publications, 1990).
321 P. R. Sarkar, "Service Psychology and Group Psychology," *A Few Problems Solved Part 8* (Kolkata: Ananda Marga Publications, 1987).
322 P. R. Sarkar, "Social Discourses: Sadvipras," *The Thoughts of P. R. Sarkar [a compilation]* (Kolkata: Ananda Marga Publications, 1981), The Electronic Edition of the Works of P. R. Sarkar, Version 1.4.0.6, Ananda Marga Publications, 2009.
323 Aashna Jain, "Can meditation heal your relationship? 5 powerful tips that will solve your problems," Pinkvilla, March 25, 2019, https://www.pinkvilla.com/lifestyle/love-relationships/can-medita-

tion-heal-your-relationship-5-powerful-tips-will-solve-your-problems-444122.
324 P. R. Sarkar, "Culture," Talks on Prout," *Prout in a Nutshell Volume 3 Part 15 [a compilation]* (Kolkata, Ananda Marga Publications, 1961), The Electronic Edition of the Works of P. R. Sarkar, Version 1.4.0.6, Ananda Marga Publications, 2009.
325 P. R. Sarkar, "Economic Self-Reliance – Excerpt F," *The Awakening of Women [a compilation]* (Kolkata: Ananda Marga Publications, 1981), The Electronic Edition of the Works of P. R. Sarkar, Version 1.4.0.6, Ananda Marga Publications, 2009.
326 P. R. Sarkar, "Shúdra Revolution and Sadvipra Society," *Human Society, Part 2* (Kolkata: Ananda Marga Publications, 1967), The Electronic Edition of the Works of P. R. Sarkar, Version 1.4.0.6, Ananda Marga Publications, 2009.
327 P. R. Sarkar, *Problems of the Day* (Kolkata: Ananda Marga Publications, 1958), The Electronic Edition of the Works of P. R. Sarkar, Version 1.4.0.6, Ananda Marga Publications, 2009.
328 P. R. Sarkar, "Nuclear Revolution," *Prout in a Nutshell, Volume 4, Part 21 [a compilation].*
329 P. R. Sarkar, "Nuclear Revolution," *Prout in a Nutshell Volume 4 Part 21 [a compilation].*
330 P. R. Sarkar, "Sadvipra Boards," *Prout in a Nutshell Volume 4 Part 18 [a compilation]* (Kolkata: Ananda Marga Publications, 1969), The Electronic Edition of the Works of P. R. Sarkar, Version 1.4.0.6, Ananda Marga Publications, 2009.
331 P. R. Sarkar, "Nuclear Revolution," *Prout in a Nutshell Volume 4 Part 21 [a compilation].*
332 P. R. Sarkar, "Social Justice," *Human Society, Part 1* (Kolkata: Ananda Marga Publications, 1959), The electronic Edition of the Works of P. R. Sarkar, Version 1.4.0.6, Ananda Marga Publications, 2009
333 "Economic Aspects of Marriage," accessed July 10, 2010, https://revelpreview.pearson.com/epubs/pearson_ember_a/OPS/xhtml/ch19_sec_11.xhtml.
334 Karina Schroeder, "How Systemic Racism Keeps Millions of Black People from Voting," Vera Institute of Justice, February 16, 2018, https://www.vera.org/blog/how-systemic-racism-keeps-millions-of-black-people-from-voting.
335 Margery Austin Turner and Lynette Rawlings, "Promot-

ing Neighborhood Diversity: Benefits, Barriers, and Strategies," The Urban Institute, 2009, https://www.urban.org/sites/default/files/publication/30631/411955-Promoting-Neighborhood-Diversity-Benefits-Barriers-and-Strategies.PDF.

336 Gene Sharp, *From Dictatorship to Democracy: A Conceptual Framework of Liberation* (New York: The New Press, 2012), 31.

337 P. R. Sarkar, "Nuclear Revolution," *Prout in a Nutshell Volume 4 Part 21 [a compilation].*

338 P.R. Sarkar, "The Four Stages of Human Progress," *Ánanda Vacanámrtam, Part 4* (Kolkata: Ananda Marga Publications, 1978), The Electronic Edition of the Works of P.R. Sarkar, Version 1.4.0.6, Ananda Marga Publications, 2009.

339 Shrii Anandamurti, "The journey towards spiritual culmination," Times of India, April 11, 2008, https://timesofindia.indiatimes.com/The-journey-towards-spiritual-culmination/articleshow/2993967.cms?

340 See the explanation of the relationship of the mind and the brain in *Universal Ideology: The Thought of P. R. Sarkar.*

341 Shrii Anandamurti, "The Path of a Genuine Sadhaka," Speaking Tree, June 11, 2014, https://www.speakingtree.in/blog/the-path-of-a-genuine-sadhaka.

342 P. R. Sarkar, "Prama – 1/ Dynamic Equilibrium and Equipoise," *A Few Problems, Solved Part 8* (Kolkata, Ananda Marga Publications, 1987), The Electronic Edition of the Works of P.R. Sarkar, Version 1.4.0.6, Ananda Marga Publications, 2009.

343 See also *Universal Ideology: The Thought of P.R. Sarkar.*

344 P. R Sarkar, "Social Justice," *Human Society, Part I* (Kolkata: Ananda Marga Publications, 1959), The Electronic Edition of the Works of P. R. Sarkar, Version 1.4.0.6, Ananda Marga Publications, 2009.

345 P. R. Sarkar, "Farmers Cooperatives," *Prout in a Nutshell, Volume 4, Part 20 [a compilation]* (Kolkata: Ananda Marga Publications, 1982), The Electronic Edition of the Works of P. R. Sarkar, Version 1.4.0.6, Ananda Marga Publications, 2009.

346 P. R Sarkar, "Cultural Revolution," *Human Society Is One and Indivisible – 3* (Kolkata: Ananda Marga Publications, 1979), The Electronic Edition of the Works of P .R. Sarkar, Version 1.4.0.6, Ananda Marga Publications, 2009.

347 P. R. Sarkar, "Taking the Opposite Stance in Battle," *Ananda Marga Ideology and Way of Life in a Nutshell, Part 11* (Kolkata: Ananda Marga Publications, 1970), The Electronic Edition of the Works of P.R. Sarkar, Version 1,4,0,6, Ananda Marga Publications, 2009.
348 P. R. Sarkar, "An exemplary life," *Ánanda Vacanámrtam, Part 7* (Kolkata: Ananda Marga Publications, 1979), The Electronic Edition of the Works of P. R. Sarkar, Version 1,4,0,6, Ananda Marga Publications, 2009.
349 P. R Sarkar, "Service Psychology and Group Psychology," *A Few Problems Solved, Part 8* (Kolkata: Ananda Marga Publications, 1987), The Electronic Edition of the Works of P. R. Sarkar, Version 1,4,0,6, Ananda Marga Publications, 2009.
350 P. R. Sarkar, "Human Society is One and Indivisible – 2," *One Human Society* (Kolkata: Ananda Marga Publications, date unknown), The Electronic Edition of the Works of P.R. Sarkar, Version 1,4,0,6, Ananda Marga Publications, 2009.
351 P. R. Sarkar, "Socio-Economic Groupification," *A Few Problems Solved, Part 9* (Kolkata: Ananda Marga Publications, 1979), The Electronic Edition of the Works of P.R. Sarkar, Version 1,4,0,6, Ananda Marga Publications, 2009.
352 P. R. Sarkar, To the Patriots, *Prout in a Nutshell, Volume 1, Part 4, 2nd edition* (Kolkata: Ananda Marga Publications, 1960), The Electronic Edition of the Works of P. R. Sarkar, Version 1,4,0,6, Ananda Marga Publications, 2009.
353 See *The Untold Story of Western Civilization*, Volume 5, Chapter 6, for an in-depth look at the education system in the United States.
354 P. R. Sarkar, "Taking the Opposite Stance in Battle," *Ananda Marga Ideology and Way of Life in a Nutshell Part 11.*
355 P. R. Sarkar, "Taking the Opposite Stance in Battle," *Ananda Marga Ideology and Way of Life in a Nutshell Part 11.*
356 P. R. Sarkar, "Social Justice," *Human Society, Part 1* (Kolkata: Ananda Marga Publications, 1959), The Electronic Edition of the Works of P.R. Sarkar, Version 1.4.0.6, Ananda Marga Publications, 2009.
357 P. R. Sarkar, "Social Justice," *Human Society, Part 1* (Kolkata: Ananda Marga Publications, 1959).
358 P. R. Sarkar, *Discourses on Prout* (Kolkata: Ananda Marga Publications, 1959), The Electronic Edition of the Works of P. R. Sarkar, Version 1.4.0.6, Ananda Marga Publications, 2009.

359 P. R. Sarkar, "Social Discourses: Problem of the Day," *The Thoughts of P. R. Sarkar [a compilation]* (Kolkata: Ananda Marga Publications, 1957 to 1981), The Electronic Edition of the Works of P. R. Sarkar, Version 1.4.0.6, Ananda Marga Publications, 2009.
360 P. R. Sarkar, "The Place of Sadvipras in the Samája Cakra," *Idea and Ideology* (Kolkata: Ananda Marga Publications, 1959), The Electronic Edition of the Works of P.R. Sarkar, Version 1.4.0.6, Ananda Marga Publications, 2009.
361 P. R. Sarkar, *The Guide to Human Conduct (*Kolkata: Ananda Marga Publications, 1979), The Electronic Edition of the Works of P. R. Sarkar, Version 1,4,0,6, Ananda Marga Publications, 2009.
362 P. R. Sarkar, The Guide to Human Conduct.
363 P. R. Sakar, "Social Value and Human Cardinal Principle," *Social Discourses: The Thoughts of P. R. Sarkar, 1957-1981*, The Electronic Edition of the Works of P.R. Sarkar, Version 1.4.0.6, Ananda Marga Publications, 2009.
364 P. R. Sakar, "Social Value and Human Cardinal Principle," Social Discourses: The Thoughts of P. R. Sarkar, 1957-1981.
365 P. R. Sarkar, "Nuclear Revolution," Prout in a Nutshell Volume 4, Part 21 (a compiliation), The Electronic Edition of the Works of P. R. Sarkar, Version 1,4,0,6, Ananda Marga Publications, 2009.
366
367 Sally Painter, Taoism Principles You Can Understand," Love to Know, accessed March 288 2021, https://feng-shui.lovetoknow.com/Taoism_Principles.
368 Sun Tzu, *Art of War*, trans. Ralph D. Sawyer, (Boulder, CO, Westview Press, 1994)
369 B. H. Liddell Hart, *Strategy* (London: Faber and Faber, 1967), 320.
370 B. H. Liddell Hart, *Strategy* (London: Faber and Faber, 1967), 320.
371 P. R. Sarkar, Human Society, Part 1.
372 P. R. Sarkar, Human Society, Part 1..
373 Prout Institute, "Strategies and Lessons for Building Localized Regional Economies," https://vimeo.com/210703545, www.proutinstitute.org.
374 Prout Institute, "Strategies and Lessons for Building Localized Regional Economies," https://vimeo.com/210703545, www.proutinstitute.org.
375 Prout Institute, "Strategies and Lessons for Building Localized

Regional Economies," https://vimeo.com/210703545, www.proutinstitute.org.

376 P. R. Sarkar, "Requirements of an Ideal Constitution," *A Few Problems Solved, Part 8* (Kolkata: Ananda Marga Publications, 1959), The Electronic Edition of the Works of P. R. Sarkar, Version 1.4.0.6, Ananda Marga Publications, 2009.

377 See P. R. Sarkar, "Requirements of an Ideal Constitution," *A Few Problems Solved, Part 8* (Kolkata: Ananda Marga Publications, 1986), Cardinal human values are those age-old traditional values that promote movement toward divinity and away from separation. Cardinal human values always represent synthesis. They are the basic tenets of a universal morality. See *Universal Ideology: The Thought of P. R. Sarkar* for an in-depth discussion of these values.

378 P. R. Sarkar, "The Cosmic Brotherhood," *Idea and Ideology* (Kolkata: Ananda Marga Publications, 1959).

379 P. R. Sarkar, "The Cosmic Brotherhood," I*dea and Ideology* (Kolkata: Ananda Marga Publications, 1959).

The Electronic Edition of the Works of P .R. Sarkar, Version 1.4.0.6, Ananda Marga Publications, 2009.

380 P. R. Sarkar, "Cult, Inference and Propensity," *Subháśita Samgraha Part 18* (Kolkata: Ananda Marga Publications, 1989), The Electronic Edition of the Works of P.R. Sarkar, Version 1.4.0.6, Ananda Marga Publications, 2009.

381 P. R. Sarkar. Parma – 1/Dynamic Equilibrium and Equipoise, *A Few Problems Solved, Part 8* (Kolkata: Ananda Marga Publications, 1987).

382 hOurworld: an international network of neighbors helping neighbors, accessed September 9, 2020, https://hourworld.org/.

383 LETS - Local Exchange Trading Systmes, accessed November 9, 2020, http://www.transaction.net/money/lets/.

384 ROCS - a Robust Comp;lementary Community Currency System, accessed September 9, 2020, http://www.transaction.net/money/rocs/#issue.

385 Wikipedia, "Barter," modified July 11, 2020, https://en.wikipedia.org/wiki/Barter.

386 See Berkshares, Inc., Local Currency for the Berkshire Region, accessed September 9, 2020, http://www.berkshares.org/ for a good example.

387 Schumacher Center for a new economics, "SHARE, Microcredit Program, accessed September 9, 2020, http://www.centerforneweconomics.org/content/share-microcredit-program.
388 Wikipedia, "Revolving Loan Fund," last modified May 4, 2020, https://en.wikipedia.org/wiki/Revolving_Loan_Fund.
389 Wikipedia, "Credit Union," last modified August 11, 2020, https://en.wikipedia.org/wiki/Credit_union.
390 Gene Sharp, From Dictatorship to Democracy: *A Conceptual Framework for Liberation*,(New York: The New Press, 2012), 124.
391 Gene Sharp, From Dictatorship to Democracy: A Conceptual Framework for Liberation, 127.
392 Gene Sharp, From Dictatorship to Democracy: A Conceptual Framework for Liberation, 133.

Glossary

agrico-industries pre-harvest industries that produce farm inputs like seeds, soil, earth worms, farm equipment, fertilizer, etc.

agro-industries post-harvesting industries which depend on agricultural products.

ahamtattva the second level of the mind; the feeling of "I do."

ahimsa one of the ten moral principles of yoga that states one should not intentionally inflict harm on any living creature by thought, word, or action.

atman equivalent to the soul in western religious philosphy.

basic needs those items and services which every individual requires for survival, including food, shelter, clothing, education, and health care.

citta the lowest level of the mind; the mental plate or field of comprehension.

command economy an economy in which production, investment, prices, and incomes are determined centrally by the national government.

counter-evolution reversing the movement of the social cycle by the application of force.

counter-revolution reversing the movement of the social cycle by the application of tremendous force.

dialectics see chart on page 46 comparing dialects to metaphysics.

disintegration the final stage of capitalism that occurs when the revolutionaries' self-direction, noncooperation, and defiance have become so complete that the capitalists lack any control over them. Under this condition, the capitalists' bureaucracy and corporations begin to break down.

economic democracy the condition in which economic and political power are bifurcated and economic power is vested with the local people. The principle goal of the administration is to remove all the impediments and obstacles which prevent the economic needs of the people being met. The universal aim of economic democracy is to guarantee the minimum requirements of life to all members of society.

economic holdings holdings where the market price of farm produce exceeds the costs of production (including capital, labor, and machinery, etc.).

ekendriya the third phase in the process of psychic development, where one establishes control over certain degrading propensities or over certain fetters related to a particular sensory or motor (indriya) organ of one's structure. When one has established control over a particular indriya, or a few indriyas, but not over all, it is call ekendriya.

evolution the mild application of force to bring about a progressive movement in the social cycle.

grand strategy the plan to execute a nuclear revolution. It calculates and develops the physical and human resources of the people and regulates the distribution of power between the different forces and campaigns.

intuitional science the philosophy that is the foundation of all yogic practices. In the East, this Intuitional Science is called Tantra, which means "expansion of mind." Tantra itself is not a religion, but the science of obtaining self-realization.

mahatattva the highest level of the mind; the feeling of "I am."

materialism philosophy that holds that matter is the fundamental substance in nature, and that all things, including ideas and consciousness, are results of material interactions

Neohumanism an extension of the spirit of humanism to everything, animate or inanimate; it includes the love of all created beings of this universe.

nonviolent coercion process by which the capitalists, even though they may initially remain in their positions and adhere to their original goals, are unable to act effectively because the people have usurped their political, social, and cultural dominance.

peoples' economy the practice of local people creating local socio-economies based upon the production of goods and services that meet everyone's basic necessities of food, shelter, clothing, health care, and education. This local self-reliant economy is not based upon the profit motive, but on meeting basic needs.

prama the concept of balance or equilibrium that is required in our physical, mental, and spiritual realms, both individually and collectively. The excellence of the social structure, culture, or civilization of a community of people is derived from the degree of prama which that community attains in its individual and collective life.

pratisaincara during process of evolution, it is the movement from crude to subtle.

progressive socialism also known as PROUT, it is the social philosophy of P. R. Sakar that allows human beings the potential for greatness beyond just meeting their basic needs. Humans are not just material beings, but also are spiritual beings, existing for the purpose of achieving Divinity. As such, social institutions must be created to assist the individuals in this pursuit. All political, economic, social, and environmental policies and institutions must be geared in this direction.

PROUT P. R. Sarkar's socio-economic theory, also called progressive socialism; it is an acronym for Progressive Utilization Theory. (See also "progressive socialism.")

Pseudo-culture the means by which one group exploits another group by forcibly imposing its own language, dress, and ideas on the exploited group, thereby psychologically paralyzing them. An example is the propagation of a consumer culture by the mass media on the part of the capitalist ruling class.

psycho-economic exploitation suppression of a people's culture in such a way that it leads to a feelings of inferiority and loss of confidence in their own potentialities, allowing them to become easily exploited.

psycho-physical parallelism condition where neither psychic waves, nor material waves exist in isolation, but continually interpenetrate each other.

revolution progressive advancement of the social cycle involing the sudden and severe application of force; in PROUT revolution results in a change from one social psychology to another.

sadvipras a new fifth social psychology (Spiritual warriors) proposed by P. R. Sarkar who, by their moral and spiritual integrity and all-round endeavor, will mobilize and guide the people on a revolutionary path; those who follow a Universal Ideology, perform spiritual practices based on the principles of self-realization; strictly adhere to the principles of morality, are ensconced in sacrificing service, and are ready to wage war against immoralists.

saincara during the process of evolution, it is the movement from subtle to crude.

samaj autonomous economic unit

self-realization act of merging (or uniting) the individual consciousness with the supreme consciousness.

sentiment - strong feeling or belief held by an individual or group of people. Common group sentiments are: geo-sentiments, cultural sentiments, and religious sentiments. Positive sentiments are those which unite society, elevate humanity, enhance collective interests, and encourage progressive development. Negative sentiments are narrow in scope and tend to divide society.

shudra revolution revolution led by the disgruntled warriors and intellectuals that have been reduced to shudrahood by circumstance in the capitalist system.

social cycle the succession of class dominance; the cycle involves the society of workers, followed by warrior dominance of the society, followed by dominance by the intellectuals, followed by dominance by the merchants; followed by a return to dominance by the warriors, etc.

social psychologies P. R. Sakars distinction of the four social classes; each psychology is determined by fundamental psychic characteristics:

Shudras: (laborers) whose minds are controlled by material waves of their physical environment

Ksattriyans: (warriors) whose minds seek to dominatel matter through physical struggle

Viprans: (intellectuals) whose minds seek to control their envi ronment through use of the intellect

Vaeshyans: (merchants/business entrepreneurs) whose minds are focused on accumulating material wealth

systaltic movement P. R. Sarkar's theory that motion rather than station is the principle characteristic of the condition of relative reality. When some phenomenon runs out of momentum, it pauses. In stopping, it gains momentum for the next pulse forward until it again runs out of energy. Thus, the movement is systaltic, characterized by periods of motion and pause.

Uneconomic holdings land holdings where the market price of the farm goods is less than the cost of their production (including capital, labor, and machinery, etc.).

universal ideology that which seeks the welfare of all while at the same time pursuing union with the Absolute Consciousness through spiritual practices like meditation, contemplation, devotion to the Supreme, etc.; it addresses our physical, mental, and spiritual needs and provides us with a vision of how to overcome our present crises by offering an optimum set of values and practices to bring us toward human unity and provides concrete solutions to our present predicament.

vikśubdha shúdras these are the warriors and intellectuals who become disgruntled workers under an exploitative capitalist system.

universal morality a morality that all people, regardless of gender, race, age, class, religion, or nation are due the same respect as human beings. Further, all plants and animals have the same existential status and require the same respect for their existence.

wave theory according to P. R. Sarkar, motion and station are alternating characteristics of all phenomena. The change from motion to station or from station to motion constitutes the wave pattern that underlies all forms, whether material or ideal. In the social cycle, a social class fights the dominant class to replace it with a new institutional form as it achieves its position as the new dominant class; as it declines it is in turn confronted by the next social class for dominance. Motion is faster at the beginning of a pulse, and at the end of the pulse during the decline. In its manifestative motion, it has more energy at the initiation and slows down as it reaches its peak.

Illustration Credits

Fig. 1: Map of Kurdistan
Description: Map showing de facto cantons of Rojava (Western Kurdistan in February 2014, as well as territories claimed by Rojava (Western Kurdistan, but not controlled by the Kurds in February 20 14
Author: Panonian
Date : 2014
This file is made available under the Creative Commons CC0 1.0 Universal Public Domain Dedication

Fig. 2: Woman Sustains Severe Eye Injury During Hong Kong Protest
Title: Hong Kong Protester Shot in Eye
Photo appears in over 22 sites; assume it is in the public domain.

Fig. 3: Where Is My Right Eye?
Title: A protester hides her right eye during an occupation of Hong Kong International Airport on Aug. 12
Photo: Jerome Favre/EPA-EFE/REX/Shutterstock
From an article entitled "An eye for an eye': How one woman's eye patch became the symbol of the Hong Kong protests from the lily.com website: https://www.thelily.com/an-eye-for-an-eye-how-one-womans-eye-patch-became-the-symbol-of-the-hong-kong-protests/.
Requested permission from Lily.com on July 13, 2020. Did not receive a response.

Fig. 4: One Woman's Eye Patch Became Symbol of Hong Kong Protests
Title: A protester at Hong Kong International Airport on Aug. 12 by Kin Cheung/AP
From an article entitled "An eye for an eye': How one woman's eye patch became the symbol of the Hong Kong protests from the lily.com website: https://www.thelily.com/an-eye-for-an-eye-how-one-womans-eye-patch-became-the-symbol-of-the-hong-kong-protests/.

Requested permission from Lily.com on July 13, 2020. Did not receive a response.

Fig. 5: Residents Sleeping at McDonald's in Cheung Sha Wan
Title: McRefugees: Homelessness in Hong Kong On the Rise
From the Blog: Wonderful World
http://macau-wonderful-world.blogspot.com/2018/08/24-hours-mc-donald.html

Index

J

K

L

M

N

O

P

Q

R

S

T

U

V

W

Y

Z

About the Authors

Charles Paprocki has spent many years working with troubled teenagers, prison inmates, welfare recipients, and migrant workers in the human services system. He also owned a graphics and advertising agency in New York City where he combined his skill and knowledge to create social marketing campaigns. He was one of the core leaders to create the Universal Pre-K program in New York State and the local food movement in Illinois. His last work was to manage an organic farm in southern Illinois. He has consulted with international NGO's on management strategies and participated in the Earth Summit in Brazil and the Social Summit in Denmark. He passed away in December of 2020.

Tom Paprocki has worked several years in social services, including starting a preschool and daycare center in rural southern Illinois and serving as an administrator for a drug education and crisis center. After receiving a Masters in Public Administration, he was hired by the NASA Goddard Space Flight Center as a Presidential Management Intern. He served thirty years at Goddard, which included positions as head of Personnel, Procurement, and Institutional Resources. He spent seven years as Director of Management Operations, which included facilities, acquisitions, environmental and health services, security, and logistics for the research and launch facilities at Greenbelt, Maryland and Wallops Island, Virginia. He is now retired living in Dunkirk Maryland.

www.ingramcontent.com/pod-product-compliance
Lightning Source LLC
LaVergne TN
LVHW100132061126
851857LV00010B/65
* 9 7 8 1 8 8 1 7 1 7 7 8 2 *